Numerical Simulation of Unsteady Flows
and Transition to Turbulence

Numerical Simulation of Unsteady Flows and Transition to Turbulence

Proceedings of the ERCOFTAC Workshop held at EPFL,
26-28 March 1990
Lausanne, Switzerland

Edited by

O. Pironneau
INRIA, France

W. Rodi
Karlsruhe Universität, Germany

I.L. Ryhming
IMHEF-EPFL, Switzerland

A.M. Savill
Cambridge University, England

T.V. Truong
IMHEF-EPFL, Switzerland

CAMBRIDGE
UNIVERSITY PRESS

Published by the Press Syndicate of the University of Cambridge
The Pitt Building, Trumpington Street, Cambridge CB2 1RP
40 West 20th Street, New York, NY 10011-4211, USA
10, Stamford Road, Oakleigh, Melbourne 3166, Australia

First published 1992

Library of Congress cataloguing in publication data available

British Library cataloguing in publication data available

ISBN   0 521 41618 3

Transferred to digital printing 2004

Table of Contents

vii

Supervisors of the Test Cases

Test Case T1
Prof. I.L. RYHMING and Dr. T.V. TRUONG
Institute of Hydraulic Machines and Fluid Mechanics IMHEF
Swiss Federal Institute of Technology-Lausanne EPFL
CH-1015 Lausanne
SWITZERLAND

Test Case T2
Prof. O. PIRONNEAU
Institut National de Recherche en Informatique et
en Automatique INRIA
Domaine de Voluceau-Rocquencourt
78153 Le Chesnay
FRANCE

Test Case T3
Dr. A.M. SAVILL
Rolls-Royce Senior Research Associate
University of Cambridge / Engineering Department
Trumpington Street
Cambridge CB2 1PZ
ENGLAND

Test Cases T4, T5
Prof. W. RODI
Institut für Hydromechanik
Universität Karlsruhe
Kaiserstrasse 12
7500 Karlsruhe 1
F.R. GERMANY

Preface

Following the formation of ERCOFTAC it was decided to organise a major ERCOFTAC workshop comparing experimental measurements of complex fluid flows with computations. The organisation of the meeting was entrusted to EPFL under the shared responsibility of Professors I.L. Ryhming (EPFL) and O. Pironneau (INRIA). The ERCOFTAC Scientific Committee was charged with the selection of suitable 'test cases' in the areas of turbulence and combustion. The committee met several times under the chairmanship of Professor B. Spalding which eventually decided that, ideally, three sessions would be needed: one exclusively concerned with turbulent flow phenomena, another involving combustion only and, lastly, to bring the communities of turbulence and combustion specialists together, to have a session dedicated to the study of combustion in strongly turbulent flow fields. However, since flows involving both turbulence and combustion have not yet been investigated in sufficient detail to enable comparisons to be made with computational models, this aspect of the workshop was postponed until the ERCOFTAC Summer School and workshop on combustion in 1992. This workshop was held on 26-28 of March 1990, at EPFL in Lausanne, Switzerland, and it was attended by more than 90 researchers representing 38 academic and industrial organisations from 12 countries (not only from Europe).

Despite the fact that industrial and academic research groups often focus on quite distinct problems the test cases at this workshop were generally acknowledged to be interesting to both these constituent parts of the ERCOFTAC research community. For the effective evaluation of the computational methods measurements of the test case flows and their documentation were undertaken especially for this workshop and the main results of these studies are published in these proceedings. Most of the computational methods that are in common use by the academic and industrial communities were applied by one or other of the 'computer' groups to the different test cases; the reader of this volume will have a unique overview of the different methods. The Workshop concentrated on the following turbulence test cases:

> T1: Boundary layer in an S-shaped duct
> T2: Periodic array of cylinders in a channel
> T3: Transition in a boundary layer under the influence of free-stream turbulence
> T4 & T5: Axisymmetric confined jet flows

In the first test case the object was to predict the flow at high Reynolds number of a three-dimensional boundary layer in an S-shaped duct, the turbulence modelling required for this flow was either a general 'Navier-Stokes' solver, or a code specific to three-dimensional boundary layers. The S-duct geometry allows one to follow the continuous development of the mean-flow and of the Reynolds stress fields over a significant downstream distance (of the order of 30-40 boundary layer thicknesses). In particular, the flow field develops

ix

'cross-over' velocity profiles along the boundary layer, a feature that has not previously been investigated in detail. In this type of three-dimensional boundary layer flow the pressure gradients and shear stress drive a secondary flow motion, while the action of the side walls further add to the secondary flows, a common feature of flows in curving ducts. Such three-dimensional flows have to be calculated in many engineering designs such as industrial aerofoils, centrifugal impellors, and complex forms of ducts. This flow is also useful for testing the prediction of viscous internal flows in ducts as well as for general three-dimensional turbulent boundary layer flows in the presence of confining side walls.

The second test case involved transition to turbulence in a complex geometry, and the ability of direct simulation, using the full Navier-Stokes equations, to predict it. The proposed problem arises in the study of the cooling of computer boards where it is desirable to increase the efficiency of air mixing induced by the introduction of metallic rods between two parallel plates. It is not obvious, without calculation or experiment, whether these will retard or enhance mixing. Hence the importance of the study. An idealised flow was considered consisting of a periodic array of cylinders between two infinite parallel walls. These studies are of considerable relevance not only to industrial applications but also to computational studies involving internal boundary conditions in infinitely long domains. In this case turbulence modelling is not relevant because in the industrial application the problem of the Reynolds number range is too low for the fully developed turbulence to exist, but high enough for the flow to be unstable and for the cylinders to have a significant effect on the growth of instabilities. Therefore conventional methods for the time mean quantities in fully developed turbulence are not appropriate; instead computational models have to solve the equations fully inducting all the unsteady effects.

The third test case dealt with the effect of free stream turbulence on the transition to turbulence in a laminar boundary layer on a flat plate and was intended to evaluate methods which can reliably predict transition and the consequential variations of the boundary structure and surface skin friction. Interest in studies of transitional boundary layers is motivated by the quest for improvements in the performance, in the handling flow qualities and in the control of boundary layer flows in engineering applications. In the complex physical processes involved in the transition phenomena, the main mechanism responsible is the Tollmien-Schlichting instability in low free-stream turbulence intensity. At higher free-stream turbulence intensities the linear stage of the transition is bypassed in favour of some other transition phenomena. This switch-over in transition mechanism presents severe problems for prediction methods since the models were primarily developed for fully turbulent flows. Two approaches were used by the 'computers', one based on the empirically adjusted use of ensemble averaged turbulence models and the other on the full simulation of the time dependent flows.

Finally the fourth and the fifth test cases concerned the prediction of confined axisymmetric jet flows with varying degrees of geometrical complexity: in T4 the jet is within a slightly divergent duct and the conditions approximate to those of the Craya-Curtet similitude theory, whereas in T5 the jet exhausts into an open space. The flow T4 occurs in many mechanical and chemical engineering flows, while the flow T5 is found in axisymmetrical burners. The difficulties of computing confined jets are caused by the presence of adverse pressure gradients, the recirculation with unfixed separation and reattachment points as well as the coexistence of both strong and weak shear regions. The prediction of all these factors depend sensitively on the turbulence models, on numerical schemes, and on the boundary conditions.

For test cases 1 and 3, the computers were not informed of the data before the workshop (though experienced groups had a shrewd idea of what they might be and adjusted their codes appropriately). Concequently the workshop was quite an exciting event as the data and the predictions were compared. In the other test cases many (but not all) of the computers were aware of the experimental data.

As will be seen in the pages that follow, the current prediction methods when used with skill by experienced users are capable of predicting well some general features of flows selected for the Workshop. Improvements in accuracy are shown to result from newer prediction methods which take more account of the flow history. Specific computational aspects, such as grid dependence and sensitivity on initial conditions, require further development. These limitations are most marked in flows affected by separation. The workshop also indicated that more detailed experiments are required to shed further light on the relative performances of various prediction methods with different turbulence models.

At the conference three general lectures were given on 'Developments in Computational Modelling of Turbulent Flows' by Professor J.C.R. Hunt, on 'Transition Description and Prediction' by Dr. D. Arnal, and on 'Recent Numerical Simulations of Compressible Navier-Stokes Flows' by Dr. J. Périaux. These lectures are included in these proceedings and provide an introduction to some of the developing techniques used in computational modelling of complex industrial flows at high Reynolds number and the assumptions and difficulties associated with their use.

Despite the success of the Workshop, much still remains to be done before reliable simulations can be carried out in any one of the areas of turbulence covered by this first ERCOFTAC Workshop. The Workshop identified both strong and weak aspects of computational modelling which was one of the main objectives of the ERCOFTAC Scientific Committee.

Prof. I. L. Ryhming

Invited Lecture

Developments in computational modelling of turbulent flows

J.C.R. Hunt – University of Cambridge.

Summary

This paper begins with a review of some of the many different approaches to computational modelling of turbulent flows in which a distinction is drawn between methods that compute, more or less approximately, individual realisations of the flow and those that compute statistics of the turbulence, such as 1 and 2-point moments; also the different levels of model are discussed. By considering the time and length scales, and eddy structure, of turbulence in different flows it is possible to justify where the statistical structure has a general form determined by the local mean velocity gradients, so that it can be simply modelled in terms of local quantities (e.g. using the mixing length approach), they are contrasted with flows in which the turbulence depends on the development of the turbulence from its initial state and on the boundary conditions. This provides a systematic method for deducing when more advanced levels of model (e.g. 2-point) are required, even for computing 1-point statistics, and what should be the appropriate levels of modelling for different statistical quantities. The principles and various techniques involved in modelling the statistics of some of the most critical aspects of turbulence such as the exchange of energy between different components, inhomogeneity and the rate of dissipation are reviewed, and related to fundamental studies of these topics based on the use of direct numerical simulation, theoretical methods and experiments.

2

Contents

§1. Introduction.

1.1. Different problems and methods for CFD.

The purpose of most practical calculations of turbulent flows is to obtain ensemble average statistics such as the mean values of velocity U_i, pressure $\bar{p}$ and temperature $\bar{\theta}$ and moments $\overline{u_i u_j}$, $\overline{p^2}$ etc, given certain boundary and initial conditions. The central assumption of most turbulence models is that there are closed sets of equations governing these statistics. Although such models are improving and providing an increasing range of practical answers, questions still remain about the equations and boundary conditions of current models such as when and where they are valid, and whether their mathematical properties are well behaved, and what are the approximate techniques for their solution by numerical methods. Usually these questions are decided by comparison between model results and experimental data and sometimes direct numerical simulations (Launder 1989).

In this review it is shown how many of such empirical comparisons can be explained and sometimes can be more focussed by a theoretical analysis of the different elements of these models, and by considering the actual eddy structure of turbulence. As we shall show, a number of factors such as the form of the initial and boundary conditions of a turbulent flow, and whether the turbulence undergoes rotational or irrotational straining by the mean velocity field, largely determine the eddy structure and thence the statistical relations between the different components of the turbulent velocity fields. Therefore, since it is the assumption about the form of these relations that chiefly distinguishes different models, considering these factors often provides answers to our questions about models (such as those listed above) and generally helps suggest the most effective use of statistical models of turbulence.

Understanding turbulence structure is not optional but essential in the study of those practical problems which depend on the local fluctuating eddy motions and therefore the individual realisations of the random field, rather than its ensemble statistics. For the case

4

of turbulent flow round bluff bodies different kinds of model for computing the turbulence statistics and individual realisations of the flow are shown in figs. 1a, 1b (from Murakami et al 1990 and Tamura et al 1990).

Another practical example where this approach is necessary is in calculating the instantaneous pattern of dispersion of pollutant using simulations of the trajectories of fluid elements.

It is also found that the individual flow fields need to be measured or computed to improve understanding and the statistical modelling of many chemical reactions in thin turbulent flows because it is not generally possible to relate the statistics of the reactions between species, e.g. A and B, in terms of low order statistics of the velocity and scalar field such as $\overline{u_i u_j}$, $\overline{u_i c_A}$ etc. [Broadwell, 1991, Leonard & Hill 1991]. Also such "realisation" modelling may be necessary for designing systems where there is real time control of flows. Various approximate methods are now available to simulate the random flow fields with an accuracy adequate to model many of the major features of the flow without having to compute the "fully resolved" 3-dimensional equations (see Fig.1b and sec. 3.1).

The main aim of this review is to discuss the principles underlying the methods for these two kinds of calculations for modelling the statistics and realisation of turbulent flows. This approach complements (and generally does not contradict) other reviews which more specifically describe and compare different models (notably that by Launder 1989).

1.2. Modelling at different computational levels of CFD.

Since there will always be limitations on computer capacity and speed, it will only be possible to calculate relatively idealised turbulent flows in complete detail; for more complex flows calculations must be performed with less detail and/or less accuracy. Of course as the speed and capacity of computers improve, our definition of "complex" also changes! (Hunt 1991). In fig. 2 schematic "graphs" are shown of contours of constant computing power, or time, for different types of calculation, ranging from

(i) direct numerical simulations (DNS) in which 100% of the flow field is defined, but these are flows confined by boundaries with simple shapes (e.g. channels) whose Reynolds numbers (for the turbulence) are less than about 200, to

(ii) approximate general methods for "realisation" modelling by simulating random fields, such as large eddy simulation (LES), approximate numerical simulation (ANS), kinematic simulation (KS) (less accurate, but more detail), or statistical modelling such as the calculation of one-point moments of the turbulence, at the second or third order level of 'closure' (Launder, Reece & Rodi 1975, Lumley 1978) (sec. 3.2).

(iii) methods that are specific to particular classes of flows, such as modelling Reynolds stress with an eddy viscosity defined by equations containing coefficients chosen for particular flows, e.g. models of boundary layers (Cebeci & Smith 1974) or integral models of jets and plumes (Turner 1973 Chap. 6), or random flight models for mean and variances of concentration fields in turbulence (Thomson 1987).

The latter models are more "expert" in that they contain ideas from complex modelling and/or experiments, but they are also available to users that are less expert in fluid mechanics and CFD! For many practical problems (e.g. from turbomachinery boundary layers to environmental flows) fast and specific methods are necessary, because they have to be incorporated into larger engineering or environmental codes. In fact there has to be a "cascade" of methods in CFD from DNS down to these specific codes!

In fig. 2 note that the contour lines for each method are fuzzy, because the methods are evolving and continually being extended. Inevitably the methods are being applied where they are not strictly applicable. As Popper explained, it is the errors in science that lead the way forward!

§2. Characterising turbulent flows.

2.1. Classification in terms of boundary and initial conditions.

The study and understanding of turbulence always has to be related to what kind of flow is being considered, because the structure of turbulence is not the same in all turbulent flows. Also, depending on the flow and its boundary conditions, the turbulence may be an intrinsic feature of the flow or may be largely determined by the turbulence on the boundary of the flow domain. Classification of turbulent flows has, surprisingly, seldom been attempted by turbulence researchers (perhaps because they are making an the ambitious attempt to describe every kind of turbulence?); but we have found it to be useful in guiding the choice of appropriate turbulence models, and for relating complex practical problems in turbulence to simpler and better understood turbulent flows. This is related to the approach, advocated by Kline (1981), of analysing complex flows in terms of "flow zones", in each of which the flow is best modelled by different methods.

We first classify turbulent flows according to three types of boundary conditions:

2.1.1. Closed domains (fig. 3):

In this case the turbulent flow domain D is confined by surfaces on which the velocity is zero or is completely specified, such as flow in a heated or stirred tank, an electromagnetic induction furnace, or by a piston in a closed cylinder. The turbulence is generated by flow instabilities (e.g. shear, buoyancy, rotational etc.), which, along with the direct effects of the boundaries, determine the large scale structure of the turbulence. (c.f. Castaing *et al* 1989, Davidson, Hunt & Moros 1988).

2.1.2. Open domains (fig. 4):

In this class, flow enters and leaves D through the bounding surface B, from or into the external domain E. The flow in D is affected by the velocity field on B, so there may be some significant correlation between the turbulence in D and E. Generally the form of these boundary conditions is specific to the particular flow problem.

(a) No turbulence enters D from E: in this case the flow entering D from E is laminar, and turbulence is generated within D. But where the flow leaves D back into E the flow may well be turbulent. An example would be a turbulent boundary layer D with no free stream turbulence; the flow leaving the layer forms a turbulent wake in E. (The transition between these regions is described by a theoretical asymptotic analysis by Neish & Smith 1988 using a simple turbulence model, and by a numerical solution using a complex turbulence model by Launder (1989).

(b) Turbulence is transported between E and D, so that turbulence in D is affected or even determined by the turbulence specified on B: there is an important physical and computational difference between flows where there is a strong mean velocity (relative to the turbulent eddy velocities), and those where there is no appreciable mean velocity across B, but turbulent energy is still being transported in or out of D by the action of turbulent eddies across B. Whether or not this transport of turbulence is significant within D, generally depends on the level of turbulent energy in D, and may also depend on the sensitivity of turbulence in D to external disturbances.

As shown in figure 4(b), a typical example of the former case is turbulence in a gas turbine approaching a row of turbine blades; outside the boundary layers on the blades the turbulence in D is largely determined by and highly correlated with that entering from E, (e.g. Goldstein & Durbin 1980), whereas the turbulence in the boundary layers depends on an interaction between the incoming turbulence and the local boundary layer instabilities (e.g. Goldstein 1981). A typical example of the latter case is turbulence near a density interface, see figure 4(c) (Carruthers & Hunt 1986).

These are both examples where the turbulence in D depends its structure at B, but there are other examples where this is not true such as fully turbulent flow in a long pipe which is independent of the structure of the turbulence at entry. The reasons are discussed in sec. 2.2.

2.1.3. Time dependent problems

Here the complete flow field $u(x, t)$ in D, involving mean and random quantities, is specified at an initial moment in time, say t_I, and boundary conditions are also applied on B. The problem is to calculate and understand how the flow changes with time. This may be very slow (in a statistical sense) if the initial form of the turbulent velocity field and the boundary conditions define a state that is close to equilibrium. On the other hand there may be a rapid rate of change if the initial turbulence does not satisfy the boundary conditions, so that the turbulence has to adjust. Examples of such flows which have been studied experimentally and theoretically include the change of isotropic turbulence under the actions of mean straining motions, or anisotropic fluctuating body forces.

2.1.4.

Most basic turbulent flows, and many practical problems, fall into one or other of the categories described in 2.1.1 to 2.1.3. Time dependent flows have proved to be useful as model problems for studying the structure of distorted turbulence in well developed or even statistically stationary flows; as Lumley remarked "turbulence is a black box that needs shaking to find out what is inside". The results of many distortions or "shakings", such as those performed in their laboratories by Corrsin, Mathieu and Townsend and their colleagues using ingenious different wind tunnel experiments, helped form the basis of much current understanding and modelling (e.g. Launder, Reece & Rodi 1975, Lumley 1978). More recently a few laboratories have focussed on the equally important flows where the turbulence is not locally homogeneous or statistically stationary (e.g. Veeravalli & Warhaft, 1989).

2.2. The time scales of the turbulence.

We have seen (in §2.1) that turbulence in a flow region D at a given time may be determined by the turbulence u_B or u_I defined by the conditions on the bounding surface B (especially where turbulent energy is transported into D) and conditions at the initial time t_I, or by turbulence generated in the interior of D. These different sources of turbulence in any given flow could only be identified by considering a number of different conditions on B and at t_I.

In order to understand and model the turbulence $u(x, t)$ in the interior of D it is essential to know the extent to which it depends on the imposed turbulence, relative to the locally generated turbulence. For example is $u(x, t)$ statistically correlated with u_B or u_I, so that the structure of turbulence in D depends sensitively on the structure of u_B or u_I, or does the variance of u ($= u_0^2$) only depend on the variance of u_B or u_I, and perhaps other factors such as the Reynolds number of the flow? An example of the former situation is the turbulence between the turbine blades which is correlated with the turbulence in the wakes of upstream blades Fig. 4b. The commonest example of the latter situation is the flow in a circular pipe when the Reynolds number is above a critical value of about 1,000 where turbulence only occurs when the inlet flow contains velocity fluctuations of finite amplitude.

As in other statistical problems in physics the criterion for the sensitivity to boundary or initial conditions must depend on whether the turbulence time scale T_L, (the intrinsic or "relaxation" time scale of the system) is of the same order as the imposed time scale T_D in the domain. T_D is the time over which the fluid elements are in the domain, where the turbulence may or may not be undergoing distortion.

One measure of the intrinsic time scale of the turbulence in the interior of D, T_L, is the "relaxation" time over which some disturbance to the turbulence structure decays. But any adjustment of a turbulence flow field always involves the transfer of energy between different scales, or "turning over" of eddies (e.g. Kellogg & Corsin 1980). Since this is also an essential part of the process of the dissipation of turbulent energy (at a rate $\mathcal{E}$ per unit mass), the time scale T_L is of the order of $(\mathcal{E}/K)^{-1}$ where K is the kinetic energy per unit mass. T_L is also of the order of the Eulerian or Lagrangian integral time scales defined by the velocity auto-correlation measured in frames of reference moving with the mean flow or with fluid particles, respectively (see Tennekes & Lumley 1971). At high R_e the quantities that determine the magnitude of T_L are the integral length scale L and the r.m.s. velocity u_0 (and not the viscosity ν); hence by dimensional arguments T_L is of order L/u_0, (Corrsin 1963) a result also consistent with nonlinear theories of turbulence (e.g. Leslie 1973, Lesieur 1991), with most observations of large eddies, and with measurements of the turbulent diffusivity of fluid particles ($D_T = u_0^2 T_L$), (Snyder & Lumley 1971). Now consider different kinds of turbulent flow as defined by different values of T_D/T_L.

Rapidly changing turbulence (RCT): this condition is defined as when the turbulence, as "seen" by fluid elements, varies over a time scale T_D that is much less than the turbulence time scale T_L, i.e. for RCT, $T_D \ll T_L$. Since T_L is the time scale over which the turbulent velocity, of a fluid element is correlated, the criterion for the turbulence **u** to be statistically correlated with the boundary or initial turbulence, u_B or u_I, is that the time which a fluid element spends in **D** must be much less than T_L; in other words this criterion is satisfied in a state of rapidly changing turbulence. For an open domain (§2.1.2.) of length H with a significant mean velocity advecting turbulence into **D** the criterion for RCT is that $H/U_0 \ll L/u_0$, and for a time dependent problem $(t - t_I) \ll L/u_0$. These are situations for which the Rapid Distortion Theory (RDT) based on linearised equations is valid and is of practical use. (e.g. Savill, 1989)

Slowly changing turbulence (SCT): this state is defined to be where the imposed timescale T_D is of the same order or greater than the turbulence timescale T_L, i.e. for SCT, $T_D \gtrsim T_L$. Therefore there can only be a weak correlation between the turbulence in **D**, and the external or initial turbulence u_B, u_I.

This point needs some amplification. In many flows the turbulence time scale at the moving location of a fluid element is changing with time, for example in a two-dimensional wake $L \propto T_{\mathbf{D}}^{\frac{1}{2}}$, and $u_0 \propto T_{\mathbf{D}}^{-\frac{1}{2}}$, so that $T_L \sim L/u_0 \propto T_{\mathbf{D}}$. The question to answer is whether T_L is increasing faster or slower than $T_{\mathbf{D}}$, in order to define whether the turbulence is rapidly or slowly changing.

It is interesting to note that *all* unconfined turbulent flows seem to adjust so that these two time scales remain of the same order i.e. $T_{\mathbf{D}} \sim T_L$. (See table 1, derived from the standard tables of results of length and velocity scales, e.g. Tennekes & Lumley 1971). This (apparently new) result implies that in all such flows the eddy structure of turbulence must display some sensitivity to boundary or initial conditions throughout the flow. The statistics in some shear flows far from the boundaries are much more sensitive to boundary conditions than others; wakes (Bevilaqua & Lykoudis 1978) are quite sensitive, whereas jets are not (Husain & Hussain 1990) (Fig.5a). Launder (1989) has shown that this sensitivity is also found in computations of the one-point statistics of wake flows, when the model equations can account for variations of the intensity of turbulence and the length scales.

The essential explanation is that in all these flows L increases as the vortices and vortical regions in the turbulence (e.g. Hussain & Clark 1983, Wray & Hunt 1990) entrain surrounding fluid, and interact and merge with each other (Fig.5b). This causes the thickening of these shear flows, which leads to diffusion of the turbulence energy, a reduction of the rate of production of turbulence and thence to a decrease in u_0. As the eddies grow and merge they preserve some dynamical properties, or "signature", so that their growing time scale T_L is always just increasing at the same rate as the time $T_{\mathbf{D}}$ that fluid elements spend in the flow. The flow visualisation experiments of Browand & Wedman (1976) and Kiya *et al* (1986) demonstrate this process. Even for boundary layers which have one confining wall it appears that $T_{\mathbf{D}} \sim T_L$ (if the external flow is not changing), presumably because the large eddies can still grow. However in pipes and channels with surrounding walls the growth of eddies is inhibited and therefore T_L is constant, so that turbulence along a pipe eventually forgets its initial conditions (over about 40 diameters according to Sabot & Comte-Bellot (1976)).

Sometimes the turbulence is rapidly changing in one region of a turbulent flow and and slowly changing in another region. In the example of flow over the turbine blades, (length H), near the surface of the blades, the turbulence scale L is small (of the order of the boundary layer thickness δ); so $\frac{T_{\mathbf{D}}}{T_L} \sim \frac{H}{U_0}\frac{u_0}{\delta}$, where U_0 is the mean flow. Thus if $\delta \lesssim H/(U_0/u_0) \lesssim H/20$, $T_{\mathbf{D}}/T_L \lesssim 1$ and so outside the boundary layer the turbulence is changing rapidly. This ratio of $T_{\mathbf{D}}/T_L$ can be used to help choose the appropriate model for any given complex turbulent flow where there are significant gradients of turbulence time scale, for example it shows why "mixing length" models can lead to the wrong sign for shear stress in rapidly changing turbulence RCT (e.g. Belcher et al 1991). (See §4). The value of $T_{\mathbf{D}}/T_L$ also indicates the level of correlation between the turbulence in $\mathbf{D}$ and on $\mathbf{B}$; if $T_{\mathbf{D}}/T_L \lesssim 1$ the boundary condition on $\mathbf{B}$ must be specified quite accurately. For example if the turbulence at $\mathbf{B}$ shown in fig 4b(i) is caused by wakes of upstream bodies it would be incorrect to assume that this is equivalent to grid turbulence, as is sometimes assumed in studies of turbomachinery. This error is now recognised (e.g. Franke & Rodi 1991) and appropriate turbulence statistical models are beginning to be used that model the unsteady effects of the wakes from the upstream flow.

2.3. Eddy Structure.

The measurements by Bevilaqua & Lykoudis (1978) (fig. 5) and Wyngnanski et al (1986) show that the profiles of mean velocity and the variance in three and two dimensional wakes are sensitive to initial conditions throughout the length of the wake. In addition their flow visualisation studies show that the structure of the large eddies also has a persistent dependence on its initial form, in the wake of a solid sphere eddies span the whole width of the wake (with length H), whereas in that of a porous disc the eddies on this scale are very weak. In shear flows these large scale eddies are determined by the initial conditions, the mean profile and how the flow evolves. Jets at very high Reynolds number such as rocket exhausts (Mungal and Hollingsworth 1989) also have significant correlated motions at low frequency and with a length scale H about equal to the jet diameter. Wakes, jets and other shear flows also have characteristic eddies on a smaller scale $L_u \ll H$.

These experiments show that understanding the eddy structure is not only necessary for calculating 2 point statistics (e.g. integral scales, spectra etc.) but also helps explain *one point statistics* (a point of view that is still controversial). If all the relevant statistical data were to be provided, discussion of the eddy structure in principle would be unnecessary; but with only limited statistical data the great value of considering eddy structure is that it indicates the spatial and temporal variation of the velocity field and provides a basis for many *statistical* models, (such as the model for the spectra in the boundary layers by Perry & Abell 1973).

Recent experimental and computational data show that "eddies" in turbulent flows can have two distinct types of dynamical form. They are "vortical" eddies, which are local regions of high vorticity (such as vortex tubes, or vortex pairs of rings) that may be advected randomly across the flow, or may induce their own transport, and on interacting with each other may merge or may disintegrate into smaller scale motions (e.g. vortex "tearing") (Hussain 1986, Falco 1977, Head & Bandyophadhay 1981). If such fluid lumps ("flussigkeit ballen") move across a mean shear flow they induce Reynolds stresses, before slowing down (or braking) by mixing with the surroundings, which defines the magnitude of the "bremsweg" or mixing length L_M (See Prandtl 1925 & 1931 and the more recent analysis of fluid lumps by Hunt 1987).

Alternatively the eddies may be "structural" with a relatively fixed pattern and location within a turbulent flow. They may be local or span the whole flow; they depend on some special feature of the flow, such as the form of the mean velocity profile, which can cause "streamline streaks" and vortices near the wall in turbulent boundary layers, or such as body forces which induce Taylor-Görtler vortices in curved flows, thermal plumes in convective turbulence or particular initial and boundary conditions such as wake vortices. These two distinct forms of eddy structure are particularly noticeable in direct numerical simulations of turbulent boundary layers (see for example Fung et al 1991; Adrian & Moin 1988) which demonstrate 'structural' eddies near the wall and 'vortical' eddies in the outer part.

In a frame moving with the local mean speed, observational studies show that most fluctuations appear as eddy motions with recirculating streamlines (such as were cited by Prandtl 1925). These and other eddy phenomena (e.g. "pairing" of vortices) may be described and analysed theoretically by a number of different methods such as vorticity dynamics, non linear waves or eigen modes, and even linear perturbations about the mean flow [compare Hussain 1986, Ho & Huerre 1984, Aubrey et al 1988, Lee et al 1990].

The existence of large-scale eddies that span the whole flow has to be considered explicitly in the study of some practical problems, such as the production of aerodynamic sound (Bridges & Hussain 1987) or the structure of temperature fluctuations (Mungal & Hollingsworth 1988). But for other problems such as the study of the mean flow or heat transfer it is of less importance. In shear flows these large "structural" eddies with scale H contain much less turbulent energy than more local eddy motions (which may be "vortical" or "structural"). Their length scale (for the normal component of velocity u_2) $L_1^{(2)}$ is determined by the mean velocity of gradient $\langle \partial U_1 / \partial x_2 \rangle$, (averaged over a distance $L_1^{(2)}$) and by the r.m.s. normal velocity u_2', so that. following Riley & Corrsin (1974),

$$L_1^{(2)} \sim \left(\frac{\langle \partial U_1 / \partial x_2 \rangle}{u_2'} \right)^{-1}, \text{ where } L_1^{(2)} \ll H \qquad (2.1)$$

In other words the peak of the energy spectrum $E(k)$ is closer to $L_1^{(2)-1}$ than to H^{-1}. A comparison of measured profiles for the atmospheric boundary layer is given by Hunt, Kaimal & Gaynor (1985) with (2.1). This hypothesis is consistent with the finding that the eddy structure on this scale has many similar features in widely different shear flows, ranging from homogeneous shear flows (as studied numerically and in wind tunnels) to those that are inhomogenous and either fully developed (such as pipeflow) or developing (such as wakes, jets and boundary layers) (e.g. Antonia & Bisset 1991, Hussain 1986, Lee et al 1990, Carruthers et al 1991).

For example in the boundary layer, the "structural eddies" near the wall are strongly linked to the local mean velocity profile and the distance from the wall (e.g. Townsend 1976, Aubrey et al 1988). But yet the "streak-like" pattern of the instantaneous contours of the longitudinal velocity and other features are also found in other shear flows. In the outer part of the layer "vortical eddies" are moving across the flow chaotically, but yet are interacting with the mean shear in a way that is found in wakes and other shear layers.

Consequently from the similarities of the statistical structure of sheared turbulence (reviewed by Townsend 1976 and Jeandel et al 1978) and the similarity of certain aspects of local eddy structure there is some physical justification for models for turbulence in shear flows being based on the assumption that certain statistical properties of turbulence (such as the relation between Reynolds stress and the mean shear) are locally determined [Launder & Spalding 1972, Lumley 1978].

On the other hand, in inhomogeneous turbulence where there are other processes of energy transport and energy transfer by pressure fluctuations between different velocity components, such as at the outer edge of a shear layer or near a free surface, turbulence does not have a universal eddy structure or common statistical form. The modelling of these processes is discussed in sec 4. Computational and analytical studies show that where there is no shearing motion or the mean motion consists of shear plus irrotational strain, the general eddy structure is sensitive to the initial conditions (such as decaying anisotropic turbulence Michard et al 1987) and/or to the boundaries of the region of turbulent flow, as in thermal convection (fig.6) where $L^{(1)} \sim H$. In turbulence approaching a bluff body all of these effects are important, the "blocking" effect of the boundaries particularly affecting the large scales, (Britter *et al* 1979). Nevertheless even in non-sheared flows statistical models can predict certain turbulence statistics of practical interest, as explained later.

2.4. Summary.

To summarise, it appears that many turbulent flows, or particular regions, within them, are sensitive to initial conditions and to external disturbances, not only when the flows are highly correlated (i.e. when $T_D < T_L$), but even when there is no direct *correlation* between the turbulence and its initial state (i.e. $T_D \sim T_L$). This sensitivity is most obvious in the structure and the dynamics of the largest eddies on the scale of H (Townsend 1956) and to the specific form of the mean flow. On the other hand, the sensitivity to the initial conditions of the energy containing scales with length scale L is much less when the turbulence has been significantly strained by a strong mean shear,

$$\text{i.e. } S^* = T_L \langle \nabla U \rangle \gtrsim 1 \qquad (\text{or } (T_D/T_L)S^* \gg 1 \text{ if } T_D \ll T_L), \qquad (2.2)$$

as has been shown most notably in D.N.S. of homogeneous shear. This situation occurs naturally in the turbulence at the bottom of a boundary layer (Lee, Moin & Kim 1990).

The tendency towards a common form of eddy structure in shear flows also indicates that it may also be reasonably accurate to model equations for the statistics in terms of local gradients of the mean flow and of the turbulence statistics. In non-shear flows this assumption is probably not valid [see sec. 4].

At very high Reynolds numbers, there are also universal features of turbulence in the structure of the small scales in the inertial range, (Kolmogorov, 1941) that are independent of the mean shear. However in many engineering flows the Reynolds number is not high enough for this universal small scale structure to exist (c.f. Mestayer, 1982; Hunt & Vassilicos, 1991). So it is clear that no one set of simplifications can encompass all the properties of interest in all the turbulent flows of interest, but for the large scales of motion in shear flows the turbulence structure may be sufficiently general that models can be constructed that are applicable over a wide (but not universal) class of turbulent flows. There seems no evidence from the eddy structure that in turbulent flows without mean shear the energy containing scales have a general form.

§3. Modelling Realisations of Turbulent Flows.

3.1. Different methods.

Methods of computational modelling of turbulence naturally divide themselves between those directed towards simulating realisations of the random flow fields and those that only calculate the statistics of the flow field. In some computations these two types of method are used simultaneously for different parts of the spectrum as in Large Eddy Simulations (LES). The former type provides direct insight into the eddies and coherent structures in turbulence which helps towards understanding and improving many kinds of turbulent models, as explained in sec. 2 and may also be of direct practical importance. In sec. 4 the form and computation of statistical models are reviewed.

In general the former "realisation methods" have to be approximate because computer capacity limits the range of Reynolds number where the computations can be exact, (sec. 1.2). Moreover, until recently approximate "realisation methods" such as LES could only be used for ideal flows (such as shear flow or wakes of circular cylinders), but now they are beginning to be applied to turbulent flows with complex boundary conditions and body forces, (Tsai et al 1991). Even more approximate, but computationally faster, realisation methods are now being developed based on aggregations of specified forms of eddy function. Either the amplitudes are computed (e.g. Aubrey et al 1988) or are specified as random functions whose statistics are determined by computations or measurements [Fung et al 1991].

3.2. Approximating and representing the flow-field.

The first step in these methods is to decide what form of approximate velocity field is to be used. It may be a vector field $\widetilde{\mathbf{u}}(\mathbf{x}, t)$ defined everywhere, or it may be defined or evaluated at discrete points $\mathbf{x}_n$, i.e. $\widetilde{\mathbf{u}}(\mathbf{x}_n, t)$, and then interpolated between these values.

In general there is some error ϵ between $\mathbf{u}$ and $\widetilde{\mathbf{u}}$ at the points $\mathbf{x}$ or $\mathbf{x}_n$. There are 3 main kinds of representation.

(a) $\widetilde{\mathbf{u}}$ may be a finite set of defined continuous functions $\mathbf{E}_n(\mathbf{x})$ with random amplitude $a_n(t)$, i.e.

$$\widetilde{\mathbf{u}}(\mathbf{x}, t) = \sum_{n=1}^{M} a_n(t)\mathbf{E}_n(\mathbf{x}) \tag{3.1}$$

where $\mathbf{E}_n(x)$ may be a Fourier component (as in rapid distortion theory Townsend 1976 or as in non-linear or spectral computations Orszag 1972) or orthogonal eigenmodes (e.g. Aubrey et al 1988). The amplitude a_n may be computed directly from the actual or approximate governing equations or may be considered as a random variable with certain statistics that are computed or assumed.

(b) $\widetilde{\mathbf{u}}(\mathbf{x}, t)$ may be the velocity field determined by a vector field $\mathbf{V}(\mathbf{x}_n, t_k)$ on a finite set of nodes $(\mathbf{x}_n, t)$ in a flow

$$\widetilde{u}_i(\mathbf{x}, t) = \sum_{m=1}^{N} V_j(\mathbf{x}_m, t)G_{ij}(\mathbf{x}, \mathbf{x}_m; t) \tag{3.2}$$

where $V_j(\mathbf{x}_m, t)$ is the velocity at the fixed nodes and G_{ij} is an interpolation function relating the velocity at $\mathbf{x}$ to its value at $\mathbf{x}_m$, as in the Direct Numerical Simulation of turbulence (e.g. Rogallo & Moin 1984), or $\widetilde{\mathbf{u}}(\mathbf{x}, t)$ may be the velocity induced by a finite set of moving vortices or vortex elements, in which case V_j is the vorticity at the moving node points and G_{ij} defines the velocity induced by the vortices [e.g. Kiya et al 1982] or vortons (Novikov 1983).

(c) The third kind of approximation involves a large scale field $\widetilde{\mathbf{u}}^{(L)}$ being computed on a number of nodes $\mathbf{x}_n$ whose spacing is larger than the smaller scales of turbulence. In order to estimate these "subgrid" scale motions a stochastic small scale field $\widetilde{\mathbf{u}}^{(s)}$ is added. Only the statistics of this small field are *determined*, but in any given realisation the small scale field can be *constructed* to have the given statistics (by a process of *stochastic simulation* denoted by braces in (3.3)).

In this case

$$\widetilde{\mathbf{u}}(\mathbf{x}, t) = \widetilde{\mathbf{u}}^{(L)}(\mathbf{x}, t) + \{\widetilde{\mathbf{u}}^{(s)}\}. \tag{3.3}$$

In Large Eddy simulation (Lesieur 1991), $\widetilde{\mathbf{u}}^{(s)}$ is generally related deterministically to $\widetilde{\mathbf{u}}^{(L)}$, while in the extension of discrete vortex modelling to allow for small scale motion (e.g. Kiya *et al* 1982), $\widetilde{\mathbf{u}}^{(s)}$ is random.

3.3. Approximating and solving the governing equations.

The next step is to consider the dynamical equations, $\mathcal{L}(\mathbf{u}^*) = 0$, which is a shorthand form for

$$\frac{\partial u_i^*}{\partial t} + u_j^* \frac{\partial u_i^*}{\partial x_j} = -\frac{1}{\rho} \frac{\partial p^*}{\partial x_i} + \nu \nabla^2 u_i^*;$$

$$\frac{\partial \rho}{\partial t} + \frac{\partial}{\partial x_i}(\rho u_i^*) = 0. \tag{3.4}$$

These equations, which are used to compute the velocity field $\widetilde{\mathbf{u}}$ as represented by one of the forms (3.1) to (3.3), are in all cases approximated and are denoted by $\widetilde{\mathcal{L}}(\widetilde{\mathbf{u}}) = 0$.

In most practical cases the approximations $\widetilde{\mathbf{u}}$ and $\widetilde{\mathcal{L}}$ differ significantly from their exact value or even their correct form; for example in the computation of the amplitude $a_n(t)$ of eigenmodes (Aubrey *et al* 1988) a certain symmetry is assumed; in discrete vortex modelling there is an assumption that the large eddies are two dimensional (Kiya *et al*

1982); linearisation of $\mathcal{L}$ is assumed in the "kinematic" simulation (KS) of flow fields using R.D.T. (Lee *et al* 1990; Carruthers *et al* 1991); or in approximate numerical simulation (ANS) the finite difference forms of $\mathcal{L}$ (third or fifth order upwind differences) are used to compute high Reynolds number flows, (e.g. $R_e \sim 10^5$) but using a grid, although very fine (e.g. $300^{(3)}$), not fine enough to resolve motions (Tamura *et al* 1990). In LES the approximate equation for the large resolved scales $\widetilde{\mathbf{u}}^{(L)}$ contains approximate non-linear terms produced by the small scales $\widetilde{\mathbf{u}}^{(s)}$. A general analysis of this "filtering" approximation has been developed by Germano (1992).

Now equation (3.4) is solved for each realisation (using whatever numerical or analytical techniques are appropriate) and therefore *boundary conditions must be applied* for each realisation. Depending on the problem (see sec. 2.1) as the flow enters the computational domain, it may be laminar or, if not, a typical turbulent flow field has to be specified (e.g. of the form of (3.1) with $a_n(t)$ taken from a random sequence). On rigid boundaries the usual kinematic conditions are applied, viz $\widetilde{\mathbf{u}} = 0$. To compute the large scale eddy motion at a shear-free interface the appropriate boundary condition is $\widetilde{\mathbf{u}} \cdot \mathbf{n} = 0$ (e.g. Biringen & Reynolds 1981). In some cases the turbulence matches with an external kind of flow field $\mathbf{u}^{(E)}$ on the interface $\mathbf{B}$ such as irrotational fluctuations, free stream turbulence or waves, in which case the velocity in the normal direction $\mathbf{n}$ and pressure (or normal stress) are continuous $[\mathbf{u} \cdot \mathbf{n}]_{\mathbf{B}} = 0$, and $[p]_{\mathbf{B}} = 0$, where $\mathbf{u}^{(E)}$ and $p^{(E)}$ are specified or computed at $\mathbf{B}$ [e.g. Phillips, 1955; Gartshore *et al* 1983, Carruthers & Hunt 1986] . The boundary $\mathbf{B}$ may be a fixed surface for analytical approximate models (as in those cited), or in computational-grid based models it can be a moving interface.

3.4. Processing the solutions.

At the completion of this stage individual flow realisations are available for examination, simply by inspection or by various kinds of objective analysis. Recently much new information about the eddy structure has been deduced from the results of D.N.S.

and L.E.S., for example by using kinematical analyses (Perry & Chong 1987; Wray & Hunt 1990) (Fig 3.1) by conditional sampling and statistical analysis [e.g. Adrian & Moin 1988], and by dynamical-system methods for investigating the significant time dependent features of the flow and the dimension of the flow [Aubrey *et al* 1988, Bergé 1987]. As we have already shown this information gives useful insight into statistical models. It can also provide designers of engineering products and processes with concepts for utilising or altering the eddy structure (Bushnell & Moore, 1991; Hunt 1991).

From the computation, by taking averages over many realisations or, if the flow is homogeneous or stationary, by averaging over the flow field in certain directions or over space time, ensemble statistics for the turbulence can be obtained, such as moments $\overline{u_i u_j}$, $\overline{u_i^3}$, spectra $\Theta_{ij}(\kappa_1)$ or probability distributions (pdf) etc. Note that second-order one-point statistics can be obtained from experiments and simulations using less data than are needed for pdfs and third moments, but the latter are of interest for the dynamical analysis of turbulence. That is one of the reasons for the hundreds of hours required for the D.N.S. for turbulent channel flow by Moser & Moin (1987). Where the results of the third stage (i.e. solutions of the approximate equation $\widetilde{\mathcal{L}}(\widetilde{u}) = 0$) can be obtained analytically, as is possible for certain simple cases in linear R.D.T., it may also be possible to calculate certain statistics, such as $\overline{u_i^2}$ or $\Theta_{ij}(\kappa_1)$, purely analytically. This has the advantage of showing very clearly how the statistical results vary as a function of initial conditions and the forms of the mean flow. (Batchelor & Proudman 1954; Hunt *et al* 1990).

§4. Models for turbulence statistics.

4.1. Different approaches and assumptions.

Model Complexity

In the simplest and computationally the fastest types of practical calculation of turbulent flows only the mean flow field is predicted (velocity $\mathbf{U}$, pressure p, concentration C)

together with those primary statistics of the turbulence that are required for the calcula-
tion of the mean fields (such as shear stress $-\rho\overline{u_1 u_2}$ in a shear flow $U_1(x_2)$, mean flux $\overline{cu_2}$
in a concentration gradient $C(x_2)$). This level of information can be derived from "eddy
diffusion" models where the eddy viscosity ν_e and diffusivity D_e are determined by the
local gradients of $\mathbf{U}(\mathbf{x}, t)$ and $C(\mathbf{x}, t)$, respectively.

At the next level of practical schemes (K-$\mathcal{E}$ or K-L methods) additional differential
equations are used in which ν_e and D_e depend on the temporal and spatial gradients of
the turbulence. These equations describe the rate of evolution of fluctuating kinetic energy
($K = \frac{1}{2}\overline{u_i u_i}$) and the length scale L of the turbulence (or the rate of dissipation $\mathcal{E}$, to be
discussed later), both to improve the calculation of the primary statistics and to estimate
the overall quantities of the turbulence field, such as might be necessary for combustion
calculations. At the third level of general practical schemes, (in 2RST and 3RST) the
assumption is discarded that the Reynolds stresses are related to the local mean velocity
gradient (i.e. that an eddy viscosity can be defined). Then the normal stresses of the
turbulence are computed, and in 3RST also third moments are computed.

In all methods currently used in practice turbulence quantities are defined at one
point (1-PM). But recently methods are beginning to be available for the estimate of
statistics at 2-points, such as second moment correlations and spectra (2-PM). So far, these
methods have only been developed for particular specialist problems, such as turbulence
in an axisymmetric internal combustion engine (Cambon 1991) or where more information
about the turbulence structure is required and where the statistics of the turbulence on the
boundaries $\mathbf{B}$ of the domain can be measured. But not enough is yet known experimentally
or theoretically on which to base a *general* method for directly computing 2-point statistics.
However this is an active area of research and such methods may emerge in the next few
years. Two current methods are briefly reviewed in sec. 4.3.

Model generality

The first models for calculating the mean flow and fluxes of momentum and heat were *specifically designed* for unidirectional shear flow, such as the "mixing length" model of Prandtl (1925). This classic example of a specific approximation for a specific class of flow or for specific regions within flows is still the basis for most practical methods, such as those for turbulent boundary layer calculations (Cebeci & Smith 1974, Johnson & King 1984). This is physically reasonable since turbulence does *not* have the same general structure in all turbulent flows, as is evident, say, from differences in ratios of normal stresses (e.g. $\overline{u_1^2}/\overline{u_3^2}$), spectra and in probability distributions. But, as explained in sec. 2.3, data from recent experimental measurements and numerical simulations of the eddy structure and statistical characteristics show that there are significant similarities in the structure of turbulence *within certain classes of flow*, such as shear flow, or flow near boundaries. This suggests that reasonably accurate approximate models for deriving basic one-point statistics can be constructed for these type of flows. Therefore constructing turbulence models specific to a particular class of flow may be the practical solution. This approach was recommended by Kline (1981) (which he called "zonal") after examining the comparisons between 1-point turbulence measurements and predictions by many types of turbulent model. Therefore a simple "mixing length" model proposal is not justifiable and also not accurate enough for many complex flows found in practice.

However a major difficulty with this "specific" approach is that there are many flows which involve several "classes" of turbulent flow or fall between classes, such as in the separated flow behind a bluff body on a plane where there are zones corresponding to (i) a free shear layer adjoining (ii) an impinging stagnation point flow whence streamlines whence streamlines bifurcate either back to (iii) a recirculating flow or (iv) into the downstream boundary layer (see for example Dianat & Castro 1991; Kiya 1989). These and other zones are shown in Fig. 7.

Another disadvantage of such specific approaches is that, although they may provide good estimates for their specific flows, it usually requires experience and empirical input to apply a sequence of such models to the different regions of a complex flow. Also there are no 'specific' models for recirculating flows.

Assumptions for general statistical models

The alternative is to develop more complex general models that can be applied to wide classes of turbulent flows, such as those that need to be understood for solving engineering and environmental problems. In many such flows the initial or the subsequent state of the turbulence is well-developed and they have a number of common features that make it possible to construct differential equation models for their moments. Launder & Spalding (1972), Lumley (1978) and others have shown that the following features are required (but are not generally satisfied):

A1 Turbulent velocity components can be considered as a perturbation about a 3-dimensional Gaussian isotropic state, so that the turbulence would tend to return to isotropy if undisturbed by anisotropic straining or forcing, or by boundary conditions. (This assumption does not apply to the *derivatives* of the velocity which are highly non-Gaussian. Examples of turbulence not satisfying this assumption are strong thermal convection and two-dimensional turbulence).

A2 Energy spectra $E(k)$ in different flows are similar, with a single maximum so that only the kinetic energy K and the integral length scale L are needed to characterise the spectrum. (Thus turbulence in a wake with vortex shedding or in a shear layer with a forced oscillation is excluded)

A3 The transfer of energy to small scales and the rate of dissipation of kinetic energy $\mathcal{E}$ by viscous processes is characterised by the large scale motions where

$$\mathcal{E} = K^{3/2}/L_{\mathcal{E}} \quad \text{and } L_{\mathcal{E}} \sim L, \tag{4.1}$$

an assumption that follows from A2 and from assuming that the Reynolds number
of the turbulence is large enough that viscosity does not explicitly enter into the
modelling, (in highly anisotropic turbulence, such as near interfaces or walls L has to
be defined in a particular way).

A4 The turbulence length scale L is small compared to the distance inhomogeneity, L_I
over which the turbulence statistics or mean strains $(\partial U_i/\partial x_j)$ change. [An example
of a flow where this assumption is not correct is where large eddies impinge on a small
obstacle, e.g. an aircraft in a gust].

One of the most intriguing aspects of turbulence research is trying to understand
why general models can give satisfactory estimates for turbulence statistics when these
assumptions are invalid. In some cases the models are extended to allow for special factors,
such as when A4 is violated or the effect of low Reynolds number (which is often modelled
using (4.1) using terms involving molecular and eddy viscosity simultaneously and when
they are of the same order, which contradicts the assumption for the validity of (4.1)!)

In the following section, the mathematical steps involved in constructing specific and
general models are described, and in sec. 4.3 their physical basis is discussed in more detail
and suggestions made about why and when they are valid.

4.2. Methods for 1 point and 2-point moments (1PM, 2PM).

There are three essential steps in the derivation of most statistical models.

4.2.1. Take moments of the Navier Stokes equation

From $\mathcal{L}(\mathbf{u}) = 0$, defined in (3.4), the equations for the joint k^{th} order moments of
velocity, or derivatives, at p points, say $M_{k,p}$ are derived without approximation for $k \geq 2$,

$$\mathcal{L}(M_{k,p}; M_{m',p'}) = 0 \text{ where } m', p' = 1, 1 \text{ and } k, p+1 \text{ and } k+1, p+1 \qquad (4.2)$$

(for the velocity moments). This connection between equations for moments k and $k+1$,
and $k+1$ and $k+2$, etc., shows that the number of equations in this set is infinite. If

the highest order moment that is to be considered, the order of truncation is k_T, then a "closure" approximation is needed to relate directly moments of order k_T to those of lower order, and equations of higher order than k_T are not needed, a point first elaborated by Millionshikov (1940). Note that any gradients in the mean velocity which may appear in (4.2) have to be calculated from the mean momentum equation

$$\mathcal{L}\big(M_{1,1}; M_{2,1}\big) = 0, \qquad\qquad (4.3)$$

which is coupled to (4.2).

A complete statistical description of turbulence requires a solution for the joint probability density function (pdf). In principle an infinite number of equations can be formulated, but they cannot be used in any way more systematically than the moment equations. Approximate equations for p.d.f.s are widely used in modelling reactions of scalar quantities in turbulent flows (Pope 1985).

4.2.2. Approximate moment equations

(a) <u>Deriving one point moments</u>

Firstly the equation (4.2) for $M_{k,p}$ involves some approximation at the k, p level of moment, and secondly appoximations are necessary in the coupled equations involving other moments $k+1$, $k+2$, ... which are truncated at the level m_T. Then to relate $M_{m_T,p}$ to $M_{m',p}$ for $m' < m_T$ further approximate relations are necessary. To be specific, equations for second moments at one point $\overline{u_i u_j}$, as used in K-$\mathcal{E}$ or Reynold Stress Transport models (Launder & Spalding 1972) are of the form:

$$L(M_{2,1}; M_{m,p}) = 0 \text{ where } m, p = 1, 1; \text{ and } 2, 2; \text{ and } 3, 2 \qquad (4.4)$$

(The reason why an equation for 1-point moments involves 2-point moments is that moments of derivatives are equal to derivatives of 2-point moments, i.e. $\overline{\left(\frac{\partial u_i}{\partial x_j}\right)^2} = -\left(\frac{\partial^2}{\partial r_j^2}\overline{\left(u_i(\mathbf{x})u_i(\mathbf{x}+r_j)\right)}\right)_{r_j=0}$ (Tennekes & Lumley 1971)).

The full equation for $M_{2,1}$ is:

$$\left(\frac{\partial}{\partial t} + U_k\frac{\partial}{\partial x_k}\right)\overline{u_iu_j} = -\left(\overline{u_iu_m}\frac{\partial U_j}{\partial x_m} + \overline{u_ju_n}\frac{\partial U_i}{\partial x_n}\right) + \overline{\left(\frac{p}{\rho}\right)\left(\frac{\partial u_i}{\partial x_j} + \frac{\partial u_j}{\partial x_i}\right)}$$

$$-\frac{\partial}{\partial x_k}\left(\overline{p(u_i\delta_{jk} + u_j\delta_{ik})} + \overline{u_iu_ju_k}\right) \underbrace{-\ \nu\left(2\overline{\frac{\partial u_i}{\partial x_k}\frac{\partial u_j}{\partial x_k}} - \frac{\partial^2}{\partial x_k^2}\left(\overline{u_iu_j}\right)\right)}_{\mathcal{E}} \quad (4.5a)$$

where

$$\frac{1}{\rho}\frac{\partial^2 p}{\partial x_k\partial x_k} = -2\frac{\partial U_i}{\partial x_j}\frac{\partial u_j}{\partial x_i} - \frac{\partial u_m}{\partial x_n}\frac{\partial u_n}{\partial x_m} \quad (4.5b)$$

The "production" terms (the first set on the right-hand-side of (4.5a)) do not need approximating. However the two-point moments $M_{2,2}$ (such as the pressure strain terms $\overline{p\partial u_i/\partial x_j}$) are approximated, usually in terms of the *local values* of one-point moments, $M_{2,1}$ and $M_{3,1}$ corresponding to particular integrals of (4.5b) for the pressure p corresponding to the first and second terms on the r.h.s. The arguments for these approximate expressions are based on tensor analysis and symmetry, and the assumption of *local* relations. Not surprisingly different authors have suggested many different forms. [Compare Launder, Reece & Rodi 1975, with the more complex expressions of Lumley 1978.] They mostly agree with the analytical results of rapid distortion theory for isotropic turbulence suddenly undergoing a rapid strain, but not if the turbulence is initially anisotropic [e.g. Maxey 1982]. There is particular disagreement about the $M_{3,1}$ terms in the expressions for $\overline{p\left(\frac{\partial u_i}{\partial x_j} + \frac{\partial u_j}{\partial x_i}\right)}$. which determine the transfer of energy between the different components when the turbulence is anisotropic (the "return to isotropy" terms) [e.g. Weinstock 1989]. The rate of dissipation $\mathcal{E} = -\nu\overline{\left(\partial u_i/\partial x_k\right)^2}$, which is another two-point moment, can either be expressed in terms of a one-point moment $K^{3/2}/L_{\mathcal{E}}$ and a "functional" length scale $L_{\mathcal{E}}$, or can be derived by another approximate equation for $\mathcal{E}$. The third order terms $M_{3,1}$ in (4.5a) are either approximated in terms of $M_{2,1}$, or else are calculated by an equation for $M_{3,2}$.

For models up to second order ($m_T = 2$) these approximations lead to a formally solvable closed set of approximate moment equations of the form:

$$\text{Mean Flow and Continuity} \qquad \widetilde{\mathcal{L}}_U^{(j)}(M_{1,1}; M_{2,1}) = 0 \qquad j = 1, 2, 3, 4, \qquad (4.6a)$$

$$\text{Moments (M)} \qquad \widetilde{\mathcal{L}}_M^{(i)}(M_{2,1}; M_{1,1}, M_{2,2}) = 0 \qquad i = 1, \ldots, I, \quad (4.6b)$$

$$\text{where } I(\lesssim 6) \text{ depends on the particular model,}$$

$$\text{Rate of dissipation} \qquad \widetilde{\mathcal{L}}_\mathcal{E}(M_{2,2}; M_{2,1}, M_{1,1}) = 0. \qquad (4.6c)$$

In the most widely used K-L and K-$\mathcal{E}$ mixing length models, the moment equation $\widetilde{\mathcal{L}}_M^{(1)}$ relates the individual one point Reynolds stresses $\overline{u_i u_j}$ ($\equiv M_{2,1}$) (or τ_{ij}) to the gradients of the mean flow in terms of an algebraic equation of the form

$$\overline{u_i u_j} - \frac{1}{3}\delta_{ij}\overline{u_k u_k} = -\nu_e\left(\frac{\partial U_i}{\partial x_j} + \frac{\partial U_j}{\partial x_i}\right) \qquad (4.7)$$

where the isotropic eddy viscosity $\nu_e = \mu\sqrt{K}L_s$ and L_s is a length scale that determines the stresses. μ is a coefficient that may differ from one class of shear flow to another.

In "mixing-length" models ν_e is defined in terms of $U_i(\mathbf{x})$ (so $I = 1$), so that (4.6b) is reduced schematically to

$$\widetilde{\mathcal{L}}_M^{(1)}(M_{2,1}; M_{1,1}) = 0. \qquad (4.8)$$

For example in boundary layer flows with mean velocity profiles $U_1(x_2)$, $\nu_e = L_s^2\left|\frac{\partial U_1}{\partial x_2}\right|$, where L_s is defined in terms of the distance from the "wall" x_2 (Prandtl 1925); in free shear layers and wakes with velocity profile $U(x_2)$, ν_e is defined in terms of the change in U across the mean flow and the scale H of the wake, $\nu_e = H^{-1}\int\left|\frac{\partial U_1}{\partial x_2}\right|dx_2$ where the integral is taken across the wake (Prandtl 1942). [This use of an integral of a modulus across a region of the flow to model an averaging effect, in this case the effect of large eddies, is a powerful technique (sometimes used in elasticity) that is often overlooked in turbulence modelling. It is discussed later in sec. 4.3]

In the K-L method, for shear flows $K = \frac{1}{2}\overline{u_i u_i}$ is calculated explicitly by an equation $\widetilde{\mathcal{L}}_M^{(2)}$ which is a contracted form of (4.5), using assumptions for $M_{3,1}$ as in (4.6). L_s is specified (so $I = 2$). Note that $\widetilde{\mathcal{L}}_M^{(1)}$ and $\widetilde{\mathcal{L}}_M^{(2)}$ are coupled, since terms in $M_{2,1}$ evaluated from (4.7) are substituted in the equation for $M_{2,1}$ in (4.6b).

In the K-$\mathcal{E}$ method, L_s is now calculated rather than being specified. It is assumed that the "functional" length scale L_s used in (4.6) has the same variation across the flow as the dissipation length scale $L_\mathcal{E} = \mathcal{E}/K^{3/2}$ (from 4.1). Thus $L_s \approx L_\mathcal{E}$, and so L_s is calculated from (4.6 (b) and (c)). [This is justifiable if the eddy structure is locally determined, as demonstrated in boundary layer flows near rigid surfaces over a wide range of Reynolds numbers ranging from those of atmosphere boundary layer experiments ($\sim 10^7$) to those of direct numerical solutions ($\sim 10^2$), e.g. Hunt et al 1989. It is not true in flows undergoing rapid acceleration or in the presence of strong buoyancy forces, as discussed below].

In the second order Reynolds Stress Transport Model (2RST), the full form of (4.6b) is computed for all 6 components $\overline{u_i u_j}$ (in general) and a seventh equation (4.6c) is required for computing $\mathcal{E}$.

Lumley (1978) and Zeman (1981) showed that the relation between third order moments and second order moments is not invariant, as is assumed in 2RST models (Launder et al 1975), especially where there are strong gradients of turbulence or strong buoyancy forces. They proposed modelling these third-order two-point moments as third-order one-point moments in a more elaborate third-order Reynolds Stress Transport Model (3RST), so (4.6) is replaced by

$$\widetilde{\mathcal{L}}_M^{(i)}(M_{2,1}; M_{m,p}) = 0, \text{ where } m, p = 1, 1; 2, 1; 3, 1 \quad (i = 1\text{–}6) \tag{4.9a}$$

$$\widetilde{\mathcal{L}}_\mathcal{E}(M_{2,2}; M_{2,1}, M_{1,1}) = 0. \tag{4.9b}$$

Then it is necessary to introduce transport equations for $M_{3,1}$, and closure approximations are used between $M_{4,1}$ and $M_{2,1}$. i.e. (4.9a) is supplemented with

$$\widetilde{\mathcal{L}}_M^{(i)}(M_{3,1}; M_{4,1}) = 0 \quad (i = 7\text{–}16)$$

and an algebraic relation connecting $M_{4,1}$ to $M_{2,1}$, namely

$$\widetilde{F}_i(M_{4,1}; M_{2,1}) = 0. \tag{4.10}$$

Note the much larger number of equations required by this higher order approximation.

(b) Deriving two-point moments

Methods for calculating cross correlations and spectra at two points have been developed for a limited range of flows. In rapidly distorted flows ($T_D \lesssim T_L$) or very low Reynolds number flows a linear analysis for the second order moments is valid (Deissler 1974, Hunt 1978, Townsend 1976), given the mean flow $U(x)$, and the linear form of body forces such as buoyancy e.g. Komori et al (1983). Unlike other models, this model can account for arbitrary inlet spectrum or anistropy of the initial or inlet turbulence (provided they are specified), and is not limited by the length scale of the turbulence (L) relative to that of the mean flow (H). The equation for $M_{2,2}$ can be written schematically as

$$\widetilde{\mathcal{L}}_M^{(i)}(M_{2,2}; M_{1,1}) = 0 \quad (i = 1\text{–}6) \tag{4.11}$$

In fact they are integro-differential equations; but the moments are more easily computed from solutions of each eigen-mode $E_n(x)$ of the velocity using the approach described in sec. 3 or Hunt (1978).

Since the conditions required for the validity of the linear theory are not satisfied in many flows of practical interest, it is necessary to use a non-linear method. As with models for 1-point moments, those for 2-point moments are also approximate; they are not restricted to the linear criterion of $T_{\mathbf{D}} \lesssim T_L$ but, at present are restricted to locally homogeneous flows. The best known is the Eddy Damped Quasi Normal Markovian (E.D.Q.N.M.; Lesieur 1990)

$$\widetilde{\mathcal{L}}_M^{(i)}\left(M_{2,2}; M_{m,p}\right) = 0; \qquad m, p = 1, 1; \text{ and } 3, 2 \tag{4.12}$$

$$\widetilde{\mathcal{L}}_M^{(i)}\left(M_{3,2}; M_{m,p}\right) = 0; \qquad m, p = 1, 1; \text{ and } 4, 2 \tag{4.13}$$

Bertoglio (1981) compared RDT and EDQNM results, which revealed which aspects of the turbulence structure are most sensitive to the non-linear effect in different flows.

The important new concept of Orzsag (1970) was to suggest that the fourth moment $M_{4,2}$ can be related to $M_{3,2}$ as well as to $M_{2,2}$ by a simple differential equation with a relaxation term i.e.

$$L_M^i\left(M_{4,2}; M_{3,2}, M_{2,2}\right) = 0 \tag{4.14}$$

(By including the $M_{3,2}$ term in (4.12), unphysical solutions where the energy tends to infinity are avoided (Proudman & Reid, 1954)). Note that all the above equations (4.6 − 4.12) are second order partial differential equations of elliptic form. They are all non-linear except for (4.11). In some cases of thin shear layers, the equations can be further approximated to parabolic form. Where the "mixing length" is expressed as an integral of $\partial U_i / \partial x_j$ as in (2.1)(and (4.21)) then they are integro-differential equations.

Another method for deriving these moments is to set up stochastic differential equations based on similar physical and mathematical assumptions. The unknown moments of the stochastic terms (e.g. variances of "white" noise acceleration terms) are defined by statistics of the turbulence, which are either estimated from measurements of Eulerian statistics or derived from the whole computation. There is some evidence that this method leads to faster computation, with less numerical error. The stochastic simulation of the velocity field statistics can readily be used for stochastic models of diffusion and reaction processes. As with models for one-point moments, stochastic models can be quite general, e.g. Haworth & Pope (1986), or they can be framed for particular classes of flows or for deriving particular statistics such as Lagrangian statistics in stably stratified flows (Pearson et al 1983).

4.2.3. Statistical boundary conditions

By these approximations and truncation procedures sets of differential equations have been obtained to calculate the flow up to the required level of multi-point statistical moments $M_{k,p}$. In order to solve the equations it is first necessary to specify initial and boundary conditions. Where a turbulent flow enters the computational domain all the moments in the equation up to order k i.e. $M_{k',p'}$ ($k' \leq k$; $p' \leq p$) need to be specified. (This can pose a restriction on the use of these models since such detailed information may not be available). If the flow becomes turbulent within the computational domain (as in flow over an aerofoil) it is necessary to specify a subdomain where and when the flow is turbulent and to specify the appropriate moments on the boundaries of this domain in space or time, (e.g. at the transition zone) [Savill, 1991].

At the boundaries of a turbulent flow where there is no inflow, such as at a rigid surface or a density interface, it is also necessary to specify the statistical moments. For example at a rigid surface all the moments of the velocity are zero, though moments involving the gradients of velocity (such as dissipation rate) and/or fluctuation of pressure are not zero. Thus the simple kinematic boundary conditions (such as $\mathbf{u} = 0$ or $\mathbf{u} \cdot \mathbf{n} = 0$) do not uniquely specify all the moments required by the models. Local analysis is necessary to specify the conditions on other variables such as $\mathcal{E}$ and p. There is also the theoretical difficulty, even where these moments are known or estimated at the boundary's surface,

that the approximations used in deriving the approximate equations for the $M_{k,p}$ moments
are not valid in the region near the boundary. Therefore the kinematic boundary conditions
cannot in general be applied to the statistical turbulence models at the boundary. Local
dynamic analysis is also necessary.

Near a rigid surface (say at $x_2 = 0$) the local dynamic analysis is simplified by the
fact that the Lagrangian time scale of the turbulence normal to the wall $T_L^{(2)}$ tends to
zero, so that $T_L/T_D \to 0$. Then the production of turbulence energy $(-\overline{u_1 u_2} \partial U_1/\partial x_2)$ is
approximately equal to the rate dissipation $\mathcal{E}$ (Townsend 1961), because eddy transport
and advection of turbulence by the mean flow are negligible. Therefore the turbulence
(except for very large scales) and mean flow should have a universal local form, that is
independent of the large scale features of the flow.

This implies that these moments should be defined by a single dynamical parameter
usually taken to be $u_* = \sqrt{\tau_0/\rho}$, where τ_0 is the surface shear stress. The other parameters
are the kinematic viscosity ν and the nature of the surface which can be parametrised by
a local length scale l_s. If the surface has small roughness elements of scale y_0, then $l_s \sim y_0$
provided $u_* y_0/\nu \gtrsim 0$. If $u_* y_0 \lesssim 1$, these elements are submerged in the viscous surface
layer of thickness $l_s \sim \nu/u_*$. If not, roughness elements determine the flow near the surface
over the scale of this "roughness length" y_0 provided $u_* y_0/\nu \gtrsim 1$. Otherwise the surface
is "smooth" and there is a layer next to the surface where $x_2 \lesssim \nu/u_*$ and the Reynolds
number of the velocity fluctuations R_e is small and viscous effects are significant. In
both cases at a height $x_2 > l_s$ above the surface the Reynolds number R_e of the velocity
fluctuations is large, but in the latter case if $x_2 < l_s \sim \nu/u_*$, the value of R_e is small.

Since statistical turbulence models are only strictly valid if $R_e \gg 1$, then boundary
conditions should only be applied far enough above the surface that the viscous processes
are weak, but close enough to the surface, relative to the overall scale of the flow, that the
statistics of the turbulence (or at least some of them) have a universal structure. Strictly
this only occurs in flows where u_* is greater than any velocity fluctuations (w_*) produced

by body forces (such as thermal convection), and any time dependence of the mean flow occurs with a frequency $\omega \ll u_*/l_s$, and the pressure gradients along the surface are weak enough (relative to $\rho u_*^2/l_s$). Both experiments (especially those at highest measured Reynolds numbers in atmospheric turbulence) and the limiting forms of the approximate moment equations show that $x_3/l_s \gg 1$ (where $l_s = \nu/u_*$ or z_0, for smoooth or rough surfaces) and $x_3/H \ll 1$,

$$\overline{u_i^2}/u_*^2 \sim \gamma_1^2, \quad \overline{u_2^2}/u_*^2 \sim \gamma_2^2, \quad \overline{u_3^2}/u_*^2 \sim \gamma_3^2 \tag{4.15}$$

$$\mathcal{E} \sim u_*^3/(\kappa x_2), \quad U_1(x_2) \sim \frac{u_*}{\kappa}\left(\ln(x_2/l_s) + B\right) \tag{4.16}$$

where γ_1, γ_2, γ_3, κ, B are approximately universal ($\simeq 2.5$, 1.3, 2.0; 0.4; $B = 5.5$ and 0 on smooth and rough surfaces respectively). (Townsend 1976 Chap. 5). These limits are applied partially (e.g. only 4.15) or completely, and in different ways in different computational models. Analytical models of turbulent boundary layers may approach these limits asymptotically, or else (4.15) may be applied as a boundary condition to a numerical solution at a particular grid point (e.g. Launder & Spalding 1972). Analytical computational and experimental studies show that the value of x_2 at which the limiting solutions are valid decreases as the external flow departs more significantly from a zero pressure gradient boundary layer (e.g. Belcher *et al* 1991). For separated flows the logarithmic profile (4.15) may not exist at all (Dengel & Fernholz 1990).

In recent computational modelling of turbulent shear flows near rigid surfaces it has become customary to postulate that, with the addition of some extra viscous terms to the $\mathcal{E}$-equation and Reynolds stress equations, the model equations for 1-PM (e.g. 4.5, 4.6) for high Reynolds number turbulence can be applied even at the surface $x_2 = 0$. Supporting evidence has been provided by the results of D.N.S. Then the equations are solved in the whole flow domain ($x_2 \geq 0$) with the boundary conditions on $x_2 = 0$,

All authors use the fact that $\overline{(u_i)^{k'}} = 0$ for $k' = 1, 2$. There have been several suggestions for the form of the conditions on ϵ, for example $\epsilon = \nu(\partial u_i/\partial x_j)^2 = \beta_1 u_*^4/\nu$ and $\partial^2 \epsilon/\partial x_2^2 = \beta_2 u_*^6/\nu^3$ or $\int_0^\delta \epsilon dx_2 = \int_c^\delta \nu_e(\partial U_1/dx_2)^2$ where β_1 and β_2 are assumed to be universal constants, and δ is of the order of H the shear layer thickness, that might be determined by experiment or from the results of D.N.S. [Mansour et al 1988, Durbin 1991].

4.3. Critical aspects of statistical models (1-PM, 2-PM).

4.3.1. Response of turbulence to homogeneous mean straining flows.

Just as in the kinetic theory of gases, the most important component of the (such as $\overline{u_i u_j}$) change when the turbulence is distorted by various kinds of mean flow $U_i(\mathbf{x})$. As explained in sec. 4.1 the models assume that over the integral scale of the turbulence, L, the gradients $\partial U_i/\partial x_j$ of mean velocity are effectively uniform, so that in a shear flow $\left| (\partial U_1/\partial x_2) \middle/ \left(\frac{\partial^2 U_1}{\partial x_2^2} \right) \right| \gg L$.

Even when this assumption is satisfied (A4 in the list of sec. 4.1), other factors also affect the ability of models to predict the values of $\overline{u_i u_j}$ (and thence the mean profiles derived from $\overline{u_i u_j}$ in (4.6(a)); to provide a measure of their significance, appropriate dimensionless quantities are introduced.

(i) First, consider all magnitudes and forms of homogeneous straining motion whose strengths, S, are normalised on the natural timescale of the turbulence T_L. S^* is defined as

$$S^* = T_L S \tag{4.17a}$$

where $S = \left[\epsilon_{ij}\epsilon_{ij} + \frac{1}{2}\Omega_k^2 \right]^{\frac{1}{2}}$ and $\epsilon_{ij} = \frac{1}{2}\left(\frac{\partial U_i}{\partial x_j} + \frac{\partial U_j}{\partial x_i} \right)$, $\Omega_k = \epsilon_{ijk}\frac{\partial U_i}{\partial x_j}$. \hfill (4.17b)

For example in a shear flow $U_1(x_2)$, $S = |\partial U_1/\partial x_2|$, while in a pure two-dimensional mean straining flow where $\mathbf{U} = (U_1(x_1), U_2(x_2))$,

$$S = \left[\left(\frac{\partial U_1}{\partial x_1}\right)^2 + \left(\frac{\partial U_2}{\partial x_2}\right)^2\right]^{\frac{1}{2}} \tag{4.17}$$

The different forms of straining correspond to rotational and/or irrotational straining functions, which can be characterised succinctly by the parameter

$$\tilde{\pi} = \left(\epsilon_{ij}\epsilon_{ji} - \frac{1}{2}\Omega_k^2\right)/S^2, \tag{4.18}$$

where $-1 < \tilde{\pi} < 1$. Note that $\tilde{\pi} = -1$ is pure rotational motion (e.g. $U_1 = \Omega x_2$, $U_2 = -\Omega x_1$), $\tilde{\pi} = +1$ is pure strain and $\tilde{\pi} = 0$ is pure shear. In the subsequent discussion it is useful to define shear flows denoted by (Sh) as occurring when $|\tilde{\pi}| \ll 1$, or specifically when $|\tilde{\pi}|$ is less than a critical value, say $|\tilde{\pi}|_{Sh} \simeq 0.1$. Then non-shear flows, denoted by (NSh), occur when $|\tilde{\pi}| > |\tilde{\pi}|_{Sh}$ [Wray & Hunt, 1990; Kevlahan (private comm)].

(ii) Second, consider the forms of the initial energy spectrum $E_0(k)$. In most well developed turbulent flows $E_0(k)$ usually has a single peak (as in fig. 6), because one set of large eddies eventually dominates.However in the turbulent flow formed from motions with different scales, such as boundary layers disturbed by wakes of upwind obstacles, $E(k)$ may contain two maxima.

(iii) Third, the initial anisotropy of the turbulence has to be considered. For the second order moments (Lumley 1978) this is most conveniently defined by the second $II_{()}$ and third invariants $III_{()}$ of the anisotropy tensors of the mean moments of *velocity* b_{ij} and of the wave number distribution in the spectrum, c_{ij}

$$\text{viz } II_{(b)} = b_{ij}b_{ji}, \qquad III_{(b)} = b_{ij}b_{jk}b_{ki}, \text{ where } b_{ij} = \frac{\overline{u_i u_j}}{\overline{u_k u_k}} - \frac{1}{3}\delta_{ij}. \tag{4.19a}$$

$$\text{and } II_{(c)} = c_{ij}c_{ji}, \qquad III_{(c)} = c_{ij}c_{jk}c_{ki}$$

$$\text{where } c_{ij} = \frac{1}{\overline{u_m^2}} \int \frac{k_i k_j \Phi_{pp}(k)\, d\kappa}{k_l^2} \quad - \frac{1}{3}\delta_{ij} \tag{4.19b}$$

[This particular tensor has been used by Kida & Hunt (1989) and Mansour et al (1991) to analyse the sensitivity of rapid distortion to the initial anisotropy of turbulence in wave-number space. Lumley (1978) used a similar measure d_{ij} defined in terms of $(\partial u_i/\partial x_j)^2$.]

For example in isotropic turbulence $II_{(b)} = III_{(b)} = 0$, in one dimensional turbulence $(\overline{u_2^2}/\overline{u_1^2} \to 0,\ \overline{u_3^2}/\overline{u_1^2} \to 0)$, $II_{(b)} = 2/3$, $III_{(b)} = 2/9$, whereas in axisymmetric two dimensional turbulence $(\overline{u_1^2}/(\overline{u_2^2} + \overline{u_3^2}) \to 0,\ \overline{u_2^2} = \overline{u_3^2})$, $II_{(b)} = 1/6$, $III_{(b)} = -1/36$ (see fig 8). Similarly for turbulence that is highly flattened in one direction (so that the length scales are small in this direction), $II_{(c)} = 2/3$, $III_{(c)} = 2/9$ and for turbulent scales that are highly elongated in one direction and equally flattened in the other two, $II_{(c)} = 1/6$, $III_{(c)} = -1/36$.

From the many experimental, theoretical and computational studies of how different kinds of turbulence responds to gradients in the mean flow, the relative performance of different models can be reviewed as a function of the ratio of the time T_D of the distortion to the time scale of the turbulence T_L (which varies as the flow develops) and in terms of the nature of the mean straining, defined by S^* and $\tilde{\pi}$. A summary is presented in table (2), for four kinds of model; EVM (based on eddy viscosity models of stresses (equations 4.6 & 4.7), 2 RST (second order Reynolds stress models (as in equation 4.6) for one-point moments), the two-point model of R.D.T. (Rapid Distortion Theory) which is linear (but can be non local) and the non-linear EDQNM (Eddy Damped Quasi Normal Markovian).

When the mean strain is *irrotational*, such as in a wind tunnel contraction, the different components of turbulent vorticity and velocity are amplified or diminished. The resultant ratio of the components is approximately independent of the form of the initial spectrum (an exact result for RDT), but depends sensitively on the initial ratio. Similarly the form of the spectrum does not change significantly (e.g. if $E(k) \propto k^{-p}$, the exponent is constant during the distortion from linear theory). This memory persists even during the dissipation of energy and the non linear transfer of energy between components. Consequently there can be no *general* relation between the *local* Reynolds stress $\overline{u_i u_j}$ and the mean strain rate $\left(\frac{\partial U_i}{\partial x_j} \right)$ of the form of (4.7). Even the *sign* of the significant change of the stresses is only correct if the same mean strain rate in time or space extends for long enough that the fluid elements experience the strain for a time $T_D \gtrsim T_L$. In many flows of practical importance such as turbulence in a double pipe-bend or flow over a hill, this condition is not satisfied and the use of E.V.M. even leads to serious errors in the mean velocity profile (4.7) (e.g. Launder, 1989).

While 2RST models are satisfactory for most irrotational strains, in the case of very strong strains where $S^* = T_L S \gg 1$, they do not lead to the same results for $\overline{u_i u_j}$ or $\overline{p \partial u_i / \partial x_j}$ as that of the exact linear RDT theory or of the results of D.N.S. There are discrepancies even for initially isotropic turbulence, but they are greater if the initial state of the turbulence is significantly anisotropic. The reason is that the approximations for $\overline{p \partial u_i / \partial x_j}$ do not allow for the *history* of the strain on the velocity gradients in the turbulence, which should be modelled over a period of order T_L during the strain (using the same form as discussed later in (4.20b)). Introducing such additional complexity may not be necessary for most flows found in practice because the strain changes slowly over a period T_L.

When the mean flow field is a pure shear flow (e.g. $U_1(x_2)$), i.e. $\widetilde{\pi} = 0$, it is found that in all models the Reynolds shear stress (e.g. $-\overline{u_1 u_2}$) has the correct sign and approximately the correct magnitude. However the change of the variances of the different velocity components $\overline{u_1^2}, \overline{u_2^2}, \overline{u_3^2}$ are not given correctly by the eddy viscosity model (4.7), whereas 2RST and 2-point (RDT or EDQNM) can all predict these quantities. These methods also have limitations, even for this idealised flow; for example there may be some inaccuracy in the modelling of the *initial* change of $\overline{u_2^3}$ and $\overline{p\partial u_3/\partial x_3}$ in shear flows ($T_{\mathbf{D}} \ll T_L$), using 2RST which is only significant if $S^* \gtrsim 1$ (Maxey 1982); linear RDT models are only accurate for these and other distorted flows when $S^* \gg 1$ (Lee *et al* 1990). However if $S^* \lesssim 1$ and $T_{\mathbf{D}} \gtrsim T_L$, the linear theory may still be useful for estimating two-point moments, especially the effects of body forces and compressibility. A semi-empirical correction for dissipation can improve the accuracy of these estimates (e.g. Townsend 1976).

In shear flows, as in irrotational straining flows, the ratios between the Reynolds stresses are not sensitive to the form of the initial spectrum. However *unlike* irrotational strains, in shear flows the ratios between Reynolds shear stress and the kinetic energy (e.g. $-\overline{u_1 u_2}/K$) tends to a value that is rather insensitive to the initial anisotropy when $T_{\mathbf{D}} S \gtrsim 1$. This is one reason why, although EVMs do not account for the history or initial anistropy, such models are more satisfactory in slowly varying shear flows than in flows with irrotational strain.

The modelling of kinetic energy K is also more satisfactory, because in shear flows even at moderate values of R_e ($\sim 10^2$) the turbulence tends towards a self-similar structure as manifested by the energy spectrum which tends to an approximately universal form $E(k) \propto k^{-p}$ where $p \dot{\sim} 2$ (Hunt & Carruthers 1990; Ho & Huerre 1984). With this spectrum the dissipation rate $\mathcal{E}$ is determined by small scale motions, which is the basis for the turbulence models.

The third reason why, in shear flows, 1-point models for moments are satisfactory is that shear tends to reduce the length scale and so the straining is effectively homogeneous over the scale of the eddies (c.f. (2.1)), one of the key assumptions (A4) of these models.

4.3.2. Energy transfer of between different velocity components

In locally homogeneous turbulence the role of pressure (expressed as joint moments with the velocity gradients e.g. $\overline{p\partial u_i/\partial x_j}$) is to change the energy of different velocity components, with no net contribution to the total energy (see equations (4.5). In the absence of mean strain$\left(\frac{\partial U_i}{\partial x_j} = 0\right)$, $\overline{p \partial u_i / \partial x_j}$ is only non zero in anisotropic turbulence.

The dynamical process affecting the anisotropy of turbulence can be understood physically in terms of the vortex lines of turbulence being stretched, rotated and advected by the velocity fluctuations induced by other vortex lines. Clearly if the vorticity is *slightly* anisotropic the chaotic motion *tends* on average to reduce this anisotropy. But there is no reason why the anisotropy should tend to zero (Townsend 1956). (A simple counter example is an anisotropic distribution of line elements evolving in a chaotic flow containing both rotation and irrotational motions; the longer fluid elements tend to be stretched out more than the shorter ones and the final results is less anisotropic than the initial state but not completely isotropic (Kida & Hunt 1989)).

Recent experimental and computational studies also show that the interaction of vortices does not lead to random scattering of their orientation; the general structure of the groups of vortices remains remarkably stable, because, even though they split up, during the interactions they re-connect with each other afterwards in a way that depends sensitively on their initial orientations and mutual strength [e.g. Melander & Hussain 1990, Kiya *et al* 1986, Aref *et al* 1991].

This suggests that in the absence of strain the anisotropy of the velocity moments, defined in (4.19a) should *decrease* at a rate that depends on b_{ij}; the larger b_{ij} the faster the decrease. The type of anisotropic motion, which can be quantified by the values of the invariants $II_{(b)}$, $III_{(b)}$ of (4.19), also affects this decay rate. If the turbulence is *locally* approximately two-dimensional (e.g. an assembly of vortex rings or tubes), $II_{(b)}$, $III_{(b)}$ have the same value as for two dimensional turbulence (i.e. $1/6$, $-1/36$). Because of the stability of these motions, the turbulence tends very slowly to three dimensional isotropy (i.e. $|db_{ij}/dt|$ is small).

On the other hand if the turbulence consists of plane vortex sheets (i.e. locally a one-dimensional flow where $II_{(b)}$, $III_{(b)} = 1/3$, $2/3$) these motions are so *unstable*, $\left(|db_{ij}/dt|\right)$

is largest [an explanation due to A.A. Townsend]. An excellent demonstration of this dependence on $II_{(b)}$, $III_{(b)}$ was provided by Michard et al (1987) in an experiment of decaying anisotropic turbulence in a wind tunnel. In principle the variation in b_{ij} also depends on the anisotropy of *vorticity* and, generally, on the gradients of velocity, which are defined by the second anisotropy tensor c_{ij} (4.19b). However the most widely used, "Rotta", model for this "inter-component energy transfer", only allows for the anisotropy of the moments viz:

$$\frac{db_{ij}}{dt} = -\frac{C_1 b_{ij}}{T_L}, \tag{4.20a}$$

where T_L is the local timescale of the turbulence ($\sim K/\mathcal{E}$) and C_1 is a coefficient (> 0). Rotta's (1951) equation was generalised by Lumley (1978) and then by others (see Launder 1989) to account for the dependence of C_1 on $II_{(b)}$, $III_{(b)}$. Note that C_1 must be zero for two dimensional turbulence.

Recently direct numerical simulations of the decay of strained anisotropic turbulence by Lee & Reynolds (1985) and theoretical calculations by Weinstock (1987) showed that C_1 also depends effectively on $II_{(c)}$ and $III_{(c)}$ A further criticism of (4.20a) is that it is based on the assumption that b_{ij} decreases as a function of the *local* value of b_{ij}, whereas it takes a certain time ($\sim T_L$) for the random chaotic motions to affect b_{ij} (by analogy with turbulent diffusion). Using the result of the *exact* model calculation of Kida & Hunt (1989) (where, for a part of the spectrum, $C_1 \propto \frac{t}{T_L}$), suggests that (4.20a) should have the form

$$db_{ij}/dt \approx -\frac{C_1'}{T_L(t)} \int_{-\infty}^{t} b_{ij}(t') \exp\left(-(t - t')/T_L(t')\right) dt' \tag{4.20b}$$

where the integral is taken along the mean streamline. In rapidly changing flows, this relaxation equation implies that because the tendency to isotropy is slower than predicted by (4.20a), which leads to calculations of the variances which are more in agreement with the R.D.T. limit. Note that if C_1 or C_1' depends on $II_{(c)}$ and $III_{(c)}$, since these invariants depend on how the wavenumbers of the turbulence are distorted by the mean straining flow, it also means that C_1' is different in different kinds of straining flow.

4.3.3. Modelling the inhomogeneous structure of turbulence

In many practical turbulent flows the assumption A4 (§4) of the local homogeneity is not satisfied. Many turbulence models wholly ignore this limitation, as in mixing length models (4.7), or partially ignore it as in K-$\mathcal{E}$ or K-L models where, although a local Reynolds stress relation (4.7) is assumed, the inhomogeneous effects of transport of K (and of $\mathcal{E}$) by the gradients of turbulence are considered (as in (4.5)). In 2RST (and 3RST) inhomogeneous effects are treated for $\overline{u_i u_j}$, K and $\mathcal{E}$ (and third moments).

It is instructive to recall some of the main phenomena that occur when turbulence is significantly inhomogeneous and some of the specific models developed for these phenomena. These may suggest ways of improving the modelling of inhomogeneous effects in general models.

When turbulence is generated locally as in a boundary layer or in a wake or above an oscillating grid in a tank, the rotational motions of the turbulence are confined within a volume with a random interface and outside this volume there are random irrotational motions. This region of fluctuating motion spreads into the quiescent region in a diffusive-like way, so that it is appropriate to model the transport terms in the K equation

$$-\frac{\partial}{\partial x_2}\left(\overline{pu_2} + \frac{1}{2}\overline{u_2 u_2^2}\right) \quad \text{as} \quad \frac{\partial}{\partial x_2}\left(\frac{K^2}{\mathcal{E}}\frac{\partial K}{\partial x_2}\right).$$

(Clearly the boundary conditions are that K and $\mathcal{E} \to 0$ as $x_2 \to \infty$, but the results of computations, using the K-$\mathcal{E}$ equations, are sensitive to the ratio of $K/\mathcal{E}$ as $x_2 \to \infty$. Consequently numerical procedures can have an important influence on the results (Stuttgen & Peters 1987)). A remarkable analysis by Lele (1985) shows that the combined non-linear K-$\mathcal{E}$ equations have an exact analytical solution for this problem, and that this solution has the correct physical properties *provided* the coefficients used in the model equation lie within a narrow range of values, which in fact correspond to those that had been established over many years by empirical adjustment. [This model does not describe the irrotational fluctuations which decay algebraically (in proportion to x_2^{-4}) (Phillips 1955) and not exponentially as assumed in the model]. Therefore a diffusion model is satisfactory for the outer region of most shear layers.

The reason why the diffusion analysis is satisfactory here is that the turbulence length scale L is smaller than the inhomogeneity length which in this case is defined by the distance over which the turbulence energy changes, because L_I is of the order of H, the thickness of the turbulence layer,

$$\text{i.e. } L_I = \left| \frac{\partial}{\partial x_2} K \right| \bigg/ \left| \frac{\partial^2}{\partial x_2^2} K \right| \sim H \ .$$

However there are many practically important flows where L is larger than L_I, which we now briefly consider. When large scale free stream turbulence (FST) ($x_2 > 0$) impinges onto a low turbulence or laminar shear layer (SL), ($x_2 < 0$), downstream of say a turbine blade leading edge at $x_1 = 0$, velocity fluctuations are induced in the SL by vortices in the FST. This leads to the transport term in the SL, $\overline{p \frac{\partial p u_2}{\partial x_2}}$ being of the order of $\frac{K^{\frac{3}{2}}}{L} f \left(\frac{x_2}{L} \right)$, where f is a decreasing function of x_2/L. This is not a diffusive term. Once these Reynolds velocity fluctuations are induced, the Reynolds stress $(-\overline{u_1 u_2})$ grows in the SL. Theory shows that the pressure fluctuations in the SL extend up into the FST and produce a small negative Reynolds stress (i.e. $\overline{u_1 u_2} > 0$). This contradicts the concept of diffusion of Reynolds stress which would lead to $-\overline{u_1 u_2} > 0$ in the FST. [Gartshore et al (1983)]. [In terms of a local eddy viscosity this implies that $\nu_e \to -\infty$ in the FST!]

Sometimes the gradients in the mean flow, such as the shear $\partial U_1/\partial x_2$, vary over a distance of order L or less. Then the Reynolds stress $(-\overline{u_1 u_2})$ is neither proportional to $\partial U_1/\partial x_2$ nor has the same sign (i.e. $\nu_e < 0$). Clearly in such flows it is also not correct to assume that the dissipation length scale $L_{\mathcal{E}}$ is related to the local value of $\partial U_1/\partial x_2$. These situations are typically found in asymmetric shear layers (Palmer & Keffer 1972) or where two kinds of shear layer merge into each other. One such example is the interaction of a wake and a boundary layer that is a notable feature of modern aircraft design [Zhou 1985].

A simple model for the shear stress in these complex shear layers, that allows for the rapid variations in $\partial U_1/\partial x_2$ over the scale L, is to generalise the mixing length model of (4.7) by relating the Reynolds stress $-\overline{u_1 u_2}$ or the dissipation length $L_{\mathcal{E}}$ to the average value of $(\partial U_1/\partial x_2)$ taken over the length scale of the turbulence, say $L_{\mathcal{E}}$ or L (see Schlichting 1960, p.481). So that in (4.7) or (4.1) $\partial U_1/\partial x_2$ is replaced by

$$\left\langle \frac{\partial U}{\partial x_2} \right\rangle (x_2) = \frac{1}{\lambda L_{\mathcal{E}}} \int_{x_2-\lambda L_{\mathcal{E}}/2}^{x_2+\lambda L_{\mathcal{E}}/2} G(x_2, x_2')(\partial U/\partial x_2)(x_2')\, dx_2' \qquad (4.21)$$

where the filter function could be $G = 1$ or, as Dr Spalart has suggested $G = \left(e^{-(x_2-x_2')^2/\mu L_{\mathcal{E}}^2} \right)/\sqrt{\pi \mu}$, where $\mu \simeq 1$. This method requires a few iterations, since the averaging distance $L_{\mathcal{E}}$ is derived implicitly. This approach has been combined with the effects of wall boundary conditons, another source of inhomogeneity that is considered next.

(b) <u>Inhomogeneity of boundary conditions.</u>

The second major cause of inhomogeneity in turbulent flows is the effect of boundaries where the normal fluctuating velocity is imposed by certain external conditions such as the presence of a rigid surface at $x_2 = 0$ or a density interface, where the turbulent flow may have to match with wave motions. Recent research has shown how the mean flow over the surface or interface, and the shape and scale of the surface geometry, both strongly affect the structure of the turbulence near the surface and interface (hereafter called surfaces).

For flow over flat or gently sloping surfaces, as in boundary layers over rigid surfaces, wind over hills or water flow beneath the surface driven by wind, the travel time $T_{\mathbf{D}}$ is large compared to the local turnover time $T_L \simeq L_1^{(2)}/\sqrt{\overline{u_2^2}}$ for the normal component of turbulence or for normal Reynolds stresses). For these flows where turbulence is produced near the surface it is found that at *high Reynolds number* $L_1^{(2)} \sim 0.4x_2$ and $\sqrt{\overline{u_2^2}} \sim 1.3u_*$ (for $H \gg x_2 \gg l_s$, where $l_s = \nu/u_*$ or y_0) and therefore the length scale and strength of the normal components of the eddies (which help determine the Reynolds stress) are only weakly affected by the inhomogeneity caused by the wall. Therefore local statistical analysis using 1-PM is appropriate. ["Wall" corrections to 2RST 1-PM have been proposed by Launder, Reece & Rodi (1975), but they are often neglected in calculations because they are ambiguous and do not invariably add to accuracy, e.g. Newley (1986).] For these flows a model for 2-PM and spectra based on conical eddies has been developed by Perry & Abell (1975) which leads to predictions for1-PM for $\overline{u_1^2}$.

However if in the turbulent flow over the surfaces the mean velocity relative to the surface is zero, then there is no production of turbulence energy by the mean shear, the energy is supplied by transport from elsewhere (as in the turbulence just below the water surface in the river flow over a bed or grid turbulence over a plate or moving wall [Thomas & Hancock 1977]) or by thermal convection (Hunt 1984). In this case it is found that $L_1^{(2)} \simeq x_2$ and $\sqrt{\overline{u_2^2}} \propto \mathcal{E}^{\frac{1}{3}} x_2^{\frac{1}{3}}$. Therefore T_L still tends to zero at the surface and $T_L \ll T_\mathbf{D}$. But in this case the eddy structure is largely determined by its interaction with the surface by "blocking" and is highly inhomogeneous. Analyses have been developed by Gibson & Launder (1978) using an adaption of 2RST 1-PM to allow for these wall effects, which, for example, have been successfully applied to model turbulence near a water surface by Gibson & Rodi (1989). For this case a 2-PM has been developed by Hunt(1984) which agrees well with measurements of spectra and of cross correlations of u_2 at two values of x_2. See also Hunt *et al* 1989.

A useful development has been made to account for the effects of both these kinds of surface boundary layer (i.e. with and without shear) on the turbulence, while retaining the simpler framework of 1-PM EVM models of turbulence statistics. The length scales

for the normal component of turbulence that determines the functional length scales for

shear stress (L_s of (4.7)) and energy dissipation of ($L_{\mathcal{E}}$ of (4.1)) can be modelled in terms

of the distance from the surface x_2 and the average mean shear over the scale of the eddy

$\langle \partial U_1 / \partial x_2 \rangle$,

$$\text{viz } L_1^{(2)-1} \simeq L_s^{-1} \simeq L_{\mathcal{E}(K)}^{-1} = \mathcal{E} / \left[\left(\tfrac{1}{4} K \right)^{\frac{3}{2}} \right] = A_B / x_2 + A_s \langle \partial U_1 / \partial x_2 \rangle / \sqrt{(-\overline{u_1 u_2})} \quad (4.22a)$$

This relation (with $A_B = 0.6$, $A_s = 0.75$) in combination with the equations (4.5) for K,

and (4.22) (Belcher *et al* 1991) has been used for computing perturbed turbulent shear

layers including separated boundary layers. Since the normal component has a smaller

length scale than K it determines L_ϵ more directly than K. This explains why (4.22a)

should be modified to become a more specific (and more general) relationship involving

the *local* velocity gradient and the variance $\overline{u_2^2}$, viz.

$$L_{\mathcal{E}(2)}^{-1} = \mathcal{E} / \left(\overline{u_2^2} \right)^{\frac{3}{2}} = A_B / x_2 + A_s \langle \partial U_1 / \partial x_2 \rangle / \sqrt{\overline{u_2^2}}, \qquad (4.22b)$$

where $A_B \simeq 0.27$, $A_s \simeq 0.46$. This has been found by Spalart (unpublished) to agree well

with the computations of $L_{\mathcal{E}}$ in direct numerical simulations of an oscillating turbulent

boundary layer (fig. 9). By using the spatial average of $\partial U_1 / \partial x_2$ (in 4.21) this model for

$L_{\mathcal{E}}$ is satisfactory even where $U_1(x_2)$ has a maximum and $\partial U_1 / \partial x_2 = 0$ (Fig. 9).

In contrast to flow over a flat surface, when a turbulent flow with mean velocity U_0

impinges on a bluff body with diameter H, the length scale of turbulence L_1 is determined

by the upwind flow and is not simply related to the distance from the surface of the

body. This is because the travel time of the turbulence around the bluff body T_D ($=$

H/U_0) is generally much less than T_L ($\sim L/\sqrt{\overline{u_1^2}}$) and so the turbulence is not in local

equilibrium. When $H \gtrsim L$, the turbulence varies from its form near the surface of the body

to its upstream form within a distance that may be significantly less than the scale of the
turbulence. This is an extreme example of inhomogeneity and 1-point models are seriously
in error; for example they all predict an *amplification* of $\overline{u_1^2}$ (and K) on the stagnation line
caused by production of energy (where x_1 is in the flow direction), whereas, when $H \lesssim L$,
$\overline{u_1^2}$ decreases because of the blocking effect [e.g. Britter *et al* 1979]. Although no 1-PM
models are correct for this case (and the model of (4.7) is not appropriate), 2-PM such as
rapid distortion theory (RDT) can be used, though the theory has still to be converted
into a practical code for a wide range of length scales and flow fields (Hunt *et al* 1990).
At present the most general method for these problems is to use Large Eddy Simulation,
which Murakami *et al* (1991) have shown compares well with experimental measurements.

It is important to note that in most flows (including those over bluff bodies) where
$T_D \lesssim T_L$, there is insufficient time for the turbulence to affect the *mean flow* and therefore
an erroneous turbulence model has little effect on the *mean* flow. Thus, fortuitously, in
most turbulent flows 1-PM models of turbulence only affect the mean flow calculations
where the models are most appropriate (namely in shear flows where $T_L \lesssim T_D$). (c.f.
Belcher *et al* 1991). Equally the calculation of the diffusion of a scalar in these RCT flows
is quite insensitive to the turbulence model provided the upwind turbulence and the mean
flow are known. (Hunt 1985). But if the pressure fluctuations caused by the interaction
of the turbulence and the mean flow are to be modelled, then calculations based on 2-PM
models or LES are required [Hunt *et al* 1990].

§5. Conclusion.

In this paper I have attempted to relate recent fundamental studies of turbulence to
the different developments in the modelling of turbulent flows. So far such studies have not
provided significantly new techniques for practical turbulence models (though there have
been some detailed improvements), because these models have usually only been applied

to flows that are essentially perturbations of shear flows, for which (as explained here) the 1-PM models can be quite satisfactory.

But when these 1-PM models are applied to turbulent flows, that are highly inhomogeneous or where the strain is not simply pure shear (e.g. highly irrotational or being pure rotation) or are rapidly changing, they begin to fail, as has been recognised in other reviews (e.g. Launder 1989). It has been proposed here that more fundamental or, at least more complex approaches are fruitful, because firstly they indicate where and why current models fail and secondly they indicate possible developments, such as using 2-PM or DNS to improve the modelling of the $M_{2,1}$, $M_{3,1}$ pressure-strain terms in 2RST (equation 4.5), or using history-dependent models, (4.20b), or spatial averages $\langle \ \rangle$, (4.21) to allow for effects that are non-local in time or space.

An important practical aspect of 1-PM turbulence models that has only been touched on here is their applicability when the Reynolds number is only just high enough for the flow to be turbulent but too low for the assumptions A1–A4 to be generally valid. Generally in those low-R_e flows that have been successfully modelled, the mean flow is *highly sheared*. The success of these *ad hoc* models is partly explained by the fact that there are only small differences between the statistics of the turbulence such as $-\overline{u_1 u_2}/\overline{u_2^2}$ and $\mathcal{E}L_1^{(2)}/(\overline{u_2^2})^{\frac{3}{2}}$ between low and high Reynolds number forms of turbulence as the results of experiments and DNS demonstrate. (Hunt & Carruthers 1990). Therefore it is reasonable to expect that turbulence models may be used reliably in shear flows at low Reynolds number (especially if judiciously corrected). But in unsheared turbulent flows, where the spectrum of turbulence is not so closely controlled by the shear, it is less likely that simple modifications to the models can be developed which are applicable in a wide class of turbulent flows.

This study has also emphasised how many kinds of turbulent flow are sensitive to the form of initial and boundary conditions, even a long distance from the region of primary energy input, (e.g. wakes), and that certain parts of a flow (e.g. the outer part of boundary layers) or certain components of a flow (e.g. large scale fluctuations parallel to the surface) are much more sensitive than others. (This is an example of how most general statements about the validity and practicality of models have to be tightly qualified). This sensitivity of *real* turbulence should mean that models of turbulence are equally sensitive to small changes in the specification of initial or boundary conditions. The results of the test cases T1 and T3 of the ERCOFTAC workshop certainly demonstrated this sensitivity, but not always correctly.

Finally it needs emphasising that in future an increasing number of practical flow problems will involve modelling individual realisations of the flow, and there is at least as much scope for devising practical methods for these kinds of turbulence as for one point moments.

Acknowledgements

I am grateful to the organisers of the ERCOFTAC workshop for inviting me to give the introductory lecture, and for help with the text from A.M.Savill & V.Saxena. P. Spalart has kindly allowed me to use the graph he derived at NASA AMES. It is a pleasure to acknowledge that many of the ideas and results here came from joint research with colleagues in Cambridge, Boulder, Stanford, Houston, Lyon, and in the Institute of Industrial Science, University of Tokyo Murakami-Kato laboratory. Other funding that is gratefully ackowledged came from Shell (Amsterdam), the Commission for European Communities (JRC Ispra) and the U.K. Ministry of Defence.

Table 1   for

	u_0	L	R_e	T_L	T_D
Plane wake	$-\frac{1}{2}$	$\frac{1}{2}$	0	1	1
Self-propelled plane wake	$\frac{3}{4}$	$\frac{1}{4}$	$-\frac{1}{2}$	1	1
Axisymmetric wake	$-\frac{2}{3}$	$\frac{1}{3}$	$-\frac{1}{3}$	1	1
Self-propelled axisymmetric wake	$\frac{4}{5}$	$\frac{1}{5}$	$\frac{3}{5}$	1	1
Mixing layer	0	1	1	1	1
Plane jet	$-\frac{1}{2}$	1	$\frac{1}{2}$	$\frac{3}{2}$	$\frac{3}{2}$
Axisymmetric jet	-1	1	0	2	2
Plane plume	0	1	1	1	1
Axisymmetric plume	$-\frac{1}{3}$	1	$\frac{2}{3}$	$\frac{4}{3}$	$\frac{4}{3}$
Boundary layer (in a zero pressure gradient)	$\frac{1}{\ln x}$	$\frac{x}{\ln x}$	$\frac{1}{\ln^2 x}$	x	x

The downstream variation of the turbulent velocity and length scales u_0, L; the local Reynolds number $R_e = L u_0 / \nu$, the turbulent time scale $T_L = L/u_0$ and the time a fluid particle spends in the flow $T_D = \int dx / \overline{u_1} \sim x/(u_0 + U_0)$, defined in powers of x (e.g. $u_0 \propto x^{-1/2}$). Adapted from Tennekes & Lumley 1971.

Table 2

Property of flow	Observations of numerical simulations	Time Scales	1-P models		2-P models	
			EVN	2RST	RDT	EDQN
(i) $\dfrac{\overline{u_i u_j}}{K}$, $\left[\dfrac{p\frac{\partial u_i}{\partial x_j}}{KS}\right]$		$(T_L S \gg 1)$	N	Y [P]	Y [G]	Y [G]
(ii) Correct sensitivity of property (i) to the initial $E(k)$	None	$T_D/T_L \ll 1$	Y	Y	Y	Y
	Some	$T_D/T_L \sim 1$	N	N	Y	Y
	None (Sh)	$T_D/T_L \gg 1$	Y (Sh)	Y (Sh)	Y	Y
	Some (NSh)	$(T_L S \gtrsim 1)$	N (NSh)	N (NSh)	Y	Y
(iii) Correct sensitivity of property (i) to the initial anisotropy b_{ij}, c_{ij},	Some	$T_D/T_L \ll 1$	N	Y (approx)	Y	Y
	Weak	$T_D/T_L \sim 1$	N	Y	Y (?)	Y
	None (Sh)	$T_D/T_L \gg 1$	Y (Sh)	Y (Sh)	N (Sh)	N
	Some (NSh)		N (NSh)	Y (NSh)	Y (NSh)	Y
(iv) Correlation of $u_i(\mathbf{x})$ with initial (or boundary) turbulence	Strong	$T_D/T_L < 1$	N	N	Y [G]	Y (?)
	Weak	$T_D/T_L \gtrsim 1$			Y [P]	Y

Comparative merits of 1-point (1-PM) and 2-point (2-PM) models for different flow properties and turbulent flows (defined by time scales) in homogeneous strains with shear (Sh) and without shear (NSh) away from rigid boundaries or other interfaces. EVM = Eddy Viscosity Models; 2RST = Second Order Reynolds Stress Transport; RDT = Rapid Distortion Theory; EDQNM = Eddy Damped Quasi Normal Markovian.

Y = Yes, can be computed or is correct in principle;

N = No.

[G] = Good Model in practice;

[P] = Poor Model in practice.

REFERENCES

Adrian, R.J. & Moin, P. 1988.
Stochastic estimation of organized turbulent structure: homogeneous shear flow. J. Fluid Mech. **190**, 531–559.

Antonia, R.A. & Bisset, D.K. 1991
3-D aspects of the organized motion in a turbulent boundary layer. In "Turbulence and coherent structures" (Ed O.Métais & M.Lesieur), pp. 141–158. Kluwer Academic Press.

Aref, H. 1991
Computations of interacting vortices. (Submitted).

Aubry, N., Holmes, P., Lumley, J.L. & Stone, E. 1988
The dynamics of coherent structures in the wall region of a turbulent boundary layer. J. Fluid Mech. **192**, 115–173

Batchelor, G.K. & Proudman, I. 1954
The effect of rapid distortion on a fluid in turbulent motion. Quart. J. Mech. & App. Math. **VII**, pt.1. pp. 83–103.

Belcher, S.E., Weng, W.S. & Hunt, J.C.R. 1991
Structure of turbulent boundary layers perturbed over short length scales. 8th Symp. on Turb. Shear Flows paper 12-2. Tech Univ. of Munich.

Bergé, P. 1987
Chaotic behaviour in a non-linear system: turbulence in Rayleigh-Benard convection. Advances in Turb. 1, pp. 56–65 (Ed. G. Comte-Bellot & J. Mathieu). Springer.

Bertoglio, J.-P. 1981
A model of three dimensional transfer in non-isotropic homogeneous turbulence. Proc. Symp. Turbulent shear flows, **3**, pp. 17.1–6, Springer-Verlag.

Bevilaqua, P.M. & Lykoudis, P.S. 1978
Turbulence memory in self-preserving wakes. J. Fluid Mech. **89**, pp. 589–606.

Biringen, S. & Reynolds, W.C. 1981.
Large eddy simulation of the shear-free turbulent boundary layer. J. Fluid Mech. **103**, 53–63.

Bridges, J. & Hussain A.K.M.F. 1987
Roles of initial condition and vortex pairing in jet noise. J. Sound Vibration. **117**, 289–312.

Broadwell, J.E. & Mungal, M.G. 1991
Large scale structures and molecular mixing. Phys. Fluids. A **3**, 1193–1206.

Britter, R.E., Hunt, J.C.R. & Mumford, J.C. 1979
The distortion of turbulence by a circular cylinder. J. Fluid Mech., **92**, 269–301.

Browand, F.K. & Weidman, P.D. 1976
Large scales in the developing mixing layer. J. Fluid Mech. **76**, 127–144.

Bushnell, D.M. & Moore, K.J. 1991
Drag reduction in nature. Ann. Rev. in Fluid Mech. **23**, 65–79.

Cambon *et al.* 1991
On the amplification of time dependent scaling to the modelling of turbulence undergoing compression. Eur. J. of Mech. (To be published.)

Carruthers, D.J. & Hunt, J.C.R. 1986
Velocity fluctuations near an interface between a turbulent region and a stably stratified layer. J. Fluid Mech. **165**, 475–501.

Carruthers, D.J., Hunt, J.C.R. & Perkins, R.J. 1991
The emergence of characteristic (coherent?) motion in homogeneous turbulent shear flows. In "Turbulence and Coherent Structures" (Ed. O.Métais & M.Lesieur) pp. 29–44. Kluwer Academic Press.

Castaing, B., Gunaratne, G., Heslot, F., Kadanoff, L., Libchaber, A., Thomne, S., Wu, X., Zaleski, S. & Zaretti, G. 1989.
Scaling of hard thermal turbulence in Rayleigh-Bénard convection. J. Fluid Mech. **204**, 1–30.

Cebeci, T. & Smith, A.M.O. 1974
Analysis of turbulent boundary layers. Academic Press.

Corrsin, S. 1963
Estimation of the relation between Eulerian & Lagrangian scales in large Reynolds number turbulence. J. Atmos. Sci. **20**, 115–119.

Davidson, P.A., Hunt, J.C.R. & Moros, P.A. 1988
Turbulent recirculating flow in liquid metal MHD. Proc. Fifth Beer Sheva Seminar, Israel, In: Liquid Metal Flows: Magnetohydrodynamics and Applications. (Ed. H. Branover, M. Mond & Y. Unger). American Inst. of Aeronautics & Astronautics, **111**, 400–420.

Deissler, R.G. 1968
Effects of combined two-dimensional shear and normal strain on weak locally homogeneous turbulence and heat tranfer. J. Math. Phys. **47**, 320–331.

Dianat, M. & Castro, I.P. 1991
Turbulence in a separated boundary layer, J. Fluid Mech. **226**, 91–124.

Durbin, P.A. 1991
Near wall closure modelling without "Damping Functions". Theoretical & Computational Fluid Dynamics. Springer-Verlag.

Dengel, P. & Fernholz, H.H. 1990
An experimental investigation of an incompressible turbulent boundary layer in the vicinity of separation. J. Fluid Mech. **212**, 615–636.

Falco, R.E. 1977
Coherent motions in the outer region of turbulent boundary layers. Phys. Fluids **20**, S124–32.

Franke R. & Rodi, W. 1991
Calculation of vortex shedding past a square cylinder with various turbulence models. 8th Symp. on Turb. Shear Flows, Munich. Paper 20-1.

Fung *et al.* 1991
   Defining the zonal structure of turbulence using the pressure and invariants of the deformation tensor. Advances in turbulence 3 (Ed. A.V.Johannsen & P.H.Alfredson), pp. 395–405. Springer-Verlag.

Gartshore, I.S., Durbin, P.A. & Hunt, J.C.R. 1983
   The production of turbulent stress in a shear flow by irrotational fluctuations. J. Fluid Mech. **137**, 307–329.

Germano, M. 1992
   Averaging invariance of the turbulence equations and similar subgrid modelling (to be published in J. Fluid Mech.)

Gibson, M.M. & Launder, B.E. 1978
   Ground effects on pressure fluctuations in the atmospheric boundary layer. J. Fluid Mech. **86**, 329–357.

Gibson, M.M. & Rodi, W. 1989
   Simulations of free effects of turbulence with a Reynolds stress model. J. Hyd. Res. **27**, 233–244.

Goldstein, M.E. & Durbin, P.A. 1980
   The effect of finite turbulence spatial scale on the amplification of turbulence by a contracting stream. J. Fluid Mech. **98**, 473–508.

Goldstein, M.E. 1981
   The coupling between flow instabilities and incident disturbances at a leading edge. J. Fluid Mech. **104**, 217–46.

Head, M.R. & Bandyopadhyay, P. 1981
   New aspects of turbulent layer structure. J. Fluid Mech. **107**, 297–338.

Haworth, D.C. & Pope, S.B. 1986
   A generalized Lagrangian model for turbulent flows. Phys. Fluids **29**(2), 387–405.

Ho, G.M. & Huerre, P. 1984
   Perturbed free shear layers, Ann. Rev. Fluid Mech. **16**, 365–424.

Hunt, J.C.R. 1978
A review of the theory of rapidly distroted turbulent flows and its applications. Proc. of XIII Biennial Fluid Dynamics Symposium, Kortowo, Poland. Fluid Dynamics Transactions. **9**, 121–152.

Hunt, J.C.R. 1984
Turbulence structure in thermal convection and shear-free boundary layers. J. Fluid Mech. **138**, 161–184.

Hunt, J.C.R. 1985
Turbulent diffusion from sources in complex flows. Ann. Review of Fluid Mechanics **17**, 447–485

Hunt, J.C.R. 1987
Vorticity and vortex dynamics in complex turbulent flows. Trans. Can. Soc. Mech. Engs. **11**, 21–35.

Hunt, J.C.R., Kawai, H., Ramsay, S.R., Pedrizetti, G. & Perkins, R.J. 1990
A review of velocity and pressure fluctuations in turbulent flows around bluff bodies. J. Wind Eng. & Ind. Aero. **35**, 49–85.

Hunt, J.C.R. & Carruthers, D.J. 1990
Rapid distortion theory and the problems of turbulence. J. Fluid Mech. **212**, 497–532.

Hunt, J.C.R. 1991
Industrial and environmental fluid mechanics. Ann. Rev. Fluid Mech. **23**, 1–41.

Hunt, J.C.R. et al. 1989
Cross correlation and length scales in turbulent flows near surfaces. Adv. in turb. 2, 128–134. (Ed. H.H. Fernholz & H.E. Fieldler). Springer.

Hunt, J.C.R. & Vassilicos, J.C. 1991
Kolmogorov's contributions to the physical and geometrical understanding of small scale turbulence and recent developments. Proc. R. Soc. Lond. A. **434**, 183–210.

Hussain, A.K.M.F. & Clark, A.R. 1983
On the coherent structure of the axisymmetric mixing layer: a flow visualization study. J. Fluid Mech. **104**, 263–294.

Husain, H.S. & Hussain A.K.M.F. 1991
    Elliptic jets. Part 2. Dynamics of coherent structures: pairing. J. Fluid Mech. **233**, 439–482.

Jeandel, D., Brison, J.F. & Mathieu, J. 1978
    Modeling methods in physical and spectral space. Phys. Fluids **21**(2), 169–82.

Johnson, D.A. & King, L.S. 1984
    A new turbulence closure model for boundary layer flows with strong adverse pressure gradients and separation. AIAA paper No. 84-0175.

Keffer J.F. 1965
    The uniform distortion of a turbulent wake, J. Fluid Mech. **22**, 135–159

Kellogg, R.M. & Corrsin, S. 1980
    Evolution of a spectrally local disturbance in grid-generated, nearly isotropic turbulence. J. Fluid Mech. **96**, 641–669

Kida, S. & Hunt, J.C.R. 1989
    Interaction between turbulence of different scales over short times. J. Fluid Mech **201**, 411–445.

Kiya, M., Ohyama, M. & Hunt, J.C.R. 1986
    Vortex pairs and rings interacting with shear layer vortices. J. Fluid Mech. **172**, 1–15.

Kiya, M., Sasaki, K. & Arie, M. 1982
    Discrete vortex simulation of a turbulent separation bubble. J. Fluid Mech. **120**, 219–244.

Kline, S Cantwell, B. & Lilley, G.K. (eds)
    Comparison of computation and experiments. Proc. 1980-81 AFOSR-HTTM — Stanford Conf. on Complex Turb. Flows. Stanford.

Kolmogorov, A.N. 1941
    The local structure of turbulence in incompressible fluid for very large Reynolds numbers. Dokl. Akad. Nauk SSSR. **30** (4), 301–305.

Komori, S., Veda, H., Ogino, F. & Mizushina, T. 1983
Turbulence structure in stably stratified open channel flow. J. Fluid Mech. **130**, 13–26.

Launder, B.E. 1989
Second moment closure: present and future. Int. J. Heat & Fluid Flow. **10**, No.4, 282–299.

Launder, B.E., Reece, G.J. & Rodi, W. 1975
Progress in the development of a Reynolds stress turbulence closure. J. Fluid Mech. **68**, 537–566.

Launder, B.E. & Spalding, D.B. 1972
Mathematical models of turbulence. Academic Press.

Lee, M.J., Kim, J. & Moin, P. 1990
Structure of turbulence at high shear rate. J. Fluid Mech.**216**, 516–583.

Lee, M. & Reynolds, W.C. 1985
On the structure of homogeneous turbulence. Proc. 5th Symp. on turbulent shear flows, 17.7–17.12. Cornell.

Lele, S.K. 1985
A consistency condition for Reynolds stress closures. Phys. Fluids **28**, 64–68.

Leonard, A.D. & Hill, J.C. 1991
Scalar dissipation and mixing in turbulent reacting flows. Phys. Fluids. **3**(5), 1286–1299.

Leslie, D.C. 1973
Developments in the theory of turbulence, Clarendon Press.

Lesieur, M. 1990
Turbulence in fluids (2nd Ed.). Kluwer Academic Press, Dordrecht.

Lumley, J. 1978
Computational modelling of turbulent flows. Adv. Appl. Mech. **26**, 183–309.

Mansour, N.N., Kim, J. & Moin, P. 1988
Reynolds-stress and dissipation rate budgets in a turbulent channel flow. J. Fluid
Mech. **194**, 15–44.

Maxey, M.R. 1982.
Distortion of turbulence in flows with parallel streamlines. J. Fluid Mech. **124**, 261–
282.

Melander, M.V. & Hussain, F. 1990
Cut & connect of anti-parallel vortex tubes. Proc. IUTAM Symp. Topological Fluid
Mechanics, 485–500. (Ed. H.K.Moffatt & A.Tsinober) Cambridge University Press.

Mestayer, P. 1982.
Local isotropy and anisotropy in a high Reynolds number turbulent layer, J. Fluid
Mech. **125**, 475–503.

Mansour, N.G., Shih, T-H & Reynolds, W.C. 1991
The effects of rotation on initially anisotropic homogeneous flows. Phys. Fluids A, **3**,
2421–2425.

Michard *et al.* 1987
Grid generated turbulence exhibiting a peak in the spectrum. Adv. in Turbulence pp.
163 (Eds. G. Comte-Bellot & J. Mathieu), Springer-Verlag.

Millionshtchikov, M. 1941
On the role of third moments in isotropic turbulence. C.R. Acad. Sci. U.R.S.S. **32**,
619.

Moser, R.D. & Moin, P. 1987
The effects of curvature on wall bounded turbulent flows. J. Fluid Mech. **175**, 479–510.

Mungal, M.G. & Hollingsworth, D.K. 1989
Organised motion in a very high Reynolds number jet. Phys. Fluids B **1**, 1615–1623

Murakami, S., Mochida, A. & Hayashi 1990
Examining the K-$\mathcal{E}$ model by means of a wind tunnel test and large eddy simulation
of the turbulence structure around a cube. J. Wind Engg. & Indust. Aerodyn. **35**,
87–100.

Neish A. & Smith, F.T. 1988
   The turbulent boundary layer and wake of an aligned flat plate. J. Engng. Maths **22**, 15.

Newley, T.M.J. 1986
   Turbulent air flow over hills. Ph.D. thesis. University of Cambridge.

Novikov, E.A. 1983
   Generalized dynamics of 3-D vortical singularities, Sov. Physics—JETP, **57**, 566.

Orszag, S.A. 1970
   Analytical theories of turbulence. J. Fluid Mech. **41**, 363–383.

Orszag, S.A. 1972
   Comparison of pseudospectral and spectral approximation. Stud. Appl. Maths. **51**, 253–259.

Palmer, H.J. & Keffer, J.F. 1972
   An experimental investigation of an asymmetrical turbulent wake. J. Fluid Mech. **53**, 593–610.

Pearson, H.J., Puttock, J.S. & Hunt J.C.R. 1983
   A statistical model of fluid element motions and vertical diffusion in a homogeneous stratified turbulent flow, J. Fluid Mech. **129**, 219–249.

Perry, A.E. & Abell, C.J. 1975
   Scaling laws for pipe flow turbulence, J. Fluid Mech. **67**, 257–271.

Perry, A.E. & Chong, M.S. 1982
   A description of eddying motions and flow patterns using critical-point concepts. Ann. Rev. Fluid Mech. **19**, 125–156.

Phillips, O.M. 1955
   The irrotational motion outside a free turbulent boundary. Proc. Camb. Phil. Soc. **51**, 220–229.

Pope, S.B. 1985
   PDF methods for turbulent reactive flows. Prog. Energy Combust. Sci. **11**, 119–192.

Prandtl, L. 1925
Bericht Über Untersuchungen zur ausgebildeten Turbulenz. ZAMM, **5**, 136–139.

Prandtl, L. 1942
Bemerkungen zur theorie der freien Turbulenz. ZAMM **22**, 241–43.

Proudman, I. & Reid, W.H. 1954
On the decay of a normally distributed and homogeneous turbulent velocity field. Phil. Trans. A Roy. Soc. **247**, 163–189.

Riley, J.J. & Corrsin, S. 1974
The relation of the turbulent diffusivities to Lagrangian velocity statistics for the simplest shear flow. J. Geophys. Res. **19**, 1768–1774.

Rogallo, R.S., Moin, P. 1984
Numerical simulation of turbulent flows. Ann. Rev. Fluid Mech. **16** 99–137.

Rotta, J.C. 1951
Statistiche Theorie nicht-homogenen Turbulenz. Z. Phys. **129**, 547-572.

Sabot, J. & Comte-Bellot, G. 1976
Intermittency of coherent structures in the core region of fully developed turbulent pipe flow. J. Fluid Mech. **74**, 767–796.

Savill, A.M. 1987
Recent developments in Rapid-Distortion-Theory. Ann. Rev. Fluid Mech. **19**, 531–521.

Savill, A.M. 1991
A synthesis of T3 test case predictions. Proc. ERCOFTAC Workshop. Cambridge Univ. Press.

Schlichting, H. 1960
Boundary layer theory, McGraw-Hill (4th ed).

Snyder, W.H. & Lumley, J.L. 1971
Some measurements of particle velocity auto correlation functions in a turbulent flow. J. Fluid Mech. **48**, 41–71.

Spalart, P. & Baldwin, B.S. 1987
Direct simulation of a turbulent oscillating boundary layer. Proc. 6th Symp. on turbulent shear flows. Toulouse.

Stuttgen, W. & Peters, N. 1987
Stability of similarity solutions by two equations models of turbulence. AIAA. Vol. 25. No. 6. pp. 829–830.

Tamura, T., Ohta, I. & Kuwahara, K. 1990
On the reliability of two dimensional simulation of unsteady flows around a cylinder type structure. J. Wind Eng. & Ind. Aero. **35**, pp. 275–298.

Tennekes, H & Lumley, J.L. 1971
A first course in turbulence. MIT Press.

Thomas, N.H. & Hancock, P.E. 1977
Grid turbulence near a moving wall. J. Fluid Mech. **82**, 481–496.

Thomson, D.J. 1987
Criteria for the selection of stochastic models of particle trajectories in turbulent flows. J. Fluid Mech. **180**, 529–556.

Townsend, A.A. 1961
Equilibrium layer and wall turbulence. J. Fluid Mech. **11** 87–120.

Townsend, A.A. 1956
Structure of turbulent shear flow. (1st Edn.) Cambridge Univ. Press.

Townsend, A.A. 1976
Structure of turbulent shear flow. (2nd Edn.) Cambridge Univ. Press.

Tsai et al. 1991.
Thermal striping in structures in interacting jets. In "Turbulence and coherent structures" pp. 125–138. (Eds. O.Métais & M.Lesieur). Kluwer Academic Press.

Turner, J.S. 1973
Buoyancy effects in fluids. Cambridge University Press.

Veeravalli, S. & Warhaft, Z. 1989
The shearless turbulent mixing layers. J. Fluid Mech. **207**, 191–229.

Weinstock, J. 1989
A theory of turbulent transport. J. Fluid Mech. **202**, 319–338.

Wray, A. & Hunt, J.C.R. 1989
Algorithms for classification of turbulent structure. Proc. IUTAM Symp. on Topological Fluid Mechanics, pp. 95–104. (Ed. H.K.Moffatt & A.Tsinober) Cambridge Univ. Press.

Wygnanski, I.J., Champagne, F.H. & Marashi, B. 1986
On the large scale structures in two dimensional, small deficit turbulent wakes. J. Fluid Mech. **168**, 31.

Zeman, O. 1981
Progress in modelling of planetary boundary layers. Ann. Rev. Fluid Mech. **13**, pp. 253–272.

Zhou, M.D. 1985
A new modelling approach to complex turbulent shear flows. Adv. in Turbulence **2**, 146–150. (Ed. H.H.Fernholz & H.E.Fiedler). Springer-Verlag.

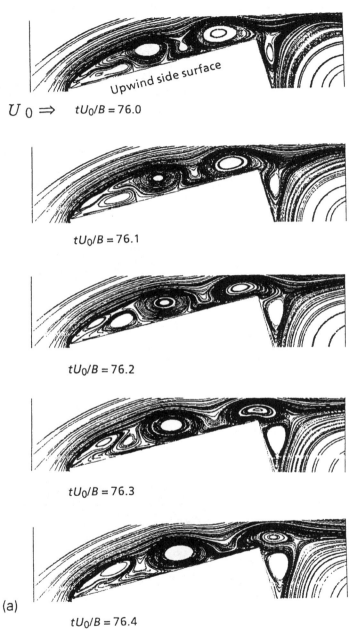

$U_0 \Rightarrow$ $tU_0/B = 76.0$

$tU_0/B = 76.1$

$tU_0/B = 76.2$

$tU_0/B = 76.3$

(a)

$tU_0/B = 76.4$

Fig.1 Comparison of "realisation" and statistical modelling of turbulent flow around a cube.

(a) A sequence of realisations of "approximate numerical simulation" of flow around a cube by Tamura *et al* 1990. ($200 \times 100 \times 850$ points at $R_E = 10^4$ for mean flow).

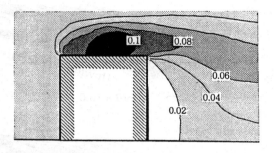

(1) **wind tunnel experiment**

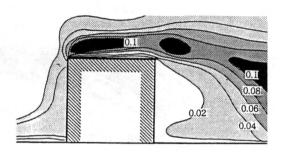

(2) LES

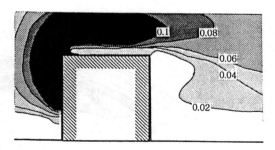

(3) $k - \varepsilon$ model

(b) Comparisons between measurements of kinetic energy around a cube in a turbulent flow ($L/H \sim 1$), and two kinds of computation using a purely statistical
model (K-$\mathcal{E}$) and using the statistics derived from a realisation model (Large
Eddy Simulation). (Murakami *et al* 1990).

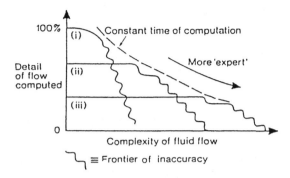

Fig.2 Schematic graph showing the degree of detail of the computation as a function of the complexity of the turbulent flow for different levels of computational model, for given computational capacity and run-time.

(i) indicates an advanced realisation model (e.g. D.N.S.) for an ideal flow obtained with a super computer (many realisations) over 10–10^2 hours.

(ii) indicates an advanced statistical model (e.g. 2RST, 3RST) for one typical engineering flow over 10 hours.

(iii) indicates simple statistical model (e.g. EVM) for many trials of complex engineering flows over 10 hours.

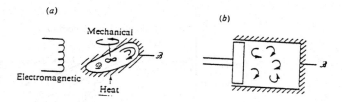

Fig.3 Example of turbulent flow in a closed domain **D** and deterministic boundary conditions on **B** e.g. driven by moving boundaries or body forces.

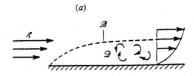

Fig.4 Open domains where the flow enters the computational domain **D** through the boundary **B**

(a) Entering flow is not turbulent, transition occurs within **D**

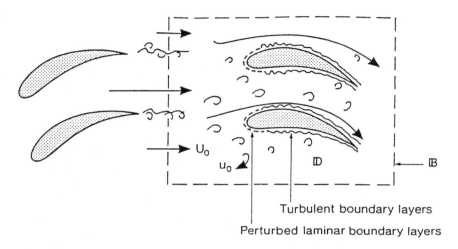

Turbulent boundary layers

Perturbed laminar boundary layers

(b) Entering flow is turbulent with a mean velocity U, significantly greater than the r.m.s. turbulence velocity u_0.

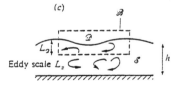

(c) Flow crossing **B** is random with no significant mean motion.

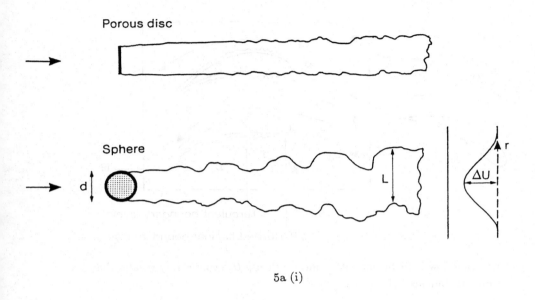

5a (i)

Fig.5 Demonstration of the sensitivity of turbulent flow to initial conditions far downstream.

  (a) Three dimensional wakes of a porous disc and solid sphere with the same drag. Note the differences in eddy structure and magnitude of turbulence, but that they have the same similarity law. From Bevilaqua & Lykoudis 1978.

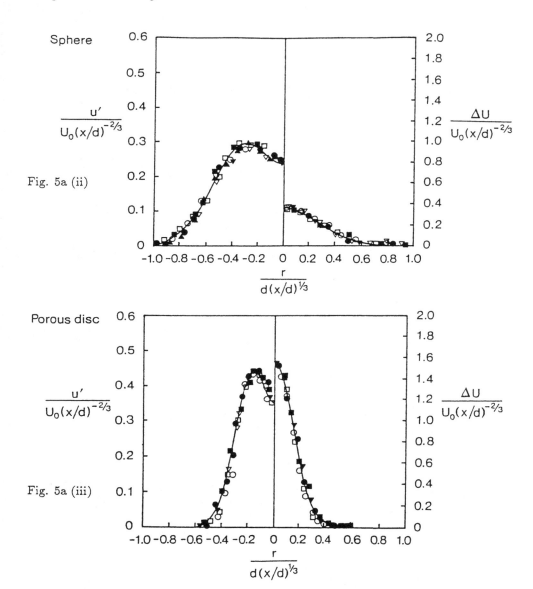

Sphere

Fig. 5a (ii)

Porous disc

Fig. 5a (iii)

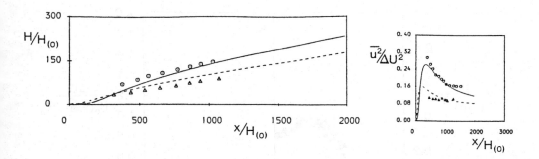

(b) Two dimensional plane wakes of an aerofoil (o) and a solid plate (△), from Wygnanski *et al* (1986). Note that the turbulence modelling accounts for the persistent difference in the wake turbulence (Launder 1989). $H(0)$ is a fixed length (=
momentum thickness). ——— denote the best model results.

Fig. 6a

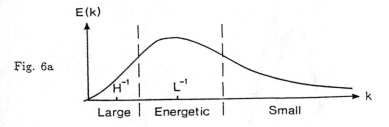

Fig.6 Schematic diagram showing the main features of the eddy motions with different
ranges of length scales.

(a) Energy of the eddy motions described by the energy spectrum $E(k)$.

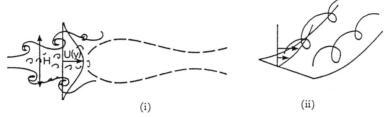

Fig. 6b

(i) (ii)

(b) "Large" eddies spanning the relevant region of the flow with length scale H. These are characteristic for the particular flow (i) oscillations on a jet or (ii) Taylor-Gortler vortices.

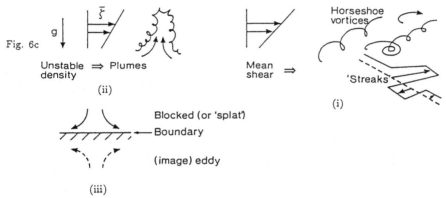

Fig. 6c

Unstable ⇒ Plumes
density

(ii)

Mean ⇒
shear

Horseshoe vortices

'Streaks'

(i)

Blocked (or 'splat')
Boundary

(image) eddy

(iii)

(c) "Energetic" eddies with length scale L, and having a general form characterised by the local straining motion, or body force, or boundary conditions, examples are (i) shear $(L_1^{(2)} \sim \frac{\partial U_1 / \partial x_2}{\sqrt{\overline{u_2^2}}})$, (ii) buoyancy $(L_1^{(2)} \sim [\text{buoyancy flux}/(\sqrt{\overline{u_2^2}})^{\frac{3}{2}}]$ this is the Monin-Obukhov length), (iii) rigid boundary $L_1^{(2)} \propto x_2$.

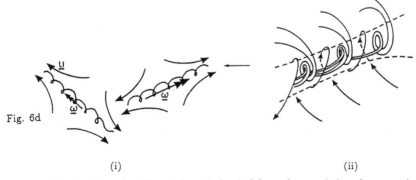

Fig. 6d

(i) (ii)

(d) "Small" scale eddies with typical spiral form characteristics of most turbulent flows at high Reynolds number or low strain rate. (Despite their non isotropic form the eddies have approximately isotropic orientation, leading to isotropic second order statistics.)

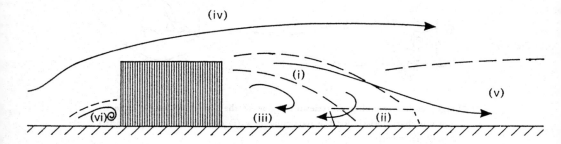

Fig.7 The different zones having distinct characteric turbulence dynamics in flow over a bluff body

   (i) Boundary layers and separated shear layer $|\pi| < \pi_{Sh}$

  (ii) Impinging shear layer (i.e. $\pi > \pi_{Sh}$) dominated by irrotational strain.

 (iii) Recirculating flow region — inhomogeneous ($L \sim L_I$).

 (iv) External flow — advection, production, large scale inhomogeneity (or blocking) ($T_D < T_L$).

  (v) Perturbed shear layer (with strong memory effects of the wake — large 3-D eddies persist ($T_D \sim T_L$; $|\pi| < \pi_{Sh}$).

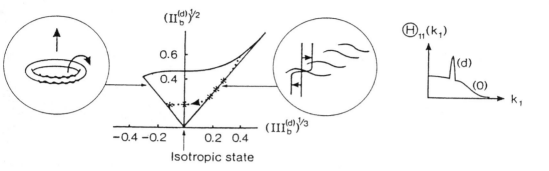

Fig.8 Results of an experimental study of the transfer of energy between velocity components of initially anisotropic turbulence as it decays. In this case an anisotropic disturbance (d) with a narrow band energy spectrum $\Theta_{11}^{(d)}(k)$ was superposed on a broad spectrum $\Theta_{11}^{(o)}(k)$. Note that as the turbulence evolves the changes of the invariants $II_{(b)}^{(d)}$, $III_{(b)}^{(d)}$ correspond to the eddies changing from an unstable form (vortex-shear-like eddies) to a more stable form (vortex-ring-like eddies). There is little evidence of tendency to a final isotropic state! (Michard *et al* 1987).

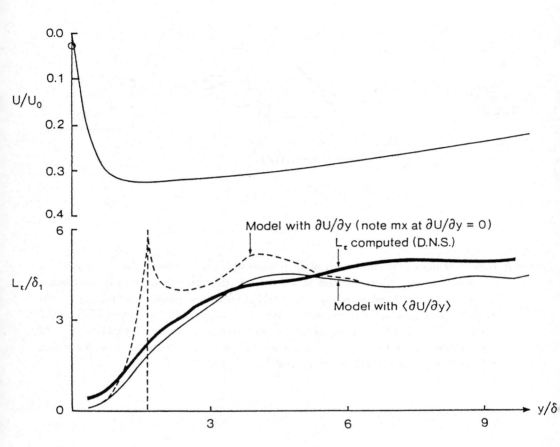

Fig.9 Graph of dissipation length scale $L_\varepsilon^{-1}(u_2)$ in an oscillating shear flow computed by P. Spalart using Direct Numerical Simulation compared with model equation (4.22). Note the improvement using the spatial averaging of $(\partial U/\partial x_2)$, defined by (4.21).

Test Case T1:
"S"-shaped duct flow

Test case T1: Boundary layer in a "S"-shaped channel

Trong-Vien Truong and Maryelle Brunet

Fluid Mechanics Laboratory
Swiss Federal Institute of Technology, Lausanne, Switzerland

1. DESCRIPTION OF THE EXPERIMENT

The present experiment has been conducted with the purpose of studying a pressure-gradient driven three-dimensional boundary layer, and to provide a comprehensive data base to examine critically different turbulence models. Additionally, the intention is to compare different measurement techniques and to develop new ones. There are a number of ways to produce a three-dimensional boundary layer: e.g. by using appropriate curved walls to generate a transverse pressure gradient, by arranging for a local transverse motion of one bounding surface, by introducing a two-dimensional body mounted normal to a flat plate, etc... As the number of measurements to be carried out to define a three-dimensional flow development in satisfactory detail is large due to the extra dimension involved, special attention must be paid to facilitate variations of different parameters governing the flow and to give easy access to different kinds of instrumentation. To achieve these goals, arguments were put forwards for an S-shaped duct geometry to produce a series of experiments with gradual variation in crossflow inducing and relaxing features to improve the understanding of the physics of turbulence in three-dimensional boundary layer flows. In this type of duct, the flow is pressure-gradient driven and the ratio of shear forces to pressure forces can be varied at will; the S-shape geometry also exhibits the interesting feature of "cross-over" of the crossflow profile since the external streamlines in the core of the duct have an inflection point [1].

1.1 Description of the experimental set-up

Accessibility plus flexibility in the tunnel hardware are therefore key factors in designing the tunnel. This points to a sandwich construction in which the vertical side walls forming the S-shape of the duct can be easily changed. However, the "sandwich" itself must have a flexible height so that, at a later stage, in testing the series of duct flows, the streamwise pressure-gradient can be changed at will. Only one of the four walls making up the duct can therefore be rigid. Consequently the tunnel roof is a rigid aluminum construction which supports the vertical side walls, the measurement plate and eventually special equipments. The bottom of the tunnel is made up of a series of plastic plates, reinforced with plastic ribs to form rigid entities; these plates are mounted in pairs across the tunnel, each plate covering approximately half the tunnel width and fastened individually to the tunnel main frame by a regulating device that allows the plate to move longitudinally and vertically and to rotate. Various slots with soft material edges are installed between the ribs of these platic plates so that different probe holders can slide easily along and across the test section. The main side walls of the tunnel are 12 millimetres cast plastic plates and allow the tunnel interior to be viewed from the access platform. In the main S-part of the duct, a pair of additional plastic side walls are inserted to shave off the side wall boundary layers building up along the side walls of the first straight portion of the tunnel.

The tunnel which is semi-closed and mounted vertically has a test section consisting of three consecutive parts: a first straight part followed by a "S"-shaped part (the "S" is horizontal like a snake on a flat horizontal surface), which in turn is followed by a second straight part. Each part is 3 metres long, 0.5 metre high and 1.2 metre wide; in the S-part the width between the internal side walls is 1 metre (Figure 1). The flow enters the first straight portion with a uniform velocity at the exit section of the tunnel contraction with a contraction ratio of 7.35, the geometrical shape of which is chosen according to a cubic equation. A free floating, polished, perfectly planar aluminum plate of thickness 6 millimetres with an elliptical leading edge facing the tunnel contraction exit, is mounted at 0.08 metre below and all along the roof of the tunnel test section. The plate is 1 metre wide with upwards curved lips of 50 millimetres wide at both sides on the first 3.5 metres from

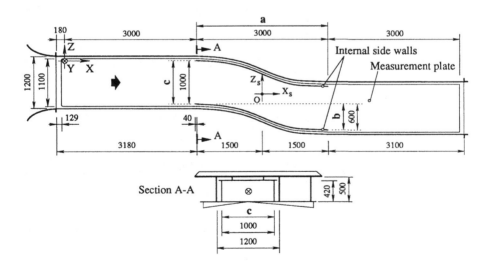

Figure 1 : General lay-out of the "S" shaped channel

the leading edge. This measurement plate has 1000 static pressure tappings to determine the pressure distribution on the first 6 metres and 15 surface plugs interchangeable and flush mounted at defined stations for surface probes. The pressure lines are assembled between the roof of the tunnel and the measurement plate; great care was paid to keep at a minimum level the flow blockage effect. However, some blockage is still evident in the measurements.

A special traversing mechanism (Figure 2) is installed under the test section to position the probe in the wind tunnel. It has 5 degrees of freedom entirely controlled by the computer: 1) linear movement X along the test section with a range of 9 metres in 3 steps, a positioning accuracy of 1 millimetre and a positioning resolution of 0.01 millimetre, 2) linear movement Y normal to the measurement plate with a range of 0.5 metre, a positioning accuracy of 0.1 millimetre and a positioning resolution of 0.01 millimetre, 3) linear movement Z across the test section with a range of 1.5 metre, a positioning accuracy of 0.25 millimetre and a positioning resolution of 0.01 millimetre, 4) rotation α around the Y axis with a range of 360°, a positioning accuracy of 0.1° and a positioning resolution of 0.02°, 5) rotation ψ around the probe axis mounted parallel to the measurement plate or with some pitch angle β, a positioning accuracy of 0.6°, a positioning resolution of 0.3°. This traversing mechanism allows to position the probe at any pitch, yaw and roll angle and at any place in the test section; it is fixed on the instrumentation platform separated from the tunnel main frame and the access platform.

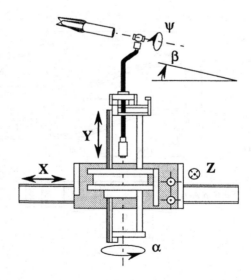

Figure 2 : Sketch of the traversing mechanism

The tunnel fan has a thyristor-regulated d.c. motor that allows for continous speed regulation up to 45 m/s and for holding constant flow parameters for long periods of testing. Installed power is 30 kW. The air temperature drift at start-up of the tunnel is 0.2° C/min at 18 m/s; stable conditions can be reached after 20 minutes of turning. The tunnel can be run with either constant flow velocity or constant Reynolds number at the exit of the contraction through the computer control.

1.2 Description of the "S" geometry

To facilitate the design of any particular duct geometry, numerical calculation schemes have been put together for calculating the flow in the duct. These calculations at the beginning have served as a guide in the layout of the duct; briefly, the scheme is built up around two basic codes, an Euler code and a three-dimensional turbulent boundary layer code. The Euler code is a three-dimensional incompressible code in curvilinear, body fitted coordinates [2]. The three-dimensional boundary layer code due to Krogstad [3] is, basically, an ADI code that has also a small crossflow and a plain two-dimensional variant as options. The coupling between the boundary layer and the inviscid core flow was considered in the simplest way: the calculated inviscid velocity-pressure distribution on the wall was used as an outer boundary condition in the boundary layer code on that same wall.

The first duct considered is designed to give a rather mild three-dimensional turbulent boundary layer behavior on the plate in order to avoid too much crossflow leading to separation and other difficulties. The geometry of this duct is chosen according to the formula $z_s = -\frac{b}{2} \cdot \frac{\tanh(x_s/d)}{\tanh(a/2d)} \pm \frac{c}{2}$ where the $x_s z_s$ system is fixed with the centre of the S-duct. The parameters chosen were: a = 3 metres (S shape length), b = 0.6 metre (vertical side wall off-set), c = 1 metre (effective tunnel width), d = 0.65 metre (steepness parameter of sidewall $\frac{dz}{dx}|_{x=0} \approx -\frac{b}{2d}$). The vertical internal and external side walls are thus parallel to each other and the maximum side wall angle is $\tan^{-1}(\frac{b}{2d}) \cong 25°$. By varying these parameters within the space available around the tunnel, one can obtain a stronger three-dimensional flow with or without separation on the side wall. The general configuration of the geometry chosen is displayed in the figure 1.

2. MEASUREMENTS IN THE S-DUCT

To facilitate the comparison between the measured and the computed results, the orthogonal coordinates system [OXYZ] is chosen with the origin fixed at the left corner of the flat surface of the measurement plate at the entrance to the test section (the contour of the test plate is curved), the X direction is pointing in the downstream flow direction, the Y direction is normal to and oriented away from the test plate. In this coordinates system, each station is defined by the following code: X1111Z2222, where 1111 corresponds to the X value and 2222 to the Z value in millimetres of that station; since all the Z values are negative, the sign "-" is deliberately omitted. As the wind tunnel is running with constant Reynolds number, the flow parameters and the measurement data are non-dimensionalized with the following set of reference values [U_∞, p_∞, T_∞] defined at the station [X= -120 mm; Y= 210 mm; Z= 0 mm] between the exit section of the wind tunnel contraction and the leading edge of the test plate (see Figure 1). In the table 1 are reported the coordinates of the measurement stations, stations defined as initial conditions for numerical simulation and stations for detailed comparison between computational and experimental results; the figure 3 displays the location of these stations on the measurement plate.

2.1 Pressure distribution on the measurement plate

Wall pressures were measured at different stations along and across the duct and given by $C_p = (p-p_\infty) / (\frac{1}{2}\rho_\infty U_\infty^2)$ as function of coordinates X and Z. The coordinates of the pressure measurement stations are reported in the figure 4 where also the plug stations on the measurement plate are displayed. The C_p-values were obtained directly by the ratio of the outputs of the two *ElecTorr* differential pressure transducers. The accuracy of the C_p measurement is less than 1% (due to the non-linearity, the hysteresis, the sensitivity-drift). The Figure 5 displays the pressure contours on the test plate calculated from the measurement data. Around the station [X= 3000 mm, Z= 0 mm], the pressure contours show a distorsion in the flow which is due to the upstream effect of the S-shaped side wall. By inclining this vertical side wall with an angle of 2^0, this distorsion is considerably reduced.

2.2 Skin friction measurement

Skin friction measurements were performed by using different techniques depending on the accessibility of the station considered: the conventional Preston tube with various diame-ters, the surface fence, the pulsed wire anemometer and the triangular block. All these tech-niques were checked against themselves at the entrance of the S-shaped part and in the centreline of the duct for different flow Reynolds numbers.

The Preston tube is used in preference principally in locations where the other techniques cannot be adopted. Aligned with the flow at the specific stations, the Preston tube measurements provide the flow angle at the wall and the magnitude of the skin friction by using the Patel calibration without any further correction [4]. The surface fence is 0.1 mm thick, 0.1 mm hight, 5 mm long and fitted in different plugs available on the measurement plate. Aligned with or across the flow, the rotation of the surface fence allows the measurement of the wall flow angle with an overall difference less than 0.5^0. The skin friction is obtained with the surface fence normal to the flow. The pulsed-wire anemometer used in the S-shaped channel is the one developed by the *Technische Universität Berlin* [5]. The probe consists of an array of 3 parallel wires with a wall distance of 0.03 mm, a separation distance between the single wires of 0.7 mm. The central wire is the pulsed wire with a diameter of 5 μm, a length of 3 mm. The sensor wires are gold plated with a total length of 2 mm, a sensing lenght of 0.5 mm and a diameter of 15 μm. The heating voltage pulse is about 5 μs, the wire temperature rises almost instantaneously to about 300°C. A heat tracer fluid element is formed and is convected with the flow; the time taken by the tracer to reach a sensor wire is a measure of the skin friction. Due to the diffusion and the electronic processing, the calibration expression of the adopted pulsed-wire anemometer follows :

$$\tau_w = A\left(\frac{1}{T}\right) + B\left(\frac{1}{T^2}\right)$$

describing the relationship between the flight time T and the wall shear stress τ_w. This log-law independant technique is preferable in this three-dimensional flow and has to be improved to measure high skin friction values with a sufficiently compact measuring volume.

The triangular block used is developed at the *Politecnico di Torino* [6]. As an extension of the McCroskey gage for 3-dimensional flows, this probe of height equal to 0.7 mm , consists of an equilateral triangular block of 9 mm side with one static pressure tap at the centre of each side and one pressure tap at the centre of the triangle. By using three functions F, G, y* = f(x*) obtained by calibration, the triangular block can provide the wall flow angle and the skin friction magnitude without rotation of the device:

$$F(\delta) = \frac{p_1 - p_2}{p_1 - p_3} \qquad\qquad G(\delta) = \frac{\left(p_1 - \frac{p_2 + p_3}{2}\right)_\delta}{\left(p_1 - \frac{p_2 + p_3}{2}\right)_{\delta=0}}$$

$$y^* = \log\frac{\tau_w h^2}{\rho\nu^2} \qquad\qquad x^* = \log\frac{\left(p_1 - \frac{p_2 + p_3}{2}\right)_{\delta=0}\cdot h^2}{\rho\nu^2}$$

where p_1, p_2, p_3, p_4 are the measured pressure values, ρ the density, ν the kinematic viscosity of the fluid, h the thickness of the triangular block and δ the angle between the flow and the normal to the side of the triangular block facing the flow. The measurement data provide the flow angle δ through the function F(δ). Subsequently, the nondimensional variable x^* is calculated with the function G(δ) and the angle δ; finally, the wall shear stress τ_w is determined with the function y* = f(x*). The measured value p_4 at the centre of the device gives an approximate value for the static pressure at the station where the triangular block is located.

Table 2 displays the experimental results, the wall shear-stress and the wall flow angles in the [OXYZ] reference system of coordinates, obtained with these different techniques at various stations in the S-shaped channel. The overall acccuracy of the skin friction measurement is estimated of about 5% due to the lack of appropriate calibration in three-dimensional flow and for the wall flow angles it is less than $\pm 1^{\circ}$.

2.3   Velocity profiles measurements

The velocity profile measurements were performed at 21 stations defined as initial conditions for numerical simulation of the flow in the S-shaped channel mainly with the *Dantec* X-wires probe [7] and at 24 stations defined as stations for comparison between computational and experimental results mainly with a 3-wires probe developed at the *RTWH, Aachen* [8]. For both of these probes, the diameter of the sensing wire is of 5 μm with a length of 1 mm.

With the X-wires probe, the conventional rotation-method was used to obtain from the measurement data, with respect to the XYZ system the mean velocity components [U,V,W] and the Reynolds stress tensor [$\overline{u^2}$, $\overline{v^2}$, $\overline{w^2}$, $\overline{uv}$, $\overline{vw}$, $\overline{uw}$]. The calibration of the X-wires probe in magnitude and in direction was performed each time by using the King's law [9] and the following functions:

$$F = \frac{U_{c1}}{U_{c2}} \qquad\qquad \varepsilon = c_0 + c_1 \left(45^{\circ} - \text{arctg}\,\frac{U_{c1}}{U_{c2}}\right)$$

with U_{c1}, U_{c2} denoting the cooling velocities of the probe wires. In the ideal case the values 1, 0, 1 apply respectively for F, c_0 and c_1.

The 3-wires probe offers the possibility to obtain directly the mean velocity components [U,V,W], the Reynolds stress tensor [$\overline{u^2}$, $\overline{v^2}$, $\overline{w^2}$, $\overline{uv}$, $\overline{vw}$, $\overline{uw}$] and the triple correlations of the fluctuating velocity $\overline{u_i u_j u_k}$. The calibration of the 3-wires probe in magnitude and in direction was performed with one function H^2 for the magnitude and two functions for the direction:

$$H^2 = (U_{c1}^2 + U_{c2}^2 + U_{c3}^2) / (2 |\vec{U}|^2)$$

$$\cos^2\varphi_2 \;=\; K_2(\varphi_3) + M_2(\varphi_3) \cdot F_2^2$$

$$\cos^2\varphi_3 \;=\; K_3(\varphi_2) + M_3(\varphi_2) \cdot F_3^2$$

with : $$F_2^2 = U_{c2}^2 / (U_{c1}^2 + U_{c2}^2 + U_{c3}^2)$$

$$F_3^2 = U_{c3}^2 / (U_{c1}^2 + U_{c2}^2 + U_{c3}^2)$$

where $|\vec{U}|$ is the magnitude of the instantaneous velocity, and U_{c1}, U_{c2}, U_{c3} the cooling velocities of the probe wires (Figure 6).

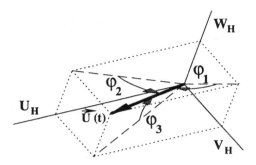

Figure 6: Definition of the angles φ_i in the probe coordinates system $[U_H V_H W_H]$

In the ideal case, H^2 has the value 1 as the coefficients M_2, M_3 and the coefficients K_2, K_3 are zero.

The X-wires probe and the 3-wires probe are positioned relatively to the measurement plate optically such that the closest wall distance is determined by measuring with a micrometer through a telescope the distance between the probe and its image on the polished test plate. The outputs of the anemometers are recorded simultaneously through a transient recorder at a frequency of 5 kHz for a time about 2 s and stored on the computer disc for later reduction. This integration time was chosen in view of keeping at a reasonable level the data reduction time for all the defined stations in which the measurements were repeated 2 or 3 times. By analyzing the shear stress data close to the wall, some scatter in the results were found; it appears that this integration time is not sufficient and has to be increased at least for the locations close to the wall. Additional checks have shown later that improvements in reducing the scatter in the data in the near wall region can be achieved by increasing 2 or 4 times the data recording time.

At the stations defined for comparison between computational results and measurement data, the following profiles in the tunnel fixed [OXYZ] coordinates system are plotted as function of y, the wall distance of the probe in millimetres, at the end of this report:

$$\frac{U}{U_\infty} , \quad \frac{W}{U_\infty} ,$$

$$\gamma = \tan^{-1}\left(\frac{W}{U}\right) , \quad \gamma_\tau = \tan^{-1}\left(\frac{\overline{vw}}{\overline{uv}}\right) , \quad \gamma_g = \tan^{-1}\left(\frac{\partial W/\partial y}{\partial U/\partial y}\right) ,$$

$$\frac{\overline{u^2}}{U_\infty^2}, \frac{\overline{v^2}}{U_\infty^2}, \frac{\overline{w^2}}{U_\infty^2}, \frac{\overline{uv}}{U_\infty^2}, \frac{\overline{vw}}{U_\infty^2}, \frac{\overline{uw}}{U_\infty^2}$$

The three angles γ, γ_τ, γ_g as defined in the Specifications of the test case T1, have the inverse sign with respect to usual convention. In addition, the profiles of the angle γ_g at different stations were calculated from the profiles of the mean velocity components U and W fitted respectively with the following functions by the least squares method:

$$\frac{U}{U_\infty} = a \cdot y^n \qquad\qquad \frac{W}{U_\infty} = [1 - \exp{(a_1 y)}] \cdot \frac{ay + b}{cy + d}$$

The reduction of the measured data with the hot-wire probes did not take into account the pressure gradients effect. The accuracy of the flow angles determined from the velocity profile measurements is less than $\pm 1^o$. The velocity data obtained with the X-wires probe are less accurate than with the 3-wires probe due to the conventional rotation-method adopted and the repeatability of the measurements with the 3-wires probe is less than $\pm 1\%$

3. ACKNOWLEDGEMENT

This test case "Boundary layer in a 'S'-shaped channel" is one of the two experiments [10] of the *Joint European Three-dimensional Turbulent Boundary Layer Experiment* initiated after the Eurovisc-Workshop-Berlin-1982 by Dr. G.Drougge (FFA,Sweden), Prof. E.Krause (RWTH, Germany) and Prof. H.Fernholz (TU-Berlin, Germany). The "Duct flow experiment" was undertaken under the responsability of Prof. I.L.Ryhming after different meetings with the participation of various research groups from RAE, Imperial College (England), ONERA (France), DFVLR, RWTH, TU-Berlin (Germany), NLR (The Netherlands), NTH ·(Norway), FFA, KTH (Sweden), EPFL, ETHZ (Switzerland). This research has been carried out with the support of CERS/KWF, Switzerland.

As this test case covers a wide range of the boundary layer behavior including the feature of "crossover" of the crossflow profile and as new measurement techniques in three-dimensional flows will be available in the near future, the data base obtained for this Workshop will be continously updated and made available to anyone expressing an interest.

4. REFERENCES

[1]   Truong T.-V., Dengel P., Moreau V., Nakkasyan A., Drotz A., Ryhming I.L.: *"The Euroexp series of duct flows at the EPFL"*, Interim Report No: 2 by Specialist Group 2, IMHEF-EPFL, 1986

[2]   Rizzi A., Eriksson L.E.: *"Computation of Flow Around Wings Based on the Euler Equations"*, Journal of Fluid Mechanics, No:148, pp. 45-71, 1984

[3]   Krogstad P.Å.: *"Investigation of a Three-Dimensional Turbulent Boundary Layer Driven by Simple Two-Dimensional Potential Flow"*, Ph.D. Thesis, Norwegian Institute of Technology, 1979

[4]   Patel V.C.: *"Calibration of the Preston Tube and Limitations on its Use in Pressure Gradients"*, Journal of Fluid Mechanics, vol 23, part 1, 185-208, 1965

[5]   Dengel P.: *"Pulsed hot-wire Anemometer TU-Berlin system"*, Technische Universität Berlin Communication, 1986

[6]   Onorato M.: *"Design, Calibration and Reduction of the triangular-block device for skin friction measurement"*, Politecnico di Torino Communication, 1989

[7]   DANTEC and DISA Information Series, Measurement and Analysis, 1982-1990

[8]   Schön Th.: *"Design, Calibration, Acquisition and Reduction Programs for 3-wires probe"*, KWTH Communication, 1989

[9]   King, L.V.: *"On the convection of heat from small cylinders in a stream of fluid: Determination of the convection constants of small platinum wires with applications to hot-wire anemometry"*, Phil.Trans.Roy.Soc. A, 214,373-432, 1914

[10] Euroexpt Meeting Report, 16-17 January 1984, Dr. H.-P. Kreplin, 22-59,AVA, Göttingen, Germany

[11] Bertelrud A., Alfredsson P.H., Bradshaw P., Landahl M.T., Pira K.: *"Euroexp-Report of the Specialist Group for Instrumentation"*, FFA TN 1983-56, Stockholm 1983

Table 1: Coordinates of the measurement stations

For initial flow conditions	X1130Z0940 X1130Z0750 X1130Z0563 X1130Z0500 X1130Z0437 X1130Z0250 X1130Z0060 X1193Z0940 X1193Z0750 X1193Z0563 X1193Z0500 X1193Z0437 X1193Z0250 X1193Z0060 X1255Z0940 X1255Z0750 X1255Z0563 X1255Z0500 X1255Z0437 X1255Z0250 X1255Z0060
For comparison	X2945Z0660 X2945Z0500 X2945Z0340 X3320Z0660 X3320Z0500 X3320Z0340 X4125Z0960 X4125Z0800 X4125Z0646 X4700Z1250 X4700Z1050 X4700Z0890 X4700Z0730 X5005Z1300 X5005Z1150 X5005Z1000 X5005Z0850 X5255Z1300 X5255Z1150 X5255Z1000 X5255Z0850 X5500Z1328 X5500Z1198 X5500Z1068

	U_∞ = 15.66 [m/s]				U_∞ = 16		U_∞ = 18		U_∞ = 18 [m/s]		
	pulsed-wire		Surface Fence		triangular block				S. Fence / Preston		
Stations	τ_w [N/m²]	γ_w [°]	τ_w [N/m²]	γ_w [°]	τ_w [N/m²]	γ_w [°]	τ_w [N/m²]	γ_w [°]	τ_w [N/m²]	γ_w [°]	100 c_f [-]
X2900Z0340		- 4.30	0.536	- 5.00					0.673	- 6.80	0.364
X2900Z0500		- 2.50	0.536	- 4.00					0.674	- 3.70	0.364
X2900Z0660		- 0.75	0.558	- 1.50					0.687	- 4.10	0.371
X3295Z0340	0.491	- 6.50							0.656	- 5.30	0.356
X3295Z0500	0.534	- 7.50			0.530	- 7.85	0.670	- 7.85	0.698	- 7.60	0.377
X3295Z0660	0.560	- 7.50							0.750	- 7.40	0.405
X4125Z0646	0.596	- 28.25	0.634	- 28.00					0.773	- 25.20	0.419
X4125Z0800					0.650	- 27.70	0.770	- 27.70	0.835	- 24.90	0.451
X4125Z0960	0.656	- 28.00							0.892	- 25.90	0.483
X4700Z0730	0.607	- 25.00			0.590	- 24.15	0.700	- 24.15	0.817	- 25.10	0.446
X4700Z0890	0.607	- 21.75	0.645	- 24.00					0.791	- 24.70	0.424
X4700Z1050	0.553	- 25.00							0.726	- 25.10	0.394
X4700Z1250	0.450	- 29.75			0.460	- 26.45	0.540	- 26.45	0.628	- 30.20	0.340
X5005Z0850									0.805	- 17.60	0.435
X5005Z1000									0.706	- 16.40	0.382
X5005Z1150									0.631	- 15.10	0.343
X5005Z1300									0.582	- 15.80	0.316
X5255Z0850									0.728	- 11.80	0.394
X5255Z1000									0.656	- 9.90	0.356
X5255Z1150									0.604	- 8.80	0.327
X5255Z1300									0.545	- 8.00	0.300
X5500Z1068	0.472	+0.75	0.520	+0.50	0.460	- 1.13	0.560	- 1.15	0.654	- 1.70	0.355
X5500Z1198	0.524	- 0.50	0.562	0.00					0.620	- 1.10	0.338
X5500Z1328	0.445	0.00	0.479	+0.50	0.420	+0.35	0.490	+0.35	0.586	-0.80	0.319

Table 2 : Wall shear-stress and wall flow angle measured data at different stations in the S-shaped channel

Figure 3: Coordinates of *the measured velocity profile stations* on the measurement plate

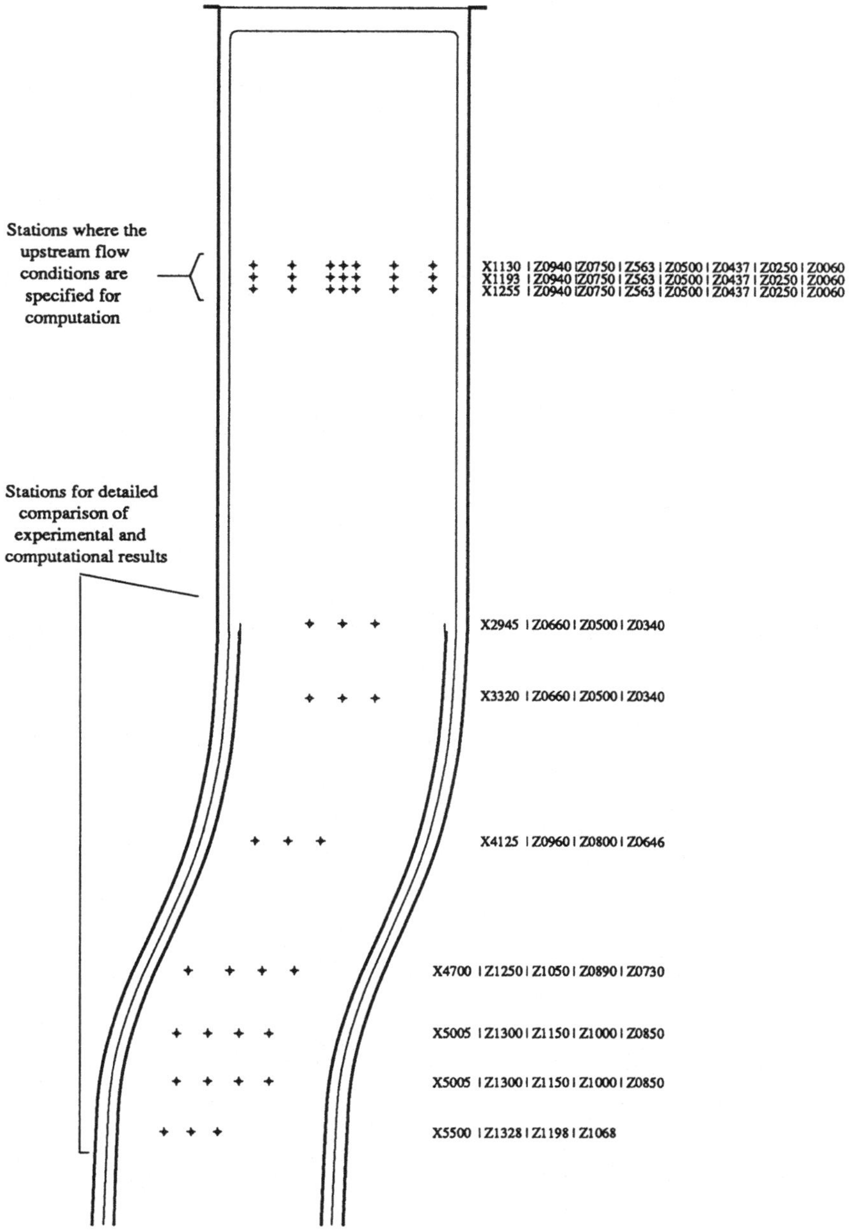

Stations where the upstream flow conditions are specified for computation

X1130 | Z0940 | Z0750 | Z563 | Z0500 | Z0437 | Z0250 | Z0060
X1193 | Z0940 | Z0750 | Z563 | Z0500 | Z0437 | Z0250 | Z0060
X1255 | Z0940 | Z0750 | Z563 | Z0500 | Z0437 | Z0250 | Z0060

Stations for detailed comparison of experimental and computational results

X2945 | Z0660 | Z0500 | Z0340

X3320 | Z0660 | Z0500 | Z0340

X4125 | Z0960 | Z0800 | Z0646

X4700 | Z1250 | Z1050 | Z0890 | Z0730

X5005 | Z1300 | Z1150 | Z1000 | Z0850

X5005 | Z1300 | Z1150 | Z1000 | Z0850

X5500 | Z1328 | Z1198 | Z1068

Figure 4: Coordinates of *the pressure taps and the plugs* on the measurement plate

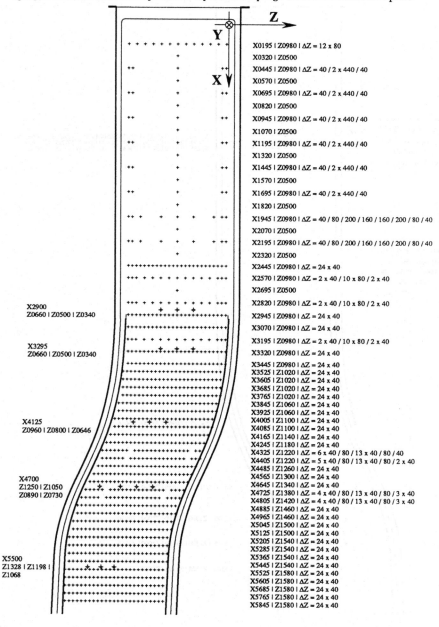

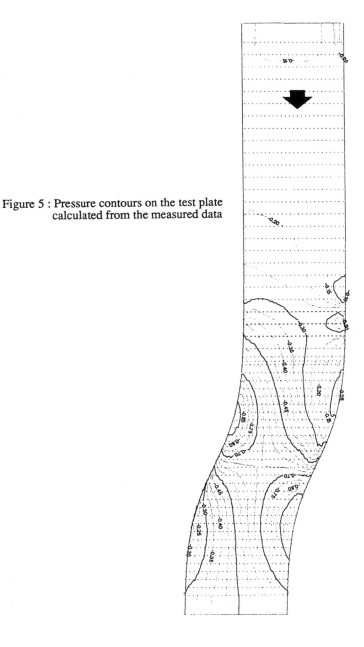

Figure 5 : Pressure contours on the test plate
calculated from the measured data

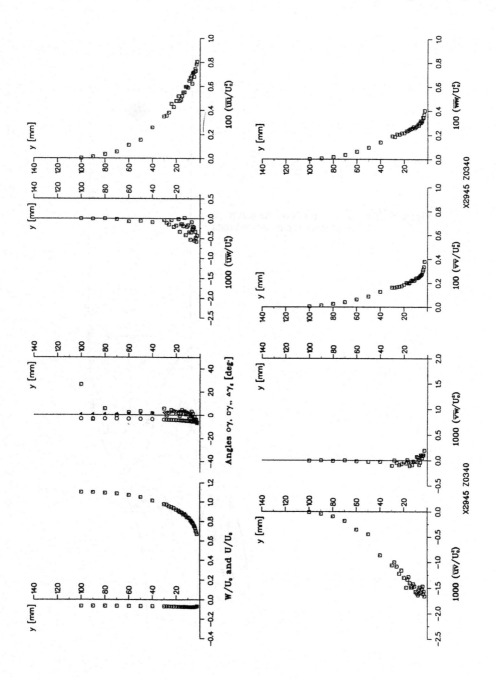

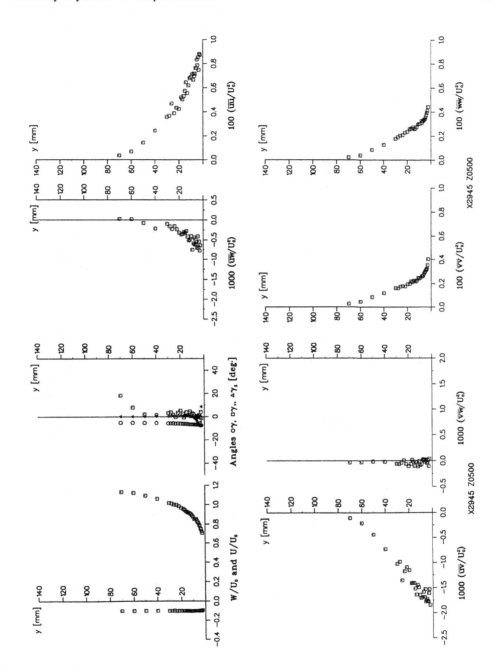

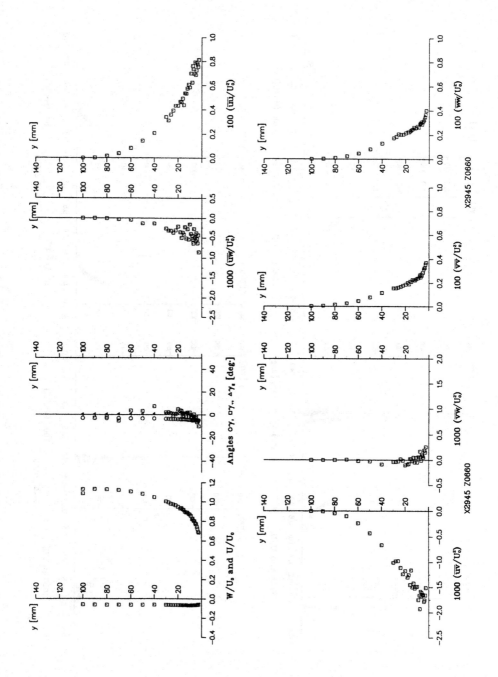

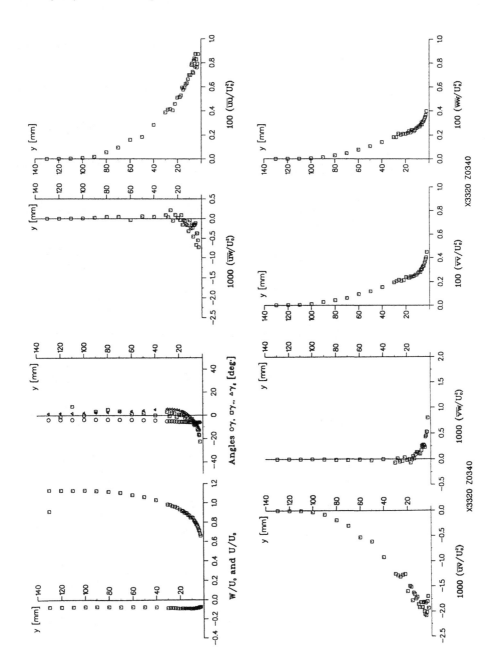

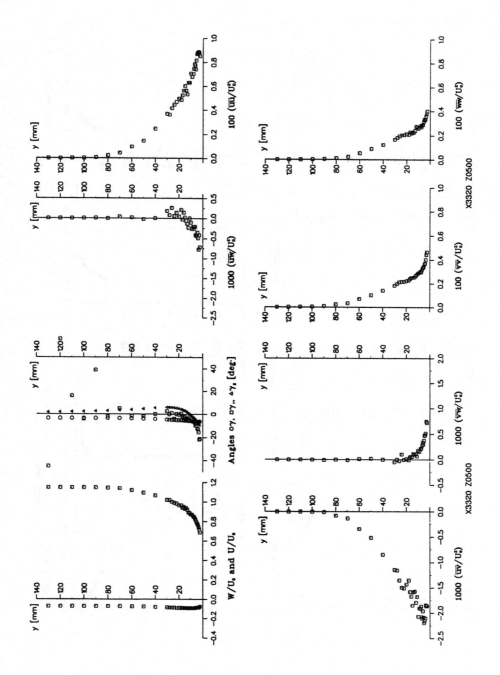

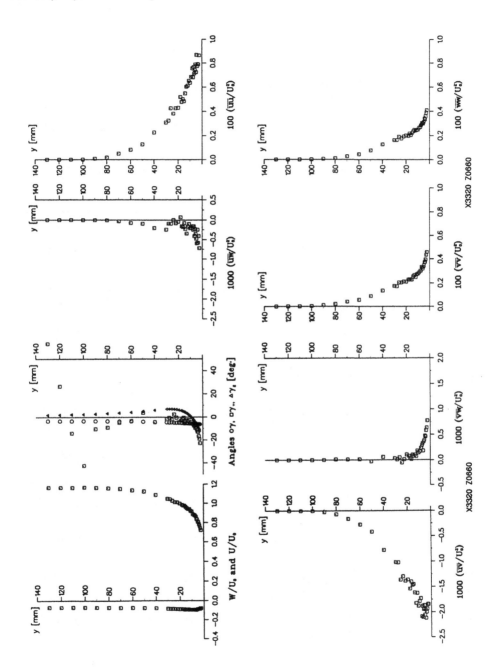

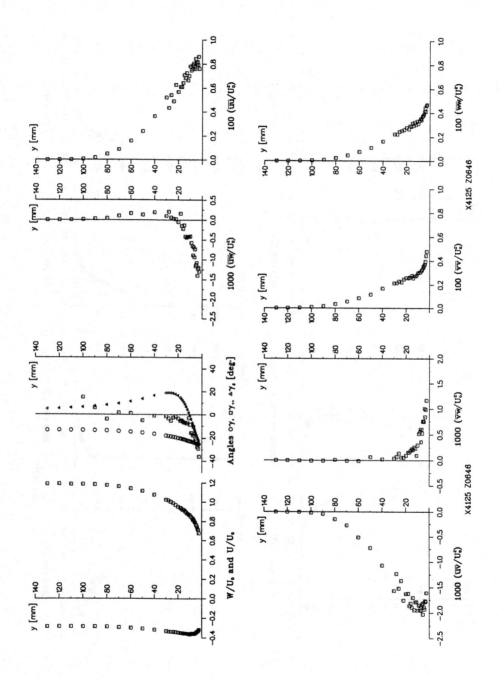

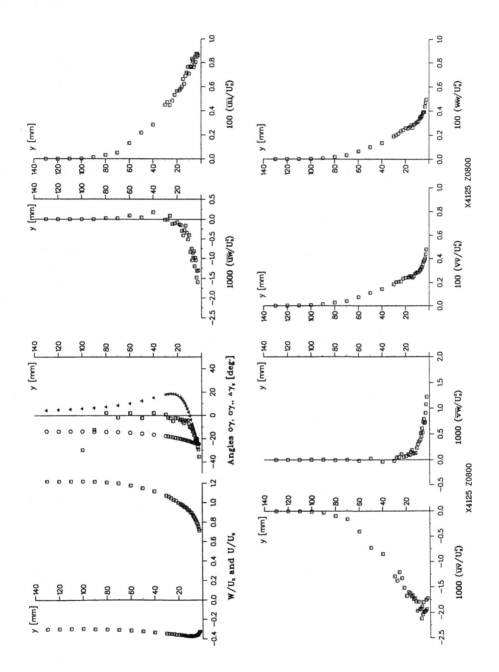

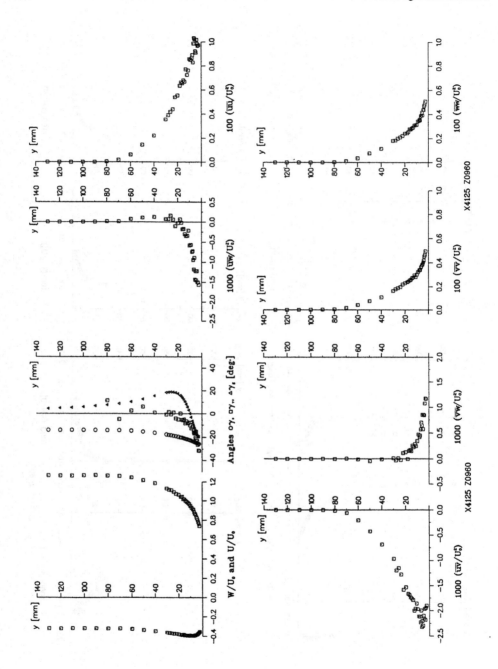

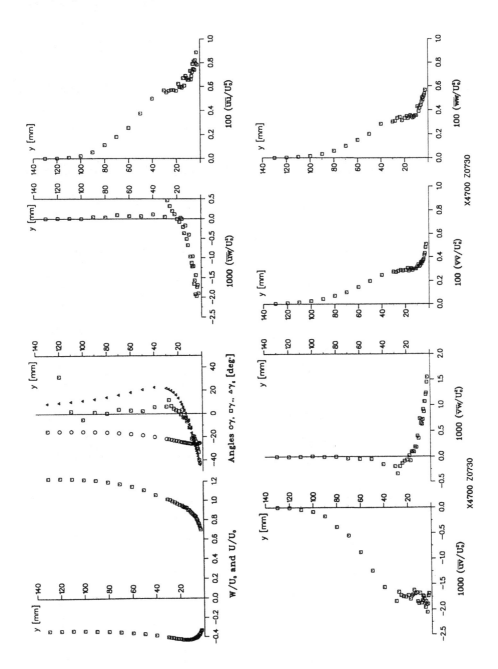

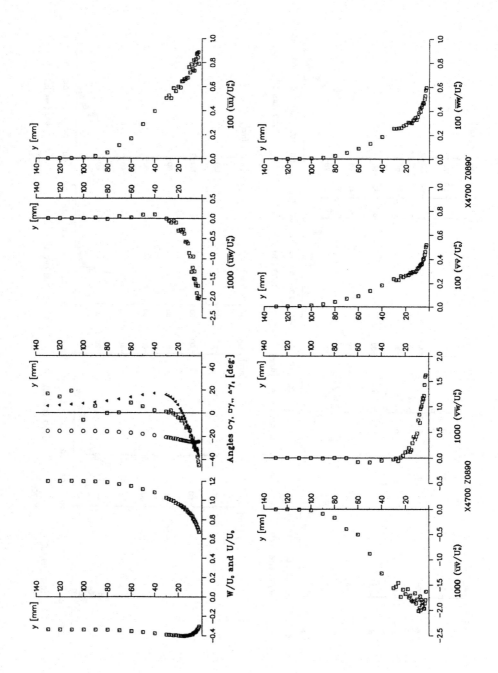

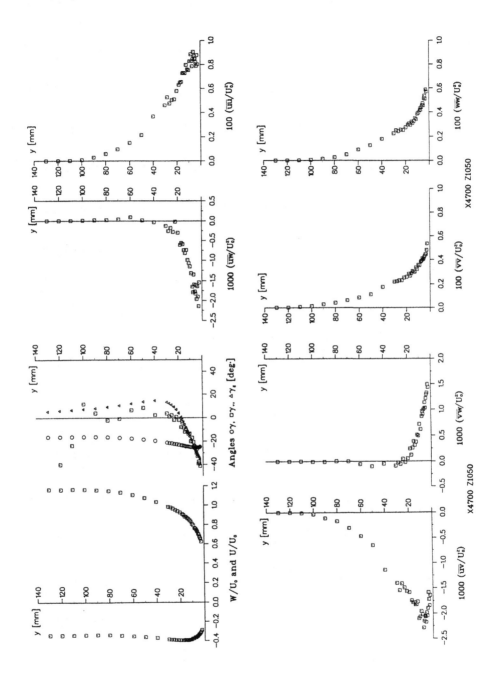

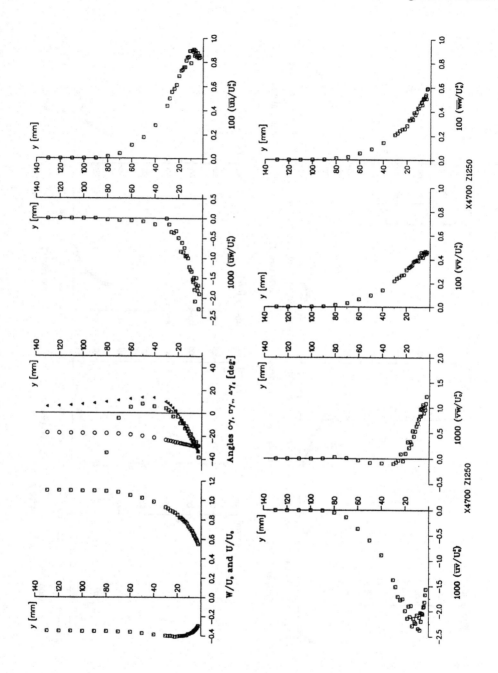

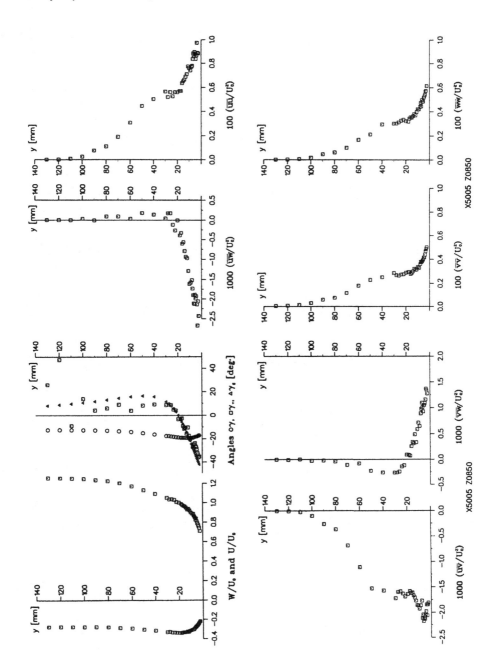

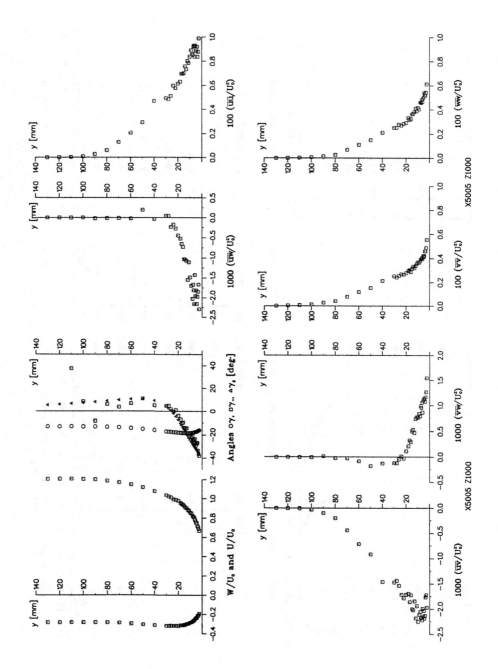

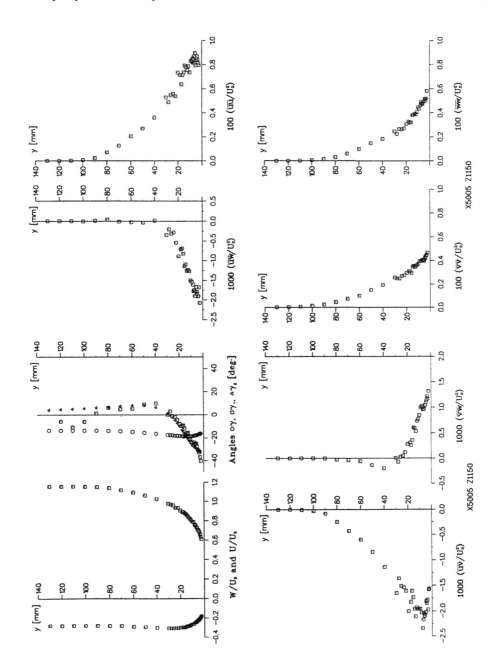

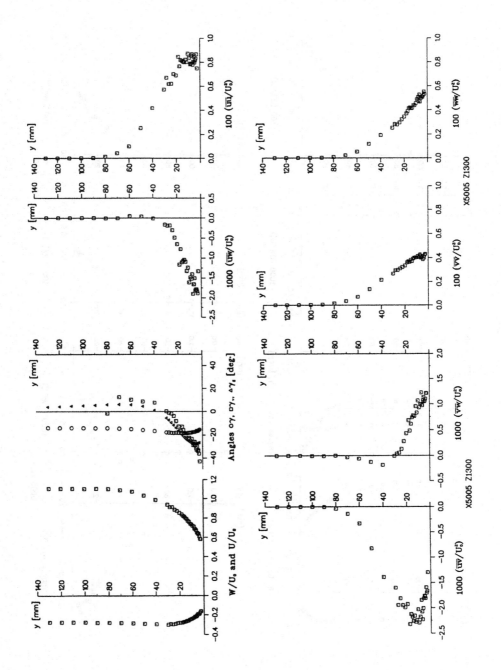

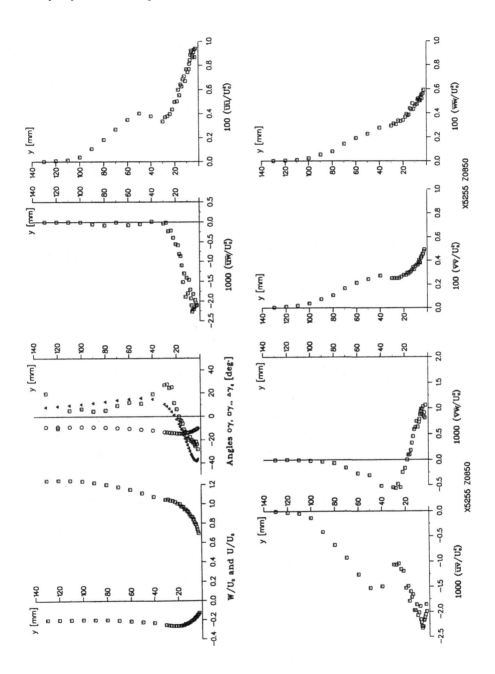

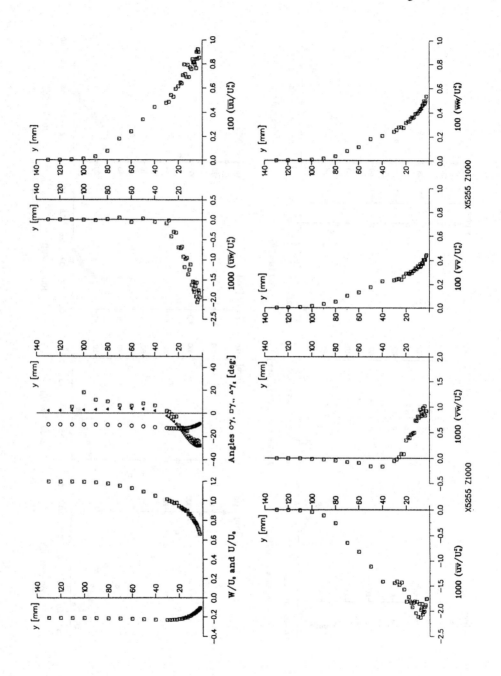

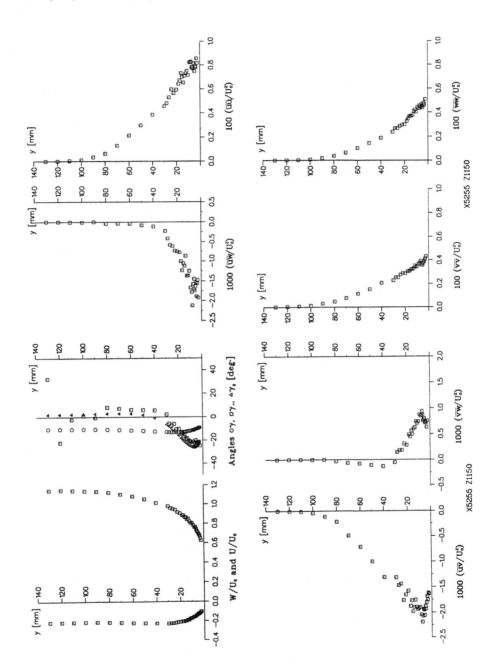

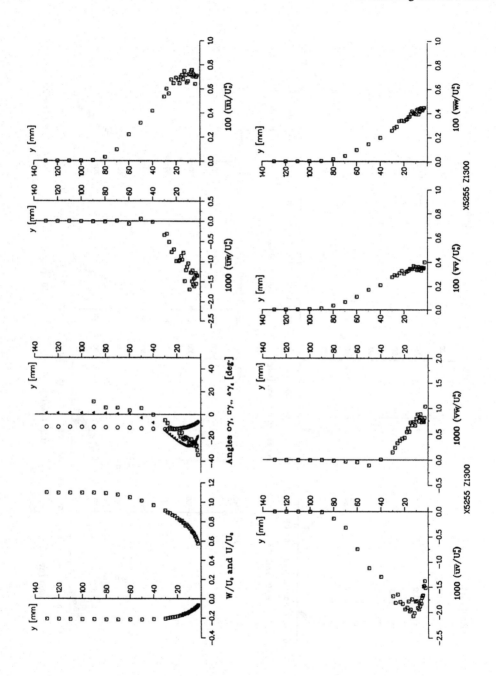

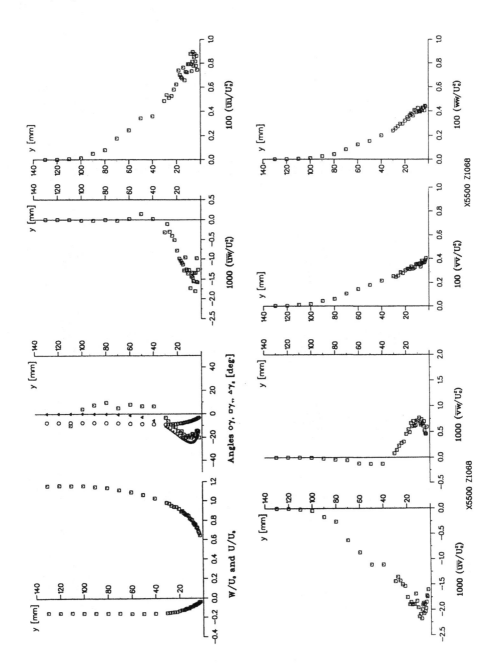

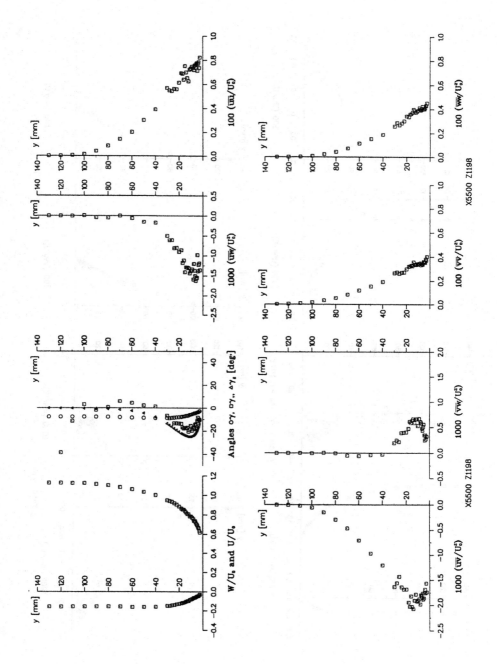

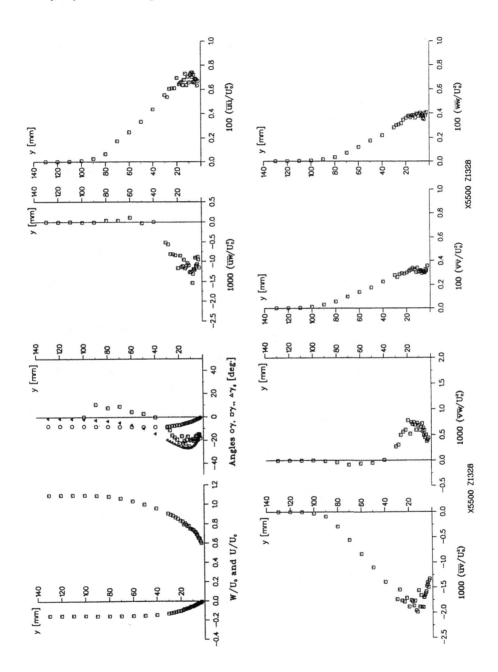

Case T1: a Boundary-Layer Computation

M. BETTELINI, T.K. FANNELØP

Division of Fluid Dynamics, ETH Zürich

1 INTRODUCTION

This paper contains a brief description of some computations carried out for the Problem T1: boundary-layer in a "S"-shaped channel.

The approach chosen is the classical one where in the first step the inviscid "external flow" is computed. Using this as boundary condition, the first-order 3D boundary-layer equations are then solved by means of a finite-difference technique. No viscous-inviscid interaction is considered. The Reynolds stresses are computed by means of algebraic turbulence models or alternate models using very simple transport equations.

The sections which follow describe the computations of the external flow and of the boundary-layer, the turbulence models used and how the initial conditions are generated For additional details the reader is referred to *Bettelini (1990)*.

2 BOUNDARY-LAYER COMPUTATION

The details of the computational method, an extension of the method developed by *Fanneløp (1968)* and *Fanneløp & Humphreys (1974)*, are described by *Bettelini (1990)*.

The most important characteristics are as follows:

- First-order boundary-layer equations in streamline coordinates
- steady, incompressible
- direct-mode integration
- finite-difference method, full implicit
- accuracy: 1. order in the streamwise direction
                2. order in the crossflow and the wall-normal directions

The dependence principle is taken fully into account.

116

3 TURBULENCE MODELLING

The models used are:

a) Algebraic models (isotropic):

    - Cebeci & Smith (1974)
    Fanneløp & Humphreys (1974)
    - Baldwin & Lomax (1978) (not presented)
    - Michel et al. (1968)

b) Models using simple transport equations (isotropic):

    - Johnson & King (1984) (not presented)
    - Johnson & King modified *(Bettelini, 1990)*

The transport equation used in the *Johnson & King* model is applied to 3D boundary-layers by introduction of an additional convective term in the crossflow direction, as done by *Abid (1988)*. In the modified version, the eddy viscosity distribution is written as

$$\mu_t = \mu_{to} \left(1 - \exp - (\mu_{ti} / \mu_{to})^n \right)^{1/n} , \quad n=2$$

By setting $n = 1$ the original formulation is recovered.

More complete results are given in *Bettelini (1990)*.

4 EXTERNAL FLOW

The external flow has been computed both from potential-flow theory and based on the experimental C_p distribution. As the boundary-layer equations are formulated and solved in streamline coordinates, the streamlines outside the boundary-layer, as well as the metric coefficients and the curvatures of the coordinate lines, must be computed very accurately.

The potential-flow computations have been carried out using:

    a) *Uchida's (1980)* method (iterative computation of streamlines and orthogonal lines using the continuity equation and the condition of vanishing vorticity)
    b) The panel method developed by *Eriksson (1975)* (vorticity distribution on the surface).

The computations of the external streamlines using the experimental C_p-distribution proceed as follows:

a) Independent integration for each individual streamline by an explicit 4.-order Runge-Kutta scheme, of the equation

$$\frac{\partial \alpha_{ext}}{\partial s} = \frac{\partial \alpha_{ext}}{\partial x} + \tan \alpha_{ext} \frac{\partial \alpha_{ext}}{\partial z} = \frac{1}{2(1-C_p)}\left[\frac{\partial C_p}{\partial x} \tan \alpha_{ext} - \frac{\partial C_p}{\partial z}\right]$$

where s denotes the arc-length along the streamlines. This expression is derived from the Euler equations, using

$$u = Q_{ext} \cos \alpha_{ext}$$

$$w = Q_{ext} \sin \alpha_{ext}$$

$$\frac{Q_{ext}}{U_\infty} = \left(1 - C_p\right)^{1/2}$$

b) Computation of the orthogonal lines and integration of the stream function along these by means of

$$\psi = \psi_0 + \int Q_{ext}\, dn$$

where n denotes the arc-length along the orthogonal lines.

c) Correction of the streamline position by means of the stream function. ψ_0 is given by the condition, that the mean location of the streamlines must be unchanged.

d) Correction of the orthogonal lines.

e) The required values for C_p and their partial derivatives in the x and z-direction are computed at each location using locally one-dimensional polynomial approximations of variable order, computed by least squares.

The difference between the pressure distribution computed using the potential-flow method and that obtained from measurements, has been found to be important. It has considerable influence on the results of the boundary-layer calculations. Therefore only results computed with the experimental external pressure distribution will be presented.

5 INITIAL CONDITIONS

Only the mean velocity profiles along a section perpendicular to the external streamlines are required. The experimental profiles given at x = 1.193 m are used here. The initial profiles for the computation are generated as follows:

a) Smoothing of the experimental profiles using cubic splines
b) Linear interpolation to compute the velocities at each z-location required.

6 RESULTS

Grid-independence of all results presented has been verified by variation of the stepsizes by a factor 1.5 to 2.

The computational domain defined by the initial conditions, according to the dependence principle, is presented in Fig. 1. Results are presented only for stations located inside this domain. The streamlines are calculated based on the experimental C_p-distribution.

The spread in the results obtained for the different turbulence models investigated is found to be small.

7 REFERENCES

Abid, R. (1988): Extension of the Johnson-King turbulence model to the 3-D flows. *AIAA Paper 88-0223.*

Baldwin, B., Lomax, H. (1978): Thin-layer approximation and algebraic model for separated turbulent flows. *AIAA Paper 78-257.*

Bettelini, M. (1990): Diss. ETH Zürich No. 9182

Cebeci, T., Smith, A.M.O. (1974): *Analysis of Turbulent Boundary Layers.* Academic Press.

Eriksson, L.E. (1975): Calculation of two-dimensional potential flow wall interference for multi-components airfoils in closed low speed wind tunnels. *FFA AU-1115.*

Fanneløp, T.K. (1968): A method for solving the 3D laminar boundary-layer equations with application to a lifting reentry body. *AIAA Journal* 6, 1075-1084.

Fanneløp, T.K., Humphreys, D.A. (1974): A simple finite-difference method for solving the threedimensional turbulent boundary layer equations. *AIAA Paper 74-13*.

Johnson, D.A., King, L.S. (1984): A new turbulence closure model for attached and separated turbulent boundary layers. *AIAA Journal* 23, 1684-1692.

Michel, R., Quémard, C., Durant, R. (1968): Hypotheses on the mixing length and application to the calculation of the turbulent boundary layers. In *Proceedings - Computation of Turbulent Boundary Layers. Vol. I: Methods, Predictions, Evaluation and Flow Structure.* 1968 AFOSR-IFP-Stanford Conference, Ed. S.J. Kline et al..

Uchida, S. (1980): Theory of streamline analysis method referred to the orthogonal curvilinear coordinates. *Memoirs of the Faculty of Engineering, Nagoya University,* Vol. 32(1).

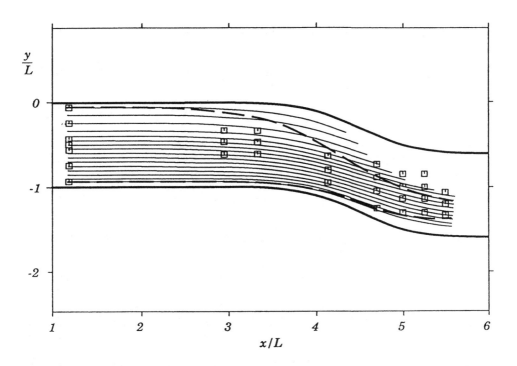

Fig. 1   *Computational domain defined by the given initial conditions, according*
to the influence principle (computed using the turbulence model CS).

Calculation of Test Case T1 at ONERA/CERT/DERAT

C. GLEYZES, R. HOUDEVILLE, C. MAZIN, J. COUSTEIX

ONERA-CERT, Toulouse (FRANCE)

1 INTRODUCTION

Calculations of the test case T1 of ERCOFTAC workshop have been performed using a new field method developed at ONERA/CERT/DERAT. This method is using three different edge conditions deduced from experimental pressure distribution through preliminary calculations. We will briefly present here the main results of these calculations. More details can be found in [4].

2 DETERMINATION OF EDGE CONDITIONS

A method providing edge velocity direction and magnitude from measured pressure distributions had been developed for fuselage type bodies [3]. For incompressible irrotationnal flows we solve, in a general surface coordinate system, the Bernoulli equation associated with $\Omega_3 = 0$, where Ω_3 is the vorticity component normal to the wall.

Starting from initial flow direction, integration is performed by x-marching, the transverse derivatives being computed according to influence-dependence rules. For an open domain, boundary conditions should be imposed in the areas where fluid enters the domain.

For the present calculations, pressure distributions were measured on a mesh not really adapted to the geometry of the channel, and consequently no more adapted to calculations. In particular, pressure measurements were not available at some stations, in the vicinity of the lateral walls. In addition, there was a strip of 60mm on both sides of the test plate where initial conditions were not provided. The calculation domain was then limited to a core of $880\,mm$ in Z, centered on the centerline of the test plate, on which pressure distribution was interpolated by cubic splines. As a consequence, it was difficult to give proper boundary conditions on the sides of the calculation domain, and three types of calculations were performed.

For the first one (A), flow was assumed to be parallel to the lateral tunnel walls, which seemed at that time to be the most logical assumption as no information was available from experiments. However, boundary layer results showed some unexpected behaviour (see paragraph 3-2), and a second one (B) was then done, with no boundary conditions imposed (allowing the flow to get in or out the calculation domain). In both cases, initial flow direction was given by boundary layer surveys at $X = 1193mm$. A third case was computed to look at the influence of initial conditions, assuming 2D flow at $x = 1193\,mm$ (which is not true in the experiments), and flow parallel to the lateral walls (C). We have plotted, on figure 1, the evolutions of edge velocity direction in some characteristic sections. Although initial measured angles are small (around $2°$ in $z < 0$ direction), their influence remains visible all along the test plate. Influence of lateral boundary conditions (using same initial con-

ditions) seems much smaller, except in the areas were flow can enter the computation domain. We will see later the consequences on boundary layer calculations.
A second important comment concerns pressure distributions : at all stations we have noted (see fig. 6 and 7), differences on magnitude of edge velocity up to 5% between boundary layer survey ($Q_e = \sqrt{U_e{}^2 + W_e{}^2}$) and pressure measurements ($Q_e = U_0\sqrt{1 - C_p}$). As our definition of edge velocity fits the second definition, this can perhaps explain some differences between computed and measured boundary layer profiles, even in the straight initial part of the test section.

3 BOUNDARY LAYER CALCULATIONS

3.1 Basic Equations
The basic equations are the continuity and momentum equations .
In these equations the pressure field is a datum of the problem. It is well known that the boundary layer equations are parabolic, due to the diffusion terms. However, if the sub-set of equations containing only the first order partial derivative terms is considered, then its nature is hyperbolic. This implies the existence of influence and dependence domains, in relation with the characteristic surfaces. This property has already been used by CEBECI for his characteristic box scheme [1]. The streamlines are characteristic lines as well as the lines normal to the wall. Discretization of the equations along these lines ensures dependence rules are automatically satisfied. Details can be found in [5].

3.2 Discretization Scheme
The discretization scheme is the simplest one can imagine, although it is fully implicit(report to Figure 2 for the following details). We suppose that the velocity field is known at x^U (line KLM). To calculate u at point Q of the downstream station x^D, the diffusion term is discretized at point Q over 3 points in the η direction. The first derivative of u in the η direction is also calculated using a centered discretization.
To calculate the term $du/ds(\eta)$, it is necessary to know the velocity at point R(η) at the upstream location. u^R is obtained by interpolating the velocity between KL or LM, depending on the direction of the velocity at point L.
After rearrangement of the unknowns, a set of linear equations is obtained for the unknown quantities $u(\eta_j)$ the coefficients of which being only functions of the velocity at point R_j. Therefore, the calculation of u at station Q does not depend on the velocity at the other points of the downstream station. The w-component of the velocity is calculated by the same way, using the momentum equation in the z-direction. v is obtained through the continuity equation.

3.3 Turbulence Model
As the code used to perform the calculations of the boundary layer in the S-shape channel is still under development, it is based on a very simple turbulence model. The molecular and turbulent stresses are assumed to be in the same direction, the eddy viscosity μ_t being calculated from the mixing length formulation proposed by Cebeci [2].

4 RESULTS

4.1 Computation conditions
All calculations were performed on a grid with 25 points in Z, 33 to 48 points in Y (self adaptative grid with first point at $y^+ = 10$), and around 900 points in X (with adaptative X steps). Computing times on a CRAY XMP14SE were around 22s.

3.2 Computation results
First calculation was performed with edge condition (A). In this case, we imposed transverse velocity to be 0 in the boundary layer along the lateral boundaries, to be consistent with the outer edge conditions. Mathematically, boundary conditions can be imposed only where characteristic lines enter the calculation domain. In fact, it is not the case along some parts of the lateral boundaries, in particular at the inflexion point of the channel, at the inner side of the bend, and this leads to a fast thickening of the boundary layer on the considered line of the mesh, until 2D separation occurs (negative velocity at the wall). This can be seen on the computed wall streamlines plotted on figure 3.

Generally speaking, the problem of the conditions to impose on the lateral boundaries is difficult. We tried then to use edge conditions (B), without any imposed conditions on the lateral boundaries. In fact, if we do not force these conditions, with the proposed discretization scheme, the velocity on the lateral boundaries is calculated exactly as at any other point. If the velocity enters the domain, we just impose $\partial/\partial z = 0$. Plot of the corresponding wall streamlines (Fig. 4) shows now computational access to the whole calculation domain, and gives very similar results compared with the previous one. Wall angles at the downstream boundary are however slightly lower.

To have an idea of the influence of initial conditions, last calculation was performed with edge conditions (C), assuming initial flow to be 2D, and imposing no transverse flow on side boundaries as for case (A). We have not plotted these results as they are quite similar to those for case (A).

We will not discuss here in details comparisons with experiments which will be done globally by the editor. We will just give, on figures 6 and 7, examples of comparisons between experiments and calculation (B) at stations $X = 4700\,mm$ and $X = 5225\,mm$, that is to say around and after the inflexion point of the S–channel. Note that $X^2 = 0$ corresponds to the most negative values of the tunnel reference coordinate Z. We have plotted on the upper graph the evolutions of edge velocity and wall shear stress τ_p, and on the lower one the evolutions of edge and wall angles (α_e, γ_w). One can find here the difference between measured edge velocity and velocity deduced from pressure measurement discussed at paragraph 2. As everybody knows the hyperbolic properties of Euler equations, these small differences amplified all along the calculation domain may have large effects at the end of the test section. This can explain part of differences between measured and computed edge velocity angles (shift of around 3° to 4°). We have also plotted, with a $(\bigtriangledown)$ the location of the most extreme wall streamline issued from initial conditions, setting the limit of the theoretical influence domain, and we can note the accidents on the results for larger values of X^2. Accidents for low X^2 values are probably due to the fact that in this area, flow is getting out of the domain, and thickens the side wall boundary layer, leading to a corner flow which is interacting with the boundary layer on the test plate. Determination of edge conditions and boundary layer calculations are consequently no more valid in this area, the size of which we cannot determine. Away from these zones, if agreement on wall angle is not too good, the change in sign of β_0 (angle between edge and wall streamlines direction) seems to be well predicted. We can also notice the fairly good agreement on magnitude of wall shear stress.

5 CONCLUSIONS

Compared to a classical discretization, the solution we adopted allows to take into account influence and dependence rules and gives access, for instance to both sides of the open separation on a prolate spheroid, for which no problem of lateral boundary conditions occurs, till negative velocities appear in X^1 direction [5]. Of course, the simple mixing length model is a bit crude for such kind of turbulence, and improvements are in progress about implementation of new turbulence models. Concerning test case T1, it seems that available experimental data did not allow precise enough definition of edge conditions for boundary layer calculations. It would perhaps have been fruitful to provide edge conditions identical for all participants. In the present case, the uncertainties so introduced may perhaps distort the conclusions about turbulence models.

6 REFERENCES

[1] T Cebeci, A A Khattab, K Stewardson : *"three-dimensional laminar boundary layers and the OK of accessibility."* J.F.M. 107, 1981.

[2] T Cebeci, A M O Smith : *"Analysis of turbulent boundary layers ."* Academic Press, 1974.

[3] C Gleyzes, J Cousteix : *"Calcul des lignes de courant à partir des pressions pariétales sur un corps fuselé."* La Recherche Aérospatiale 1984-3.

[4] C Gleyzes : *"Application de la méthode des caractéristiques au calcul de l'écoulement dans le canal en S de l'EPFL."* R.T. DERAT 38/5625.32 oct 1990

[5] R Houdeville, C Mazin :*"Calcul de couches limites tridimensionnelles par une méthode de caractéristiques."* R.S. ONERA/CERT/DERAT 36/5025.31, 1990.

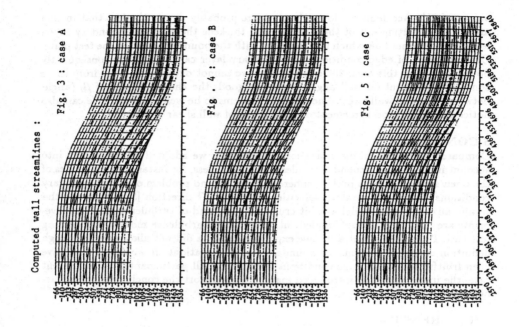

Computed wall streamlines :

Fig. 3 : case A

Fig. 4 : case B

Fig. 5 : case C

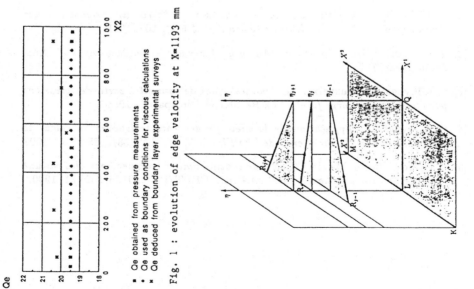

Fig. 1 : evolution of edge velocity at X=1193 mm

■ Qe obtained from pressure measurements
♦ Qe used as boundary conditions for viscous calculations
× Qe deduced from boundary layer experimental surveys

Fig. 2 : discretization of momentum equations

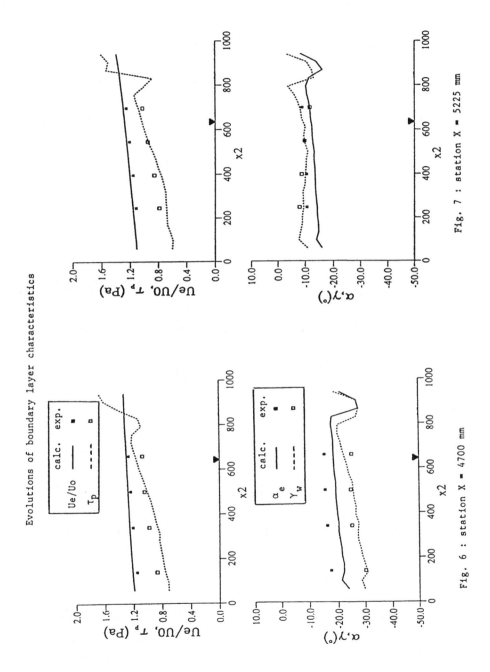

Evolutions of boundary layer characteristics

Fig. 6 : station X = 4700 mm

Fig. 7 : station X = 5225 mm

"S"- CHANNEL CALCULATIONS

Per-Åge Krogstad & Per Egil Skåre

NTH, Trondheim, Norway

1 NUMERICAL CALCULATIONS

1.1 General

Our method uses the boundary layer equations written in streamline coordinates as found in textbooks such as ref. [1]. The method which has different options for solving the equations is fully documented in ref. [2]. In addition to 2D equations, solutions along lines of symmetry, small crossflow approximations and the fully 3D boundary layer equations may be solved. In all cases the equations were transformed using a Lees-Levy type transformation in the streamwise-wall perpendicular plane and a coordinate scaling in the lateral direction. The calculations for the workshop have been performed using a mixing length formulation for the shear stress, but transport equations may also be used.

1.2 Numerical method

For all calculation a stretched grid was used normal to the surface. The stretching was performed using a geometric progression, where each vertical interval was increased with respect to the lower by 20%.

Calculations have been performed using different approximations to the equations of motion. In the small crossflow approximations all the lateral derivatives disappear. Thus calculations may be performed along one line at a time. The lines were placed so that they passed through the measured stations. The equations were solved by a three point centred difference vertically and a two point upwind difference in the streamwise direction. The discretized equations are implicit second order accurate in the vertical and first order accurate in the streamwise direction.

For the fully 3D equations the Alternative-Direction-Implicit method (ADI) was used. Here both the wall perpendicular and lateral differences are made as three point centred differences, making a strong lateral coupling between streamlines. The discretized equations are implicit second order accurate, both in the vertical and lateral direction. The solution is obtained by alterning between vertical sweeps from the wall to the freestream and lateral sweeps from one side to the other.

1.3 Turbulence model

The mixing length formulation of Michel & al. [3] was used with the low Reynolds number and pressure gradient corrections of Cebeci [4]. The eddy viscosity was assumed to be isotropic.

1.4 External velocity field

Originally the specified pressure field was used to obtain the external field. However, some scatter in the data produced wiggly streamlines and external velocity field. It was therefore necessary to smooth the measured pressure field. This was done by using BSplines to approximate the CP derivatives. Unfortunately the new distribution made it impossible for us to reach two of the specified measurements stations. From the external field a streamline coordinate system was constructed and the metrics and geodesic curvatures calculated. Calculations using the small crossflow approximations were also performed using the Euler flow external solution which, however, deviate somewhat from the measured distribution.

1.5 Grid dependence

This was checked by varying the number of grid points and observing the change in the skin friction using the ADI-method at a fixed position. This was taken to be the measurement stations at $X=5500$ and $Z=1198$, near the exit of the duct. As a reference we used 70 grid points in the wall perpendicular direction, 22 streamlines and 220 steps downstream. This is also the values we used for the data presented. Changing to 100 vertical grid points gave an increase in Cf of 0.8%. Reducing the number of streamlines to 11 gave a 9,4% increase and cutting the number of streamwise steps by 2 to 110, gave 0.3% reduction.

1.6 Initial profiles

The calculations were started by generating a profile from the law of the wall and Cole's wake function. To do this Cf and θ were specified. Originally we wanted to use the measured Cf together with the θ obtained by integrating the given velocity profile. However, no profile could be specified by this combination. We then used our own data reduction program on the velocity profile and obtained a 25% lower value of the skin friction, which gave us no problem in generating initial profiles.

2 CALCULATED RESULTS

2.1 Zone of influence

For all calculations the zone of influence was mapped after the calculations had been finished. Although the accessible zone extends all the way through the specified part of the duct, this zone becomes very narrow in the downstream part. This affects the predicted solution for most of the stations in the downstream part of the "S".

2.2 Case identification
We have supplied three sets of calculations, named Case 1, 2, 3:

Case 1.   This calculation used Euler solution as the external field. Case 1 was computed using the small crossflow approximation.

Case2.   In this calculation the external field was derived from the smoothed measured pressure field. Again the small crossflow approximation was used. The difference between Case 1 and 2 demonstrates the strong deviation between the idealized Euler field and the real external field.

Case 3.   Here we used the ADI method. This is the calculation that compared best the measurements. In this case the complete set of equation of motion is solved subject to the external conditions obtained from the smoothed measured pressure field. Unlike the small crossflow calculations, where the zone of influence degenerates to the streamline itself, the ADI solution is restricted to the flow field given by the zone of influence of the initial stations. Although the calculation domain was wider than this zone of influence, the computations did not blow up when the influence principle was violated. However, the results were found to be strongly affected. We therefore do not present any results that appear outside the computed domain of influence, shown in figure 1.

2.3 CPU-time
Using the gridsize described in 1.5, the calculation on a MicroVAX 3100 used 11.5 seconds per streamline for the two first cases. Using the ADI method as in Case 3, it took 19.1 seconds per line.

2.4 Results
We have chosen the station at X=4700 and Z=1050 to show the results of the different cases. The results are shown in figures 2 to 4. The parameters plotted are:

$$U/U_\infty \, , \, W/U_\infty \, , \, uv/U_\infty^2 \, , \, vw/U_\infty^2 \, , \, \gamma = \tan^{-1}\left(\frac{W}{U}\right) \, , \, \gamma_t = \tan^{-1}\left(\frac{\tau_z}{\tau_x}\right)$$

It was observed from this and the other stations that Case 1, using the Euler flow solution, deviates considerably from the measurements, while Case 2 and 3 gave a reasonable agreements with the measurements. We also noticed that there is only a slight difference between the small crossflow approximation and the ADI-method, the latter being as expected generally slightly better.

References

[1] White, F.M.   "Viscous fluid flows" McGraw-Hill, 1974
[2] Krogstad, P.Å.   "A method for calculating three dimensional boundary layers
                        on arbitrary surfaces" IMHEF rep. T-2-85, EPFL, 1985
[3] Michel, R. & al.   "Hypotheses on the mixing length and the application to the
                        calculation of the turbulent boundary layers".
                        Proc. Compuation of Turbulent boundary layers,
                        AFOSR-IFP Stanford Conference 1968.
[4] Cebeci, T.   "Kinematic eddy viscosity at low Reynolds numbers".
                        AIAA J. Vol. 11, 1973

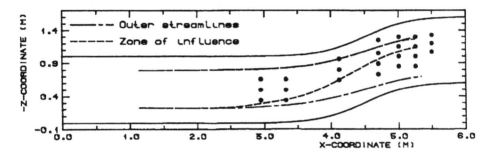

Figure 1. The domain of influence for the ADI-method.

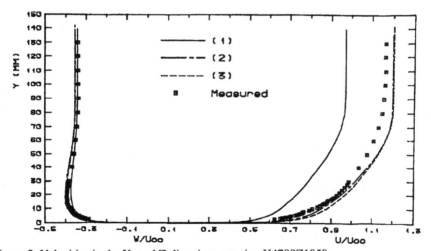

Figure 2. Velocities in the X- and Z-direction at station X4700Z1050

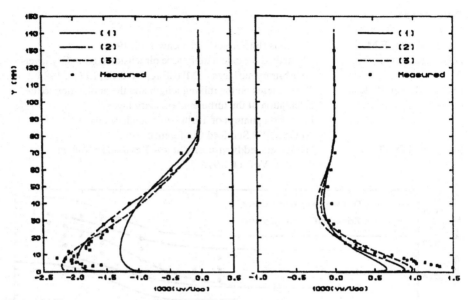

Figure 3. Shear stresses at station X4700Z1050

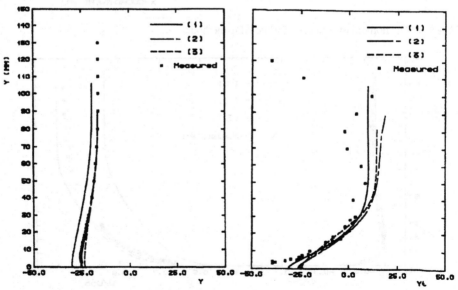

Figure 4. γ and γ_t (deg,) at station X4700Z1050

An anisotropic eddy viscosity model, using the k-ε model or an extended Johnson & King model for solving the boundary layer equations

Per-Åke Lindberg

Fluid Mechanics Laboratory
Swiss Federal Institute of Technology, Lausanne, Switzerland

1. INTRODUCTION

In the following is presented the turbulence models and solution procedures used in this contribution. Due to the fact that the space is very limited, this presentation is very brief. A more detailed description of the turbulence methods used is found in the listed references, especially Lindberg (1991).

To be able to calculate the complex velocity field in the S-duct, an anisotropic eddy viscosity is used. The anisotropy is accounted for by the parameter Tg, defined as the tangent of the angle between the Reynolds stress vector and the gradient vector of the mean velocity as proposed by Moreau (1988). With the Boussinesq approximation this gives for the two main components of the Reynolds stresses (the use of the name Reynolds stresses is not entirely correct here as it is the Reynolds stresses divided by the density usually)

$$- \overline{uv} = \frac{\nu_t}{\sqrt{1 + T_g^2}} \left(\frac{\partial U}{\partial y} - T_g \frac{\partial W}{\partial y} \right) \qquad - \overline{wv} = \frac{\nu_t}{\sqrt{1 + T_g^2}} \left(\frac{\partial W}{\partial y} + T_g \frac{\partial U}{\partial y} \right)$$

If T_g is set to zero the isotropic eddy viscosity is obtained, the Reynolds stresses are in this case parallel with the mean velocity gradient components. The value of Tg was obtained by solving a transport equation for the ratio of the Reynolds stresses in the spanwise and streamwise direction. This equation is derived from the model equations for $\overline{u_i u_j}$ suggested by Launder, Reece & Rodi (1975). The eddy viscosity is obtained from either the eddy viscosity model proposed by Johnson & King (1985) or from the k-ε model proposed by Chen & Patel (1988). The values of the modelling constants used here are those proposed by the different authors respectively, if nothing else is stated.

133

The calculations were made by using the boundary layer equations written in a curvilinear coordinate system where the x-axis is in the local direction of the external flow, the y-axis in the direction normal to the wall and the z-axis in the spanwise direction. The metrics are in x and z direction denoted by h_1 and h_3 and h_2 is taken equal to unity. The geodesic curvatures K_{13} and K_{31} are defined as

$$K_{ij} = \frac{1}{h_i h_j} \frac{\partial h_i}{\partial x_j}$$

2. TURBULENCE MODELS

2.1 The Johnson - King Model

As a simple model for the eddy viscosity, the model proposed by Johnson & King (1985) is chosen. It is extended to be valid in a 3D curvilinear coordinate system, by replacing the 2D Reynolds stress - $\overline{uv}$ with its 3D counterpart in the x-z plane. Hence, the assumption that the ratio of the Reynolds stress and the turbulent kinetic energy is constant is modified to the form

$$\frac{\sqrt{\overline{uv}^2 + \overline{wv}^2}}{k} = a_1$$

Also the mean velocity in the streamwise direction is replaced by the magnitude of the mean velocity in certain places. The production term in the original k-equation, written in a curvilinear coordinate system, will cause new terms to appear due to the curvature of the coordinate system. Hence, new terms will also be introduced in the equation for the maximum value of the Reynolds stresses, which is solved in terms of the variable

$$g(x,z) = \left(\sqrt{\overline{uv}^2 + \overline{wv}^2} \right)_{max}^{-1/2}$$

The final equation then becomes

$$\frac{U}{h_1} \frac{\partial g}{\partial x} + \frac{W}{h_3} \frac{\partial g}{\partial z} = \frac{a_1}{2L} \left(1 - \frac{g}{g_{eq}} + D \right)$$

$$- \frac{g}{2} \left(- \frac{\overline{u^2}}{k} W K_{13} - \frac{\overline{w^2}}{k} U K_{31} + \frac{\overline{uw}}{k} (U K_{13} + W K_{31}) \right)$$

The eddy viscosity is then obtained by using the 2D solution procedure for the g-equation proposed by Johnson & King (1985).

2.2 The k - ε Model

As a more complex model for the the eddy viscosity a standard k-ε model is chosen, where two transport equations for the turbulent kinetic energy, k, and dissipation, ε, are solved. The equations are written in a standard form using a curvilinear coordinate system, see Ekander & Johansson (1989), the only term that is affected by the curvature is the production term P which becomes

$$P = - \overline{uv} \, \frac{\partial U}{\partial y} - \overline{vw} \, \frac{\partial W}{\partial y} - \overline{u^2} \, WK_{13} - \overline{w^2} \, UK_{31} + \overline{uw} \left(UK_{13} + WK_{31} \right)$$

The boundary conditions for k and ε are taken from Chen & Patel (1988). In their approach no wall function is used and the k-equation solved for the entire flow field. The ε-equation is not solved in the region close to the wall, instead the dissipation is calculated from an algebraic expression using k and an algebraically length scale is obtained.

$$\varepsilon(R_y \le 250) = \frac{k^{3/2}}{l_\varepsilon} \qquad l_\varepsilon = C_1 \, y \left(1 - e^{- \, R_y/A_\varepsilon} \right) \qquad R_y = \frac{y\sqrt{k}}{\nu}$$

The eddy viscosity, ν_t, is finally calculated in the standard fashion, with a damping function close to the wall.

2.3 Model for the ratio of the Reynolds stresses

To calculate the anisotropy parameter Tg a model for the ratio of the Reynolds stresses is used which is based on the model equations for the Reynolds stresses proposed by Launder, Reece & Rodi (1975). From their model equations for $\overline{uv}$ and $\overline{vw}$ an equation for the ratio of the two Reynolds stresses is derived as follows. The equations are first written in tensor notation i.e.

$$\frac{D \, \overline{u^k u_n}}{Dt} = P^k_{\ n} + \Pi^k_{\ n} - \varepsilon^k_{\ n} + D^k_{\ n}$$

where the terms on the right hand side are the production term, the pressure-strain term, the dissipation term and the diffusion term respectively. The left-hand side and the production term are exact and the extension to a curvilinear coordinate system is therefore straight forward, see e.g. Ekander et al. (1989). In the exact expression of the pressure-strain term there are not any explicit terms due to curvature in boundary layer approximation. However, the curvature has an indirect effect through the the pressure and the fluctuating velocities which are directly affected by the curvature. Hence, the model for this term should also include curvature terms. Therefore it is advantageous to write the modeled pressure-strain term in tensor form and subsequently transform it to curvilinear coordinates. Finally, the curvature terms in the dissipation and diffusion terms were neglected. From the two equations for $\overline{uv}$ and $\overline{vw}$ the equation for the ratio of the two Reynolds stresses, $R = \overline{vw} / \overline{uv}$, becomes then

$$\frac{U}{h_1}\frac{\partial R}{\partial x_1} + V\frac{\partial R}{\partial x_2} + \frac{W}{h_3}\frac{\partial R}{\partial x_3} = -\left(\frac{3 - C_2}{11}\frac{\overline{v^2}}{\overline{uv}} - \frac{30C_2 - 2}{55}\frac{k}{\overline{uv}}\right)\left(\frac{\partial W}{\partial y} - R\frac{\partial U}{\partial y}\right)$$

$$+ \frac{14 - C_2}{11}(1 + R^2)(K_{13}U - K_{31}W) - R\frac{10 + 7C_2}{11}(K_{31}U - K_{13}W)$$

$$+ C_s\left(\frac{\partial}{\partial y}\left(2\frac{k}{\varepsilon}\overline{v^2}\right)\frac{\partial R}{\partial y} + \frac{k}{\varepsilon}2\overline{v}^2\left(\frac{1}{\overline{uv}}\frac{\partial^2\overline{wv}}{\partial y^2} - \frac{\overline{wv}}{\overline{uv}^2}\frac{\partial^2\overline{uv}}{\partial y^2}\right)\right)$$

The first terms on the first line of the right hand side is forcing the Reynolds stresses to be aligned with the gradient of the mean velocity. The terms on the second line are due to the curvature of the streamlines and the term on the last line represents the vertical diffusion. Launder et al. (1975) suggest $C_2 = 0.4$. However, for the present equation this seems to be too large, because the first term will then dominate the right hand side and the Reynolds stresses will therefore be almost aligned with the mean velocity gradient. The results presented here are calculated with $C_2 = 0.0$, however, a correct value can not be judged until comparison with the experiments has been carried out. As boundary condition for R at the wall, the Reynolds stresses are assumed to be parallel with the gradient vector of the mean velocity, other boundary conditions has been used which showed that the influence of the boundary condition is very small.

2.4 The Algebraic Reynolds stress model
To obtain the four remaining Reynolds stresses the algebraic Reynolds stress model suggested by Ekander et al. (1989) is used. In the algebraic Reynolds stress model suggested by Rodi (1980) the ratio of a u_iu_j and k is assumed to be equal to the ratio of the respective convective time derivative minus the diffusion term. If this is generalized to a curvilinear coordinate system the right hand side will contain terms due to the curvature of the coordinate system. These terms will not take part in the production-destruction process, as they vanish upon contraction of the tensors. Therefore Ekander et al. (1989) suggested that these terms should not be kept on the right hand side and one then obtain the following expression for the Reynolds stresses,

$$\frac{R^i_{\ j}}{k} = \frac{\dfrac{\partial R^i_{\ j}}{\partial t} + U^m \dfrac{\partial R^i_{\ j}}{\partial x^m} - c_s \left(\dfrac{k}{\varepsilon} R^{rm} \dfrac{\partial R^{mi}_{\ \ j}}{\partial x^m} \right)_{,r}}{\dfrac{Dk}{Dt} - D}$$

$$= \frac{U^m \Gamma^i_{lm} R^l_{\ j} - U^m \Gamma^l_{jm} R^i_{\ l} + P^i_{\ j} + \Pi^i_{\ j} - \varepsilon^i_{\ j}}{P - \varepsilon}$$

where, Γ^i_{lm} is the Christoffel symbol of the second kind.

3 NUMERICAL METHOD

The calculations were made with an existing boundary layer code originating from Krogstad (1985) and modified by Moreau (1988). The flow outside the boundary layer was obtained by integrating the Euler equations using the measured pressure distribution.

The equations are transformed to a self similar type of coordinates so that the boundary layer thickness is approximately constant in the new coordinate system. The equations are discretized using second order schemes in all directions. The equations were solved by marching in the x-direction and iterating the nonlinear equations at each x value in the y-z plane.

The step length in the x-direction was chosen small enough to give stable solutions, which means 400-800 steps, depending on which turbulence model was used. In the z-direction about 20 streamlines were used, and in the y-direction an uneven space was used with 120 points across the boundary layer. If there was any inflow at the outermost streamlines the small crossflow approximation was used, i.e. the z - derivatives were set to zero, otherwise the z-derivative for the outermost streamline was calculated using the three outermost streamlines. To avoid problems in the far end region of the calculated domain on the inflow side, one stream line was dropped every half meter after x = 3.5m. The calculations required about one to five minutes CPU-time on a Cray-2.

As inlet conditions, the measured velocity profiles in the streamwise direction were used, extended with a linear part close to the wall and a constant value above the measurements. For k and ε the shape of a calculated profile from a 2D boundary layer were used for the grid points between the wall and the measuring point closest to the wall.

4 CONCLUSIONS

Only a few comments on the results obtained with the different turbulence models will be given here, instead the reader is referred to the synthesis of the test case in this volume, or in Lindberg (1991). In both these references the sensitivity of streamline geometry, inlet conditions etc are discussed. The relatively poor agreement of the Johnson - King Model with the experimental results was analyzed after the Workshop. It showed that the calculation of the position and the values of the derivative of the mean flow etc. at the position where - $\overline{uv}$ has a maximum is critical. In the contribution to the Workshop all required values were calculated as a weighted mean with the weight $\left(\overline{uv} \;/\; \overline{uv}_{max} \right)^{10}$ in order to avoid stability problems as the maximum value changed from one grid point to another. After the workshop another method was used where only the location was calculated as described above and the other values were obtained by using that location. This latter method seems more appropriate and also give much better agreement with the experiments.

5 REFERENCES

Anderson D.A., Tannehill J.C. & Pletcher R.H. (1984) *Computational Fluid Mechanics and Heat Transfer*. McGraw Hill

Chen H.C. & Patel V.C. (1988) *Near-wall turbulence models for complex flows including separation*. AIAA Journal, Vol 26, No 6

Ekander H. & Johansson A.V. (1989) *An Improved algebraic Reynolds stress model and application to curved and rotating channel flows*. Seventh symposium on turbulent shear flows, Stanford University, August 1989

Johnson D. A. & King L. S. (1985) *A Mathematical Simple Turbulence Closure Model for Attached and Separated Turbulent Boundary Layers*. AIAA Journal , Vol 23, No 11, pp. 1684-1692

Krogstad P.-Å. (1985) *Progress in the development of a Reynolds-stress turbulence closure*. Internal Report No T-2-85, IMHEF-DME, EPFL, CH-1015 Lausanne, Switzerland.

Launder B.E., Reece G.J. & Rodi W. (1975) *A Method for Calculating Three-Dimensional Boundary Layers on Arbitrary Surfaces*. Journal of Fluid Mechanics, Vol 68, Part 3, pp. 537-566

Lindberg P.-Å. (1991) *Near-Wall trubulence models for 3D boundary layers*. Internal report no T-91-9, IMHEF-DME, EPFL, CH-1015 Lausanne, Switzerland.

Moreau V. (1988) *A study of anisotropic effects in three-dimensional, incompressible turbulent boundary layers driven by pressure gradients*. : Thesis No 755 (1988) EPF Lausanne, Switzerland.

Rodi W. (1980) *Turbulence models and their application in hydraulics*, International association for hydraulic research, Delft

Calculation results of the NLR BOLA Method for the Ercoftac test case T1: 3D turbulent boundary layer in an S-shaped duct

J.P.F.Lindhout and B. van den Berg

NLR, The Netherlands

1. INTRODUCTION

Detailed mean flow and turbulence measurements have been carried out at EPFL, Lausanne, in an S-shaped duct. The duct consists a straight part, an S-shaped part and a second straight part. Each part is 3 m long, while the duct cross section is 1.2m * 0.5m The 3D boundary layer development on one of the flat duct surfaces was measured. The flow development is especially interesting because of the presence of cross-over boundary layer velocity profiles in the downstream part of the S-shaped duct. Cross-over profiles occur when the cross-wise pressure gradients change sign, which is the case in many practical flows. In the present test case the duct width is sufficiently large for cross-over profiles to occur outside the region of influence of the duct corners. More details about the experiment and a definition of the test case may be found in ref. 1.

2. CALCULATION METHOD

The calculation method BOLA is a differential method for solving the laminar or turbulent boundary layer equations in arbitrary wall-normal coordinates. The program traces the lateral free boundaries and steps around 'forbidden' zones (characterized in practice by low skin friction or large wall flow angle), keeping within the region determined by the initial data. The lateral boundaries are constructed such that no flow can enter the re-gion to be calculated. Because the flow angles are varying along a normal, this actually means that the region calculated generally shrinks marching downstream. From the given surface pressure distribution the velocity components at the edge of the boundary are calculated employing the Euler equations, to which the boundary layer equations reduce at the edge. The inviscid flow angles at the upstream boundary are used as initial conditions.

To counteract growth of the boundary layer, a transformation depending on the displacement thickness and Reynoldsnumber is applied explicitly to the wall normal coordinate before discretization. For the standard calculation 41 points are used. Along a

139

normal the points are uniformly distributed in the transformed coordinate. More than half of the points are situated in the near-wall region, where the strongest normal gradients occur.

The stepsizes in the X and Z direction are kept constant except near the lateral boundaries where, depending on the shape, stations are created or removed. Also the Courant-Friedrich-Levy criterion can force a reduction of the X-step, but that situation did not occur in the present calculations. The stepsizes were .025m and .01m respectively for the calculation results shown. Stepsizes were varied in preliminary calculations to establish requirements for grid independence.

Off the surface, second order central differences are used for the wall normal derivatives whereas in-plane derivatives are given by Hermitian-type formulae connecting the function value and its first derivative at both points. This leads to a stable almost second order discretization. The Newton-Raphson procedure is employed to linearize the momentum equations. Four finite difference molecules are used. The choice depends on the sequence in which the calculations sweeps along a Z-coordinate line. The sequence is optimized such that the stepsize restrictions are as small as possible. No special problems occured for this type of calculation.

The Reynolds stresses are represented with an isotropic eddy viscosity model based on the mixing length formulation of Michel. The constant in the Van Driest damping factor is modified for the pressure gradient in the skin friction direction following Cebeci. For further details about the calculation method and the turbulence model used, one is referred to e.g. ref. 2.

3. BOUNDARY AND INITIAL CONDITIONS

The measured turbulent boundary layer along the initial line is very nearly collateral and its thickness varies little in spanwise direction. Therefore one and the same Coles profile at a flow angle equal to the inviscid flow angle has been used to represent the initial boundary. To fit the measurements the initial profile momentum thickness has been taken 5.0 mm and the skin friction coefficient .00344.

As only the surface pressure distribution is given, the flow angles at the boundary layer edge must be calculated enploying the Euler equations and the known flow angles at the initial line. The calculated downstream flow angles depend strongly on the initial angles. Therefore the experimental initial flow angle distribution has been smoothed assuming a linear variation. The assumed magnitude and sign of initial angles are indicated in fig. 1 together with the other initial conditions. The computed external flow angle development along the duct has been plotted in fig. 2 against the X-coordinate at the values of Z-coordinate indicated. A comparison with measured flow angles at the boundary layer edge at a few stations is also shown.

The agreement between experimental and calculation in fig.2 appears to be reasonable, apart from the measured angle at the first station, which deviates also from the general trend of both experimental and computational data. However, it should be noted that this reasonable agreement in the main part of the flow only exists because the initial flow angle in the calculations has been assumed negative, while the test case definition actually suggests the angle to be positive. It is not quite clear what the correct starting procedure is and whether the flow development in the straight duct preceding the S-shaped section is sufficiently well defined for accurate calculations. In the present calculations the external flow angle development as shown has been employed, as the interesting part of the 3D boundary layer flow occurs in the S-shaped section and acceptable agreement with the measured angles only exists with the assumed initial conditions. The good agreement between the calculations and experiment at the second measurement station suggests that this station, located a convenient short distance upstream of the S-shaped duct, may be considered as an initial station for the present calculations.

4. RESULTS
Some of the computed wall streamlines are sketched in fig.3. The borders of the region of determinacy of the initial conditions according to the calculations are also indicated. The local disturbance in the wall streamlines, which is visible at around x=3 m, is induced by the presence of a side-wall in the test set-up. In the sketch a few measurement stations are indicated. Comparisons between calculation and experiment will be restricted here to these stations. A full comparison will be made by the ERCOFTAC Workshop organizers.

As noted in the preceding chapter, some uncertainty exists about the correct external flow angles. As the flow angle at the boundary layer edge is normally an "a priori" data in boundary layer calculations, it seems sensible to focus comparisons on the difference between the flow angle in the boundary layer and that at the edge, gamma - alpha. Note that the sign change of gamma (and not of alpha), as used in the Workshop notation, has not been applied here.

Further, the measured external velocity magnitude at several stations appers to differ essentially from that derived from the measured surface pressures. If the boundary layer approximation holds and the measurements are accurate, evidently both values should be equal. The velocity magnitude at the boundary layer edge is an "a priori" data in boundary calculations based on a surface pressure distribution. For useful comparisons it is necessary in the given circumstances, therefore, to apply a correction factor to make the measured external velocity equal to the velocity derived from the surface pressure.

Mean velocity direction and magnitude are compared on this basis for a few stations in fig.4 to 7.. At the position x = 3.32 m, which could be considered as an initial station for the S-shaped part, the measured and computed velocity magnitudes correspond well. The velocity profile, twist is seen to be slighly larger in the computations. Further downstream the agreement improves rather than decreases. Even at the last station, at the downstream end of the S-shaped duct, the agreement is very satisfactory. The cross-over velocity profile, characterized by the change in sign of the velocity twist, appears to be well predicted (fig.7). For this station the measured and computed turbulent shear stresses are compared in fig.8. Surprisingly the shear stress ratio magnitude seems reasonably well predicted with the simple turbulence model used. However, deviations between measured and computed shear stress directions amount up to 10 degrees.

5. CONCLUSIONS
The agreement between measurements and computations is very reasonable considering the complex 3D turbulent boundary layer flow. To draw more precise conclusions about the still existing discrepancies, some questions about the initial and boundary conditions should first be resolved.

6. REFERENCES
[1] I.L.Ryhming, T.V.Truong: "Specifications for the test case: Boundary layer in a S-shaped channel (Problem T1)". Ercoftac Workshop 26-28 March 1990, EPFL, Lausanne
[2] J.P.F.Lindhout, G.Moek, E.de Boer, B.van den Berg: "A method for the calculation of 3D boundary layers on practical configurations". J.Fluid Eng., Vol.103, p.104, 1981

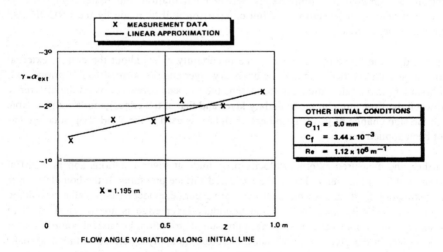

Fig. 1 Initial conditions for boundary layer and external flow calculations

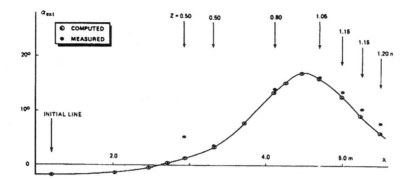

Fig. 2: Comparison of measured external flow angle variation and computations based on measured surface pressures

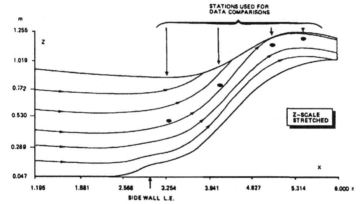

Fig. 3: Computed wall streamlines in region of determinacy

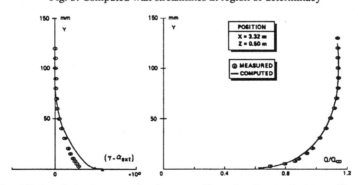

Fig. 4: Comparison of measured and computed boundary layer velocity profile twist and magnitude at $X = 3.32$

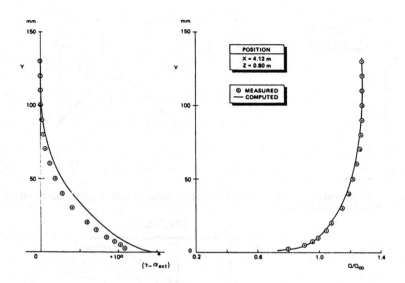

Fig. 5: Comparison of measured and computed boundary layer velocity profile
twist and magnitude at X = 4.12 m

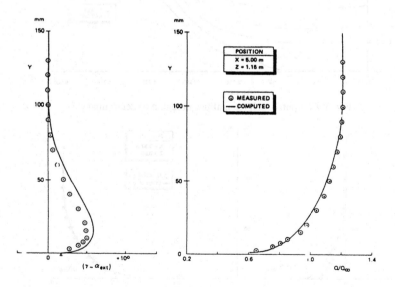

Fig. 6: Comparison of measured and computed boundary layer velocity profile
twist and magnitude at X = 5.00 m

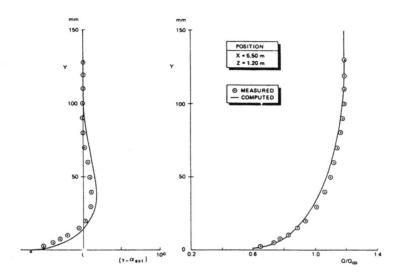

Fig. 7: Comparison of measured and computed boundary layer velocity profile
twist and magnitude at X = 5.50 m

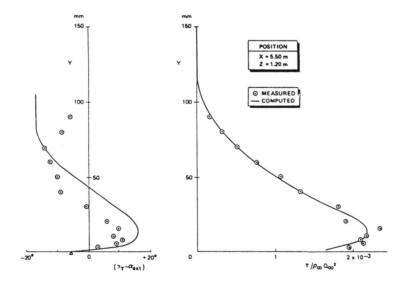

Fig. 8: Comparison of measured and computed turbulent shear stress direction
and magnitude at X = 5.50 m

An algebraic Model based on the Velocity-Profile Skewing

A. NAKKASYAN, T.S. PRAHLAD *, P. GILQUIN

IMHEF/Ecole Polytechnique Fédérale de Lausanne , 1015 Lausanne, Switzerland
*Present address: Aeronautical Develop. Agency, Bangalore-560-037, India

1. INTRODUCTION

The test-case T1 is calculated with a slightly modified version of Krogstad's (1979) "fully 3D, –implicit normal to the wall, –explicit laterally"–method[1], which is similar to Fannelop and Humphrey's (1974) finite-difference method[2].

In our algebraic, two-layer eddy-viscosity model, the locally observed difference in the directions of the resultant turbulent-stress $\vec{\tau}$ and the rate of strain $\frac{d\vec{q}}{\partial y}$ of a 3D–TBL is approximated by an anisotropy parameter $T = \varepsilon_n(y)/\varepsilon_s(y)$. Following Ryhming and Fannelop[3], T is directly related to the skewing of the velocity vector in the three-dimensional boundary layer. This idea was based on Rotta's arguments on the pressure-strain vector[4]. The model was further developed by Prahlad[5]. Accordingly, our T is defined in a *coordinate system* (s,n) *aligned with and normal to* the local mean velocity vector $\vec{q}(y)$.

The free-stream flow characteristics $[U_e; h_1, h_3$ (metrics); ψ=cte lines (stream lines); ϕ =cte lines (equipotentials)] and the initial conditions   - all are required by our 3D-TBL computer-program - are *interpolated* from the C_p-measurements and the measurements of the mean-velocity profiles, respectively, and *stored* as input data, by Gilquin[6]

2. THE 3D–TBL COMPUTER PROGRAM

In the curvilinear coordinate system $(x_1, x_2, x_3) \equiv (\psi, y, \phi)$ which follow a free-streamline, the mean-velocity vectors ($\vec{q}(y)$; $0 \le y \le \delta$) have the components $(u_1, u_2, u_3) \equiv (u, v, w)$ and satisfy the continuity-equation and ϕ- & ψ-momentum conservation-equations :

$$\frac{\partial}{\partial \phi}(h_3\, u_1) + \frac{\partial}{\partial \psi}(h_1\, u_3) + \frac{\partial}{\partial y}(h_1\, h_3\, u_2) = 0 \tag{1}$$

$$\frac{u_1}{h_1}\frac{\partial u_1}{\partial \phi} + u_2\frac{\partial u_1}{\partial y} + \frac{u_3}{h_3}\frac{\partial u_1}{\partial \psi} - u_1 u_3 K_{13} + u_3^2\, K_{31} =$$

$$-\frac{1}{\rho\, h_1}\frac{\partial P}{\partial \phi} + v\,\frac{\partial}{\partial y}\left[\, V_1\frac{\partial u_1}{\partial y} + \varepsilon_{xz}\frac{\partial u_3}{\partial y}\,\right] \tag{2}$$

$$\frac{u_1}{h_1}\frac{\partial u_3}{\partial \phi} + u_2\frac{\partial u_3}{\partial y} + \frac{u_3}{h_3}\frac{\partial u_3}{\partial \psi} - u_1\, u_3\, K_{31} + u_1^2\, K_{13} =$$

$$-\frac{1}{\rho\, h_3}\frac{\partial P}{\partial \psi} + v\,\frac{\partial}{\partial y}\left[\, V_3\frac{\partial u_3}{\partial y} + \varepsilon_{zx}\frac{\partial u_1}{\partial y}\,\right] \tag{3}$$

where h_2 is set to 1 ; $V_1 \equiv 1 + (\varepsilon_{xx}/v)$ & $V_3 \equiv 1 + (\varepsilon_{zz}/v)$.

The two coupled equations to be solved for $F \equiv \dfrac{u_1}{U_e}$ and $W \equiv \dfrac{u_3}{U_e}$ are of the form

$$G_1\, H'' + G_2\, H' + G_3\, H = G_4 \quad, \quad \text{where } H \text{ is either } F \text{ or } W \, . \tag{4}$$

The original iteration flow-chart, in which F and W are updated at (ϕ_I, ψ_J) in one pass, is slightly improved[7] by strengthening the coupling between F and W. Thus, the discretized and/or linearized terms $(F^2, W^2, 2\xi F(\partial F/\partial \xi), FW, WF_\alpha, WW_\alpha$, etc.) are updated in two passes along the same equipotential ϕ_I, following the path $[(\phi_I, \psi_J) \rightarrow (\phi_I, \psi_{J+1}) ; J = 1...to$ *approx 5)*]. The number of iterations necessary for local convergence is 10 or less, for the first pass. The second time, with updated $F_\alpha W_\alpha$, convergence is faster. The program can handle between 3 to 15 free streamlines. In practice we have taken less (see section 5).

3. THE TURBULENCE MODEL

The skewing of the local mean-velocity profile and the *definition* of the anisotropy parameter $T \equiv \varepsilon_n/\varepsilon_s$ in the (s,n)-coordinate system are the starting points of the eddy viscosity model:

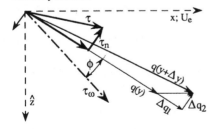

$$\tau = \tau_s + \tau_n$$

$$\tau_\omega = \tau(y{=}0)$$

$$q(y{+}\Delta y) = q(y) + \Delta q$$

$$\Delta q_1 = \left.\frac{\partial q}{\partial y}\right|_\phi \; ; \quad \Delta q_2 = |q|\,\frac{\partial \phi}{\partial s}$$

$$\Delta q = \Delta q_1 + \Delta q_2$$

In the external-streamline coordinate system where calculations are performed, T becomes

$T=T(\gamma, \gamma_g, \gamma_\tau)=\tan(\gamma-\gamma_\tau)/\tan(\gamma-\gamma_g)$, i.e. identical to Rotta's expression[4] for anisotropy ($\varepsilon \leftrightarrow \nu_{Rotta}$). And, the algebraic model is defined as

$$\begin{pmatrix} -\rho\ \overline{u'v'} \\ -\rho\ \overline{w'v'} \end{pmatrix} = \begin{pmatrix} \tau_x \\ \tau_z \end{pmatrix} = \begin{pmatrix} \varepsilon_{xx} & \varepsilon_{xz} \\ \varepsilon_{zx} & \varepsilon_{zz} \end{pmatrix} \begin{pmatrix} \dfrac{\partial u}{\partial y} \\ \dfrac{\partial w}{\partial y} \end{pmatrix} ; \quad u \equiv u_1\ ; \quad w \equiv u_3 \quad \text{- see eq. (2)\&(3) -}\ , \quad (5)$$

$$with \quad \varepsilon_{xx} = \varepsilon_s \frac{u^2+Tw^2}{u^2+w^2}\ ; \quad \varepsilon_{xz} = \varepsilon_{zx} = \varepsilon_s \frac{u\ w\ (1-T)}{u^2+w^2}\ ; \quad \varepsilon_{zz} = \varepsilon_s \frac{w^2+T\ u^2}{u^2+w^2} \quad (6)$$

in which T is approximated by a *locally constant-profile* as suggested by Prahlad[5]

$$T(y) \cong \left[\frac{w(y_m)/Ue}{1-u(y_m)/Ue}\right]^{\frac{1}{2}}\ , \quad \text{with} \quad y \cong y_m \equiv \delta/2 \quad (7)$$

and the eddy viscosity $\varepsilon_s(y)$ is approximated as a two-layer profile

$$\varepsilon_s = \varepsilon_{s,out} \tanh\left\{\frac{\varepsilon_{s,in}}{\varepsilon_{s,out}}\right\}\ , \quad (8)$$

composed of an outer-layer expression which is constant (no intermittency; -see section 7-)

$$\varepsilon_{s,out}=0,0168\rho Ue\delta^*, \text{ with } \delta^*=\int_0^\delta (1.-u(y)/Ue)dy\ ,$$

and an inner-layer expression which is derived from a mixing-length (l) hypothesis

$$\varepsilon_{s,in} = \rho\ l^2 \sqrt{\left(\frac{\partial u}{\partial y}\right)^2 + \left(\frac{\partial w}{\partial y}\right)^2}\ f(T)\ , \quad \text{with} \quad l = 0,41y\left(1 - e^{\frac{-y^+}{26}}\right)$$

$$f(T) = \sqrt{\frac{(u^2+w^2)\ [(\partial u/\partial y)^2 + (\partial w/\partial y)^2]}{[u\ (\partial u/\partial y) + w\ (\partial w/\partial y)]^2 + T^2\ [w\ (\partial u/\partial y) - u\ (\partial w/\partial y)]^2}}$$

4. INPUT DATA-FILE FOR THE 3D-TBL CALCULATIONS

First, the measurements $Cp(Xm,Zm)$ of the S-duct are completed by interpolation (Subr. SURF in IMSL/LIB). Next, derivatives of $Cp(X,Z)$ are calculated with a least-squared approximation using a B-spline function (Subr.BSLS2 in IMSL/LIB). For each free-streamline [given the initial conditions $X_{oo}=0.3; 0.2\leq Z_{oo}\leq 0.8; u_{oo}\equiv Ue(Cp(X_{oo},Z_{oo}))$; $w_{oo}=0$], a system of 3 equations corresponding to the Euler equations is solved for $Z(u,w)$ at $X=X+\Delta X$ with the method of characteristics, until convergence is obtained:

$$dZ/dX= w/u \; ; \quad u \, du= -0.5 \, (U_{ref})^2 \, (\partial C_p/\partial X) \, dX \; ; \quad udw= -0.5 \, (U_{ref})^2 \, (\partial C_p/\partial Z) \, dX \qquad (9)$$

The results are stored on a *grid* with (18x56) points at $[\phi(x_g,z_g); \psi(x_g,z_g)]$, located inside the S-shaped sub-domain $[(0.3m \le x_g \le 5.6m); (0.2m \le z_g \le 1.4m)]$.

For the interpolation of the initial profiles $[u_0(y); w_0(y)]$ at (X_0,Z_0), the complete set of twenty-one, measured mean-velocity profiles near $X_0=1.1193m$. are used.

5. COMPUTATIONAL INFORMATIONS

The *stability* of calculated solutions is primarily determined by the closure-equations of the algebraic model and by the coupling incorporated in the resolution of equations (4).

Care is taken to protect the first and last free-streamlines by adding two dummy streamlines at both edges; because, the " fully 3D, –implicit normal to the wall, –explicit laterally "–method has relatively poor *precision* for the first and last ψ. Thus, five separate jobs are run, each job calculating about five, *closely* located free-streamlines.

Initially, the *step size* in the x-direction is set to 5 mm. Later, the program reduces it if the local boundary-layer thickness varies too fast. For the variable y/δ , we use 120 vertical grid-points in geometric progression, identical at each computation-station.

The 3D-TBL calculations are done on the VAX-3600/VMS computer of EPFL/DME. The computation-time for the 3D-TBL calculations along a single free-stream line depends on the iteration/convergence flow-chart; but a good guess is 5 CPU-sec.

6. THE RESULTS OF THE ALGEBRAIC MODEL

Our comparisons between the *calc*ulated- vs measured-profiles for all 24 measurement-stations (2.9m<X≤5.5m) are summarized according to the quality of our calculations:

Class	tot.stations	location in the S-duct	*calc.*-profiles	remarks	figure included
I	16	2.9<X≤5.5; middle-Z	"good"	------	X5500Z1198
II	3	4.7≤X<5.3; highest-Z	bad	model(?)	X4700Z0730
III	5	4.1<X≤5.5; lowest-Z	bad	C_p (?)	none

The column labelled "remarks" is reconsidered under section 7 .

The predictions of the algebraic model for the stations X4700Z0730 (top) and
X5500Z1198 (bottom) are compared with measurements in the following figures:

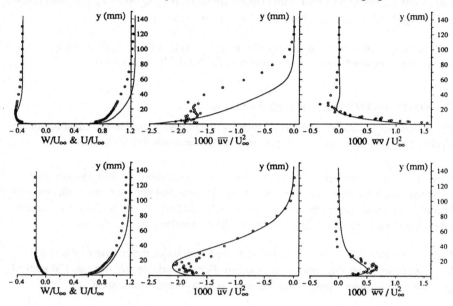

7. POST ERCOFTAC REMARKS

We have tested the closure equations (5) to (8) of our algebraic model, as directly as
possible, with the help of the experimental database $\mathcal{D} \equiv \{(u, w, \overline{u'v'}, \overline{w'v'})_{mes}; (\partial u / \partial y,$
$\partial w / \partial y, \gamma, \gamma_g, \gamma_\tau,)_{IMSL}; \varepsilon_{s,exp} \}$, where ("mes"; "IMSL" ; "exp") mean ("raw measurements";
"data smoothed with IMSL/LIBRARY"; "reference predictions -see next"), respectively.

Our conclusions are the following:

a) the predictions of the equations (5)+(6) *with* $T = T(\gamma, \gamma_g, \gamma_\tau)$ are in excellent agreement
with measured τ_x and τ_z, at all measurement stations. As a consequence, we define an
"experimental" eddy-viscosity $\varepsilon_{s,exp} = \varepsilon_s (\mathcal{D})$ using $\tau_{tot} = (\tau_x \cdot \tau_x + \tau_z \cdot \tau_z)^{1/2} = \tau_{tot}(\varepsilon_{s,exp})$;

b) the quality of the approximation $T = T(u,w,Ue)$ -see eq.(7)- varies from good to bad.
As a general rule, the largest disagreements with $T = T(\gamma, \gamma_g, \gamma_\tau)$ occur at y small. Note,
eq. (7) was not necessarily expected to be valid if the external streamline curvature
changed sign;

c) equation (8) compared with $\varepsilon_{s,exp}$ shows the need for a smoothing function other than "tanh"-type. 'Any' intermittency-function which satisfies the condition $(y \rightarrow \delta; \varepsilon_{s,exp} \rightarrow 0)$ is an improvement as compared to our $\varepsilon_{s,out}$ which is defined locally constant. *But,* the fact that the smoothing function "tanh" often hinders the evolution of $\varepsilon_{s,in}$ (which otherwise roughly follows $\varepsilon_{s,exp}$ as far as its maximum value $\varepsilon_{s,exp}(\text{max})$) has major consequences .

This post ERCOFTAC90 study indicates that one of the origins of our failure with Class II calculations is the profile of ε_s near its maximum. Also, it proves that the algebraic model may not be at the origin of the problem with our Class III calculations.

8. REFERENCES

[1]   KROGSTAD P.A. (1979) *Investigation of a Three-Dimensional Turbulent Boundary Layer Driven by Simple Two-Dimensional Potential Flow,* Thesis, NTH,Trondheim.

[2]   FANNELOP T.K. & HUMPHREYS D.A. (1974) *The solution of the Laminar and Turbulent 3-D BL Equations with a simple Finite Difference Technique .* FFA Report 126, Bromma, Sweden .

[3]   RYHMING I.L. & FANNELOP T.K. (1982) *Three Dimensional Turbulent Boundary Layers.* IUTAM Symposium, Berlin.

[4]   ROTTA J.C. (1979) *A Family of Turbulence Models for Three-Dimensional Boundary Layers .* First Symp."Turbulent shear flows I ", Springer-Verlag.

[5]   PRAHLAD T.S. (1983) *Modelling Considerations for Eddy Viscosity in 3-D Turbulent Boundary Layers .* EPFL Report T-83-9.

[6]   GILQUIN P. (1990) *Calcul des lignes de courant extérieures à partir des Cp mesurés dans le canal en S.* EPFL Report T-90-24.

[7]   NAKKASYAN A., PRAHLAD T.S. & RYHMING I.L. (1986) *A 3-D Turbulent Boundary Layer Model Based on Local Velocity Profile Skewing with parametric investigations.* EPFL Report T-86-6 .

Calculation of Three-Dimensional Turbulent Flow in an S-Shaped Channel

W. RODI and J. ZHU

Institute for Hydromechanics, University of Karlsruhe, D-7500 Karlsruhe 1
F.R.Germany

1. INTRODUCTION

The present authors have performed predictions for the test problems T1, T2 and T4 for the ERCOFTAC Workshop on Numerical Simulation of Unsteady Flows, Transition to Turbulence and Combustion held March 26-28, 1990 in Lausanne. The predictions were all carried out with a recently developed general finite-volume method for calculating incompressible elliptic flows with complex boundaries. This paper describes the essential features of the method and the computational details for the test problem T1 - three-dimensional turbulent flow in an S-shaped channel. The details relating to the test problems T2 and T4 are given separately in companion papers in these proceedings.

2. NUMERICAL SOLUTION PROCEDURE

2.1 Governing Equations
The equations for three-dimensional steady flows in non-orthogonal coordinates (x_i) using Cartesian velocity components (V_i with V_1=U, V_2=V and V_3=W) may be written in the following general form:

$$\frac{\partial}{\partial x_i}(C_i\phi + D_{i\phi}) = JS_\phi, \qquad i = 1, 2, 3 \tag{1}$$

where for different variables ϕ, the convective coefficients C_i, the diffusion terms $D_{i\phi}$ and the source terms S_ϕ are given in Table 1. J is the Jacobian of coordinate transformation between the curvilinear system (x_i) and a reference Cartesian system (y_i with y_1=x, y_2=y and y_3=z). The eddy viscosity μ_t appearing in the diffusion terms of the different transport equations (1) is calculated by using the standard k-ϵ model. The transport equations for the turbulent kinetic energy k and its dissipation rate ϵ are also of the general form (1), and the source terms for k and ϵ as well as the various turbulence model constants are also given in Table 1.

2.2 Boundary Conditions

Figure 1 shows the geometry of the flow and the coordinate system, as defined in the problem specification for test case T1. Although only the boundary layer on the top (or bottom) wall is of interest in this case, the flow in the whole S-shaped channel was calculated with an elliptic solution procedure. Because of symmetry, the calculation was however restricted to the top half of the channel, with y=0 and 210mm being the top wall and symmetry plane, respectively. At the symmetry plane, the normal velocity component and the normal gradients of the other variables were set to zero. At the outlet, zero streamwise gradient conditions were assumed. The wall-function approach (Launder and Spalding, 1974) was used to simulate the near-wall flow. In this, the resultant wall shear stress was calculated by

$$\vec{\tau}_w = -\lambda_w \vec{V} \tag{2}$$

where

$$\lambda_w = \begin{cases} \mu/y_P & \text{if } y_P^+ < 11.6 \\ \rho C_\mu^{1/4} k^{1/2} \kappa / \ln(E y_P^+) & \text{otherwise} \end{cases}$$

$$y_P^+ = \rho C_\mu^{1/4} k^{1/2} y_P/\mu, \qquad \kappa = 0.4187, \qquad E = 9.0,$$

and y_P is the normal distance of the first control-volume center from the wall. Further, the diffusive flux of k was set to zero at the wall, and the value of ϵ at the first grid point away from the wall was determined from

$$\epsilon = \frac{C_\mu^{3/4} k^{3/2}}{\kappa y_P} \tag{3}$$

The pressure on the boundary nodes was evaluated by linear extrapolation from values at interior nodes.

The flow in the upstream straight channel was first calculated with the assumption that the flow in the S-shaped channel has no influence on it (zero gradient condition at exit). The inlet boundary plane was placed at the station x=1255mm where the experimental data for U, V, W, $\overline{uv}$ and k were available, while ϵ was calculated from the relationship

$$\epsilon = C_\mu k^2 \left| \frac{\partial U/\partial y}{\overline{uv}} \right| \tag{4}$$

The flow in the S-shaped channel was then calculated using the outlet results obtained from the straight channel calculation as the inlet boundary values. The boundary layers developing along the two side walls of the straight channel have no influence on the inlet boundary conditions for the S-shaped channel due to the reduction of the channel width removing the boundary layers (see Figure 1). The outflow boundary was located at the section x=6500mm which is 500mm downstream of the end of the S-shaped channel.

2.3 Solution Algorithm

The numerical method used to solve the system of equations (1) is a finite-volume procedure designed for calculating incompressible elliptic flows with complex boundaries. It uses a non-staggered grid with all the dependent variables being stored at the same geometric centre of the control volumes. The normal-derivative diffusion terms were approximated by the central differencing scheme and the cross-derivative diffusion terms were treated explicitly as an additional source. The standard hybrid central/upwind scheme was used to approximate the convection terms. A special interpolation practice developed by Rhie and Chow (1983) for calculating the mass fluxes at the control volume faces was used to avoid pressure oscillations due to the non-staggered arrangement. Pressure-velocity coupling was achieved via the SIMPLE algorithm. The resulting set of algebraic difference equations was solved with the strongly implicit solution algorithm of Stone (1968). Convergence was declared when the maximum normalised residue of the variables was below 5×10^{-4}. The details of the present numerical procedure are given in Majumdar et al. (1990).

The calculations were performed on the Siemens/Fujitsu VP400-EX vector computer of the University of Karlsruhe. The computer code has been vectorized to a major extent so that high computational efficiency can be achieved. 26 iterations and 53 seconds of CPU-time were required for the straight channel calculation on a $30\times36\times30$ grid, and 139 iterations and 5.7 minutes for the S-shaped channel calculation on the $30\times36\times30$ grid shown in Figure 2.

3. TYPICAL RESULTS

Calculated results are shown and compared with the experimental data in Figure 3 for the wall-pressure coefficient C_p, in Figure 4 for the skin-friction coefficient C_f, and in Figures 5(a) and 5(b) for various boundary-layer profiles at two selected stations, one with x=4125mm and z=-800mm, and the other with x=5005mm and z=-1000mm. It can be seen that the calculation reproduces quite well the essential features of the flow observed in the experiment; only the shear stress $\overline{uw}$ near the wall is not well simulated.

4. REFERENCES

Launder, B.E., and Spalding, D.B., 1974, "The numerical computation of turbulent flows," *Comput. Methods Appl. Mech. Eng.*, **3**, 269-289.

Majumdar, S., Rodi, W., and Zhu, J., 1990, "Three-dimensional finite-volume method for incompressible flows with complex boundaries," *ASME Symposium on Advances and Applications in Computational Fluid Dynamics*, Dallas, U.S.A., Nov. 25-30.

Rhie, C.M., and Chow, W.L., 1983, "A numerical study of the turbulent flow past an isolated airfoil with trailing edge separation," *AIAA Journal*, **21**, 1525-1532.

Stone, H.L., 1968, "Iterative solution of implicit approximations of multidimensional partial differential equations," *SIAM J. Num. Anal.*, **5**, 530-558.

Table 1:   Form of terms in the individual equations

ϕ	$D_{i\phi}$	S_ϕ
1	0	0
V_1	$-\dfrac{\mu_e}{J}(B_j^i\dfrac{\partial V_1}{\partial x_j}+\beta_j^i\omega_1^j)$	$-\dfrac{1}{J}\dfrac{\partial}{\partial x_j}(\beta_1^j p)$
V_2	$-\dfrac{\mu_e}{J}(B_j^i\dfrac{\partial V_2}{\partial x_j}+\beta_j^i\omega_2^j)$	$-\dfrac{1}{J}\dfrac{\partial}{\partial x_j}(\beta_2^j p)$
V_3	$-\dfrac{\mu_e}{J}(B_j^i\dfrac{\partial V_3}{\partial x_j}+\beta_j^i\omega_3^j)$	$-\dfrac{1}{J}\dfrac{\partial}{\partial x_j}(\beta_3^j p)$
k	$-\dfrac{\mu_e}{J\sigma_k}B_j^i\dfrac{\partial k}{\partial x_j}$	$G-\rho\varepsilon$
ε	$-\dfrac{\mu_e}{J\sigma_\varepsilon}B_j^i\dfrac{\partial\varepsilon}{\partial x_j}$	$(c_{1\varepsilon}G-c_{2\varepsilon}\rho\varepsilon)\dfrac{\varepsilon}{k}$

$$C_i=\rho\beta_j^i V_j, \quad \omega_j^i=\beta_j^n\frac{\partial V_i}{\partial x_n},$$

$$\beta_j^i=\text{cofactor of }\partial y_j/\partial x_i \text{ in } J, \qquad J=\begin{vmatrix}\dfrac{\partial y_1}{\partial x_1}&\dfrac{\partial y_2}{\partial x_1}&\dfrac{\partial y_3}{\partial x_1}\\[6pt]\dfrac{\partial y_1}{\partial x_2}&\dfrac{\partial y_2}{\partial x_2}&\dfrac{\partial y_3}{\partial x_2}\\[6pt]\dfrac{\partial y_1}{\partial x_3}&\dfrac{\partial y_2}{\partial x_3}&\dfrac{\partial y_3}{\partial x_3}\end{vmatrix}$$

$$B_j^i=\beta_n^i\beta_n^j,$$

$$\mu_e=\mu+\mu_t, \quad \mu_t=\rho C_\mu k^2/\varepsilon,$$

$$G=\frac{\mu_t}{2J^2}(\frac{\partial V_i}{\partial x_n}\beta_j^n+\frac{\partial V_j}{\partial x_n}\beta_i^n)^2,$$

$$C_\mu=0.09, \quad c_{1\varepsilon}=1.44, \quad c_{2\varepsilon}=1.92, \quad \sigma_k=1.0, \quad \sigma_\varepsilon=1.32.$$

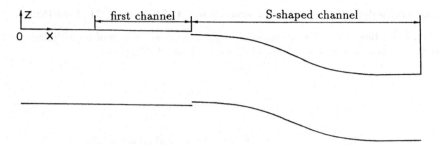

Figure 1. Geometry of solution domain

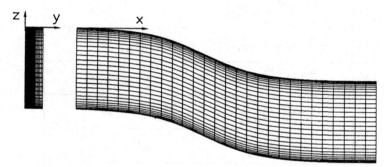

Figure 2. Computational grid for S-shaped channel

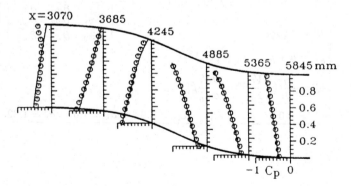

Figure 3. Pressure distribution on top wall of S-shaped channel
—— Calculation; o Experiment

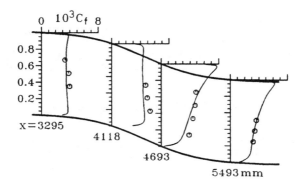

Figure 4. Skin-friction distribution on top wall of S-shaped channel
—— Calculation; o Experiment

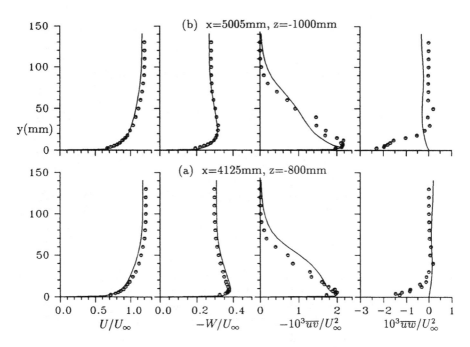

Figure 5. Boundary-layer profiles at two stations
—— Calculation; o Experiment

A Pseudo-3D Extension to the Johnson & King Model and Its Application to the EuroExpt S-Duct Geometry

A.M. Savill (1) , T.B.Gatski (2) & P.-A. Lindberg (3)

1: Rolls-Royce Senior Research Associate, University of Cambridge, England.
2: Senior Research Scientist, NASA Langley Research Centre, USA.
3: Visiting Research Scientist, EPF Lausanne Pilot Centre, Switzerland.

1 INTRODUCTION

This paper presents a submission to the T1 Test Case problem of the Lausanne ERCOFTAC Workshop on Numerical Simulation of Unsteady Flows, Transition to Turbulence and Combustion. The model adopted for these calculations is a new 3-D version of that originally proposed, by Johnson & King [1], for non-equilibrium 2-D flows approaching separation, referred to hereafter as the J&K model. Being, in the taxonomy of the 1980-81 Stanford Conference, only a 'half-equation' model (because it employs just an ordinary differential equation (ODE) for the maximum shear stress, τ_m and a prescribed mixing length, l) this closure scheme is of considerable interest to those attempting to improve current Navier-Stokes solutions for compressible flow, both through and around complex geometries, which employ the simpler well-known Baldwin-Lomax or Cebici-Smith prescriptions for l alone. In fact a number of attempts have already been made to extend the original 2-D J&K model to handle 3-D flows. In particular Abid [2] has described a scheme in which the principal shear stress $\overline{uv}$, mean strain, and mean velocity were replaced by the corresponding resultant quantities for 3-D flow. In addition an allowance for anisotropic effects was introduced by taking into account the modifications to the mixing length and eddy viscosity implied by Rotta's T model [3] (a_1 constant T value of 0.7 appeared optimum). However the additional correction factors for a (the ratio of the shear stress to the turbulence energy, k), and for τ itself, were omitted; and the only generalisation incorporated in the ODE was the inclusion of the additional Convective derivative (and not additional Production) terms. These defficiencies may explain why Abid found that the resulting 'anisotropic' J&K model failed to produce superior results for incompressible infinite swept wing and pressure-driven 3-D boundary layer test problems, compared to the 'isotropic' version without the Rotta corrections, except in the case of cross-flow properties. However, adopting the J&K model approach did significantly

158

improve the baseline Cebeci-Smith predictions, so he concluded that the inclusion of non-equilibrium effects was at least as important as the modelling of anisotropic effects, and in addition showed that a simple generalisation of the normal mixing length expression for the inner boundary layer region to: $l = \kappa (\tau / \tau_m \tau_w)^{0.5} y$ (1)
produced larger improvements in predictions than the T-model corrections he introduced.

A similar type of 'isotropic' extension, but employing the Baldwin-Lomax length scale description, was subsequently used by Abid et al.[4] and Marx [5] for 3-D Navier-Stokes computations of transonic wing flows and again considerable improvements in predictions for cases with separation were found compared to the results obtained with the simpler mixing length approach. Essentially the same approach, but again employing the Cebeci-Smith model, has also been adopted by Shirazi & Truman [6] for computations of the supersonic and hypersonic flow around bodies at incidence, including an axisymmetric compression corner. However for these attached flow cases the non-equilibrium J&K model showed no advantage over the baseline Cebeci-Smith model. Marx noted similar consistency between the J&K and Baldwin-Lomax models for an attached transonic aerofoil flow, although in that case both agreed well with experiment. Indeed it seems generally established that the original J & K formulation does not perform so well, even on 2-D flows, away from separation and this has led Bettelini [7] and Johnson & Coakley [8] to suggest different refinements to the blending of the inner and outer layer eddy viscosities (see equation (16) below). Johnson & Coakley, and more recently Granville [9], have also proposed alternative prescriptions for the inner layer viscosity (equations (10 & 11)) to ensure this satisfies the law of the wall under adverse pressure gradient condition.

Despite the above comments it was evident from [2] that isotropic versions of the model do tend to underpredict regions of strong cross-flow due to the lack of an adequate anisotropic eddy viscosity. In an attempt to remedy this Abid [10] has more recently proposed an alternative 'anisotropic' 3-D extension to the J&K model in which the same isotropic description is applied to the inner boundary layer region, but two separate ODEs, and hence outer layer eddy viscosities, are prescribed for $\overline{uv}$ and $\overline{vw}$. The resulting scheme appeared to perform very much better on the infinite swept wing test case than the earlier version or other eddy viscosity appraoches; correctly predicting both the magnitude and lag of the shear stress vector. However the only difference in this version between the outer layer viscosities and their isotropic counterparts was the independence of the values attributed to the non-equilibrium parameters σ(x) and σ(z) assigned to these, and the ODE for $\overline{vw}$ was derived directly from that for $\overline{uv}$ by simple substitution so that the same a parameter was used in each.Such approximations limit the generality and range of applicability of this approach and prevent it from being considered to be a 'full' 3-D extension of the J&K scheme. Bettelini [7] put forward a similar anisotropic model, with separate ODEs for $\overline{uv}$ and the ratio of $\overline{uv}$ to $\overline{vw}$, at the Workshop (although no results were presented for T1). However this suffers the same defficiencies since both ODEs are derived in an analagous manner from the isotropic modelled transport equation for k, using again the same a $_1$.

As a first step towards a true 'full-3D' extension the present work has returned to the simpler 'pseudo-3D' approach originally adopted by Abid [2], but the additional production terms have been included in the ODE and an attempt has been made to introduce all of the anisotropic correction factors implied by the Rotta T-model (or alternative Tg-model of Moreau [11]) in a self-consistent manner. The resulting model specification has then been introduced into an established 3-D boundary layer code, also due originally to Moreau [11], in order to evaluate its performance on the T1 Test Case prior to implementing it in a 3-D Navier-Stokes code. In a separate submission Lindberg [12] has tested a similar 3D J&K model which also includes the curvilinear production terms, but no T or Tg factors.

2 PROPOSED MODEL EXTENSION TO 3D

The present proposed extension to the J&K model follows a pseudo-3D approach whereby, like Abid [2], we work in terms of resolved shear stress, mean strain and mean velocity vectors:

$$\tau = ((\overline{uv})^2 + (\overline{vw})^2)^{0.5} \tag{2}$$

and in the notation of Moreau [11]: $\quad S = ((\partial U/\partial y)^2 + (\partial W/\partial y)^2)^{0.5}$

$$q = (U^2 + W^2)^{0.5} \tag{3}$$

Only one ODE, is retained for τ_m, but the appropriate 3-D correction factors for the mixing length 1, the structure parameter a_1 (= ratio τ to tubulence energy, k), eddy viscosity ν_t, and τ are introduced. As derived from the T model of Rotta [3] these are:

$$l_{3D} = 1.826 (T T ')^{0.5} 1 \quad : \quad \sqrt{a_1} = 1.826 \tag{4}$$

$$a_{3D} = (T' T^{-1}) a_1 \tag{5}$$

$$\nu_t = (T) 1^2 S \tag{6}$$

$$\tau = (T ') \nu_t \ \partial U/\partial y \tag{7}$$

where: $\qquad T = \left\{ 1 + (\frac{T - 1}{q^2})[\frac{(W\partial U/\partial y - U\partial W/\partial y)^2}{S^2}] \right\}^{0.5} \tag{8}$

and $\qquad T' = \left\{ 1 + (\frac{T^2 - 1}{q^2})[\frac{(W\partial U/\partial y - U\partial W/\partial y)^2}{S^2}] \right\}^{0.5} \tag{9}$

Then, in the inner boundary layer: $\qquad \nu_{t_i} = D^2 l_{3D} | \tau_m |^{0.5} \tag{10}$

For consistency, and with A+=15 [1], $\quad D = (1 - \exp (\frac{-y | \tau_m |^{0.5}}{\nu A^+})) \tag{11}$

which retains the ability to handle separated flows.

And, in the outer boundary layer: $\qquad \nu_{t_o} = \sigma (x,z) \ \gamma_{3D} | \int_0^\infty (q_e - q) | \tag{12}$

where, following Moreau [11] $\qquad \gamma_{3D} = \frac{| q | - U_e}{U - U_e} \ \gamma_{2D} \tag{13}$

with $\qquad\qquad\qquad \gamma_{2D} = [\ 1 + 5.5\ (y/\delta)^6\ ]^{-1}$ $\qquad\qquad$ (14)

and $\sigma(x,z)$ is determined, as in the basic 2D and earlier J&K 3D extensions, from an iterative procedure such that: $\qquad\qquad \nu_{\tau_\mu}\ \partial U/\partial y = -\ \tau_m$ $\qquad\qquad$ (15)

where ν_{t_m} is evaluated from the above relations assuming a blending law of the form:

$$\nu_t = \nu_{t_o}\ (\ 1 - \exp\ (-\ \nu_{t_i}/\nu_{t_o})^n\ )^{\ 1/n} \quad \text{with } n = 1 \qquad (16)$$

[Note: Alternative n=2 [7], and tanh function in place of exponent [8], for attached flow] and τ_m is evaluated from an ordinary differential equation derived from the k transport equation, but with a Production term:

$$\frac{1}{|t|}\frac{3D}{}\ [\ -\ \overline{uv}\ (\partial U/\partial y) - \overline{vw}\ (\partial W/\partial y)\ ] = (\ T^2 - T^{-1})\ l\ \partial u/\partial y \equiv (\tau_{eq})^{0.5}\ T^2\,T^{-1} \qquad (17$$

whence: $\qquad l_{3D}\dfrac{U_m(TT^{-1})}{a_1\tau_m}\dfrac{d|\tau_m|}{ds} = (T^2\,T^{-1})\ |\tau_{eq}\,|^{0.5} - l_{3D}\,[\text{Diffusion}] - |\tau_m.|^{0.5} \qquad (18)$

and $\qquad \dfrac{dg}{ds} = \dfrac{a_1}{2|\tau_m\,|}(\dfrac{T'}{T})\ \{[1 - (\dfrac{T^2}{T'})\ (\dfrac{g}{g_{eq}})] + \dfrac{C_D\ l_{3D}}{a_1\delta[0.7 - (\frac{y}{\delta})_m]}\,|1 - (\dfrac{\nu_{to}}{\nu_{toeav}})^{\,0.5}\,| \qquad (19)$

where $\qquad g = |\tau_m\,|^{0.5} \qquad a_1 = 0.3 \qquad C_D = 0.5$

Note that the present version of the model is based on the Cebeci-Smith approach because it has been applied within the framework of a boundary layer code. However the expression for the outer layer eddy viscosity could equally well be based on the corresponding Baldwin Lomax prescription, which is better suited to Navier-Stokes code implementation and has therefore been the choice adopted for many 2-D applications of the J&K model, in order to avoid difficulties in evaluating δ. In addition the original 2D Diffusion approximation has been retained in the present model extension, because it is anticipated this will only have a second-order influence on the predictions, and indeed this term has sometimes even been omitted when applying the model to 2D flows (eg. by Johnson & Coakley for σ less than 1). If the a factor were replaced by that given by (5), the T'/T factor would cancel out.

3 IMPLEMENTATION

The model specification outlined above (as derived by Savill & Gatski) has been written in the notation adopted by Moreau so that it could be implemented (by Lindberg) directly within his own improved version of the general K-curvature 3-D boundary layer code developed by Moreau and others (see [11]). This was done so that the performance of the model on the T1 S-Duct Test Case could be evaluated in direct comparison to the computations performed separately by Lindberg himself [12] using his own J & K model extension, and also a k-ε/algebraic stress model (ASM) scheme allied to an equation for Tg, with exactly the same initial/boundary conditions and numerical scheme. The T and T' factors were evaluated either using a constant value of T (=0.7), but $\overline{uv}$ and $\overline{vw}$

calculated using the value for Tg (as determined from a new transport equation for $\overline{vw}\,/\,\overline{uv}$ based on the Launder, Reece & Rodi Reynolds stress transport model equations for each of these - see [11,12]). The mixing length prescription used in every case was:

$$l \;=\; 0.085\,\delta\,\tanh(\kappa y/0.085\,\delta) \tag{20}$$

The normal stresses were evaluated assuming $\overline{u^2} = 4/7\,k\,;\; \overline{v^2} = 1/7\,k;\; \overline{w^2} = 2/7\,k \tag{21}$

For further details of the methods adopted, as well as the initial and boundary conditions, and an indication of the sensitivity of the results to refinement of the grid or model constants see [12].

4 RESULTS

The results obtained with the present pseudo-3D J&K model extension can be compared directly with those achieved by Lindberg using his alternative pseudo-3D version of the J&K model and k-ε scheme, with identical numerics, and those obtained by Bettelini [7] using an isotropic-3D J&K model, but different numerics, as well as with the experimental data from EPFL. Initial results presented at the Workshop suggested that both the anisotropic J&K models were worse than their isotropic counterpart in the sense that in both cases too high a cross-flow was predicted (Indeed when $\overline{uv}$ and $\overline{vw}$ were computed using T=0.7 the Savill-Gatski model resulted in an even higher cross-flow). As a result there was only a very short region of the flow, up to x=3.4m (including only Station 29 of the three experimental Stations selected for detailed comparison by the Test Case Supervisor, Ryhming [13]) for which the results were valid (see Fig.1a). By comparison both the k-ε model and the isotropic J&K predicted a much lower crossflow and their results were valid for all three Stations 29, 33 & 44. It was clear that such a discrepacy had to be related to the manner in which the anisotropic J&K models were implemented, since this was the same for both despite their different formulations, and following discussions at the Workshop with Bettelini it emerged that the results are very sensitive to the manner in which the maximum shear stress and related values are computed. In particular it is essential to compute τ_{eq} and the mean flow derivatives precisely at the exact location of maximum shear stress and not to interploate these using weighted residuals as Lindberg originally chose to do. When the same procedure as employed by Bettelini [7], was adopted the cross-flow predictions were greatly improved and resulted in a similar domain of validity to the k-ε model (see Fig.1b). Even before this correction was made the present model produced predictions for Station 29 which were more closely in line with the k-ε model results and the experimental data than the other two J&K model versions. In particular γ_p, $\overline{vw}$, and also $\overline{v^2}$ and $\overline{w^2}$, were better predicted while similar results were obtained for U and γ_t, but worse for W and the other Reynolds stresses (although the k-ε scheme adopted by Lindberg made use of generalised ASM relationships - see [12] - rather than assuming simple constant ratios such as those given in (20)).The present J&K model was also a clear improvement over the basic Moreau Tg and simpler Rotta T models.

After correction the predictions for Station 29 were improved (see Fig.2a&b) and the same general observations made above also held true for Stations 33 (Fig.3a&b) and 44 (Fig.4a&b), although in all cases $\overline{uv}$ and $\overline{uw}$ were much better predicted by the other two J&K models; the present model considerably over-predicting these shear stresses. This discrepancy suggests that too large an allowance was made for the 3D anisotropic influence through the T and T' factors, particularly with regard to the resolved length scale l which does not reduce to the 2D value for T,T'=1. Indeed it seems generally that some double-accounting for 3D effects has been introduced inadvertently by allowing separately for the anisotropy of l, a_1, v_t & τ when these are all inter-related. Almost certainly the a factor in (4) is incorrect. In addition (as noted above) the T'/T factor should have been omitted from the Diffusion term in (19), and although this is likely to have a small influence, a similar case can be made for reducing the T^2 factor to T in the more important Production term of (17). [**NB**: Subsequent tests revealed that omitting the a produces far better predictions for all quantities (almost equivalent to k-ε). Altering Production and Diffusion factors had a progressively far smaller effect.]

Acknowledgements

This collaborative project has been supported in part by Rolls-Royce plc through Research Brochure PVA3-100D. Additional funds were made available though the EPFL ERCOFTAC Pilot Centre who also provided access to all of the necessary Vax and CRAY 2 computing resources. Special thanks are due to Dr.V.Moreau for helpful discussions concerning his 3D boundary layer code.

References

[1] D.A.Johnson & L.S.King (1984) A New Turbulence Closure Model for Boundary Layer Flow With Strong Adverse Pressure Gradients and Separation. AIAA-84-0175.

[2] R.Abid (1988) Extension of the Johnson & King Turbulence Model to the 3-D Flows. AIAA-88-0223.

[3] J.C.Rotta (1979) A Family of Turbulence Models For Three Dimensional Thin Shear Layers. Proc. 1st Turbulent Shear Flows Symp., TSF1 (Springer-Verlag).

[4] R.Abid, V.N.Vatsa, D.A.Johnson & B.W.Wedan (1989) Prediction of Separated Transonic Wing Flows With a Non-Equilibrium Algebraic Model. AIAA-89-0558 and AIAA J. 28 (8), 1426-1431 (1990).

[5] Y.P.Marx (1989) Numerical Simulation of Turbulent Flows around Airfoil and Wing. Proc. 8th GAMM Conf. on Numerical Methods in Fluid Mechanics, Delft (Vieweg-Verlag).

[6] S.A.Shirazi & C.R.Truman (1989) Simple Turbulence Models For Supersonic and Hypersonic Flows: Bodies at Incidence and Compression Corners. AIAA-80-0669.

[7] M.S.Bettelini (1990) Numerical Study of Some Engineering Turbulence Models for Three-Dimensional Boundary Layers. Technical Science Doctoral Thesis ETH Zurich. (See also separate submission to T1 Test Case in these Proceedings).

[8] D.A.Johnson & T.J.Coakley (1990) Improvements to a Nonequilibrium Algebraic turbulence Model. AIAA Paper to appear.
[9] P.S.Granville (1990) A Near-Wall Eddy Viscosity Formula for Turbulent Boundary Layers in Pressure Gradients Suitable for Momentum, Heat, or Mass Transfer. JFE 112, 240-243.
[10] R.Abid (1988) An Anisotropic Eddy Viscosity for 3-D Separated Boundary Layer Flows. SAE Technical Paper 881544.
[11] V.Moreau (1988) A Study of Anisotropic Effects in Three-Dimensional Incompressible Turbulent Boundary Layers Driven By Pressure Gradients. Ph.D. These No.755 EPF Lausanne.
[12] P.-A. Lindberg & I.L.Ryhming (1990) An anisotropic eddy viscosity model, using the k-ε model or an extended Johnson & King model for solving the boundary layer equations. Separate T1 submission to this Workshop Proceedings Volume.
[13] I.L.Ryhming (1990) Synthesis of T1 submissions for this Workshop Proceedings Volume

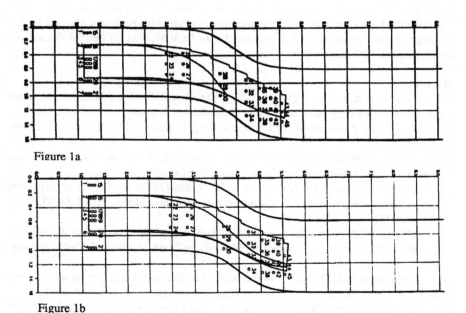

Figure 1a

Figure 1b

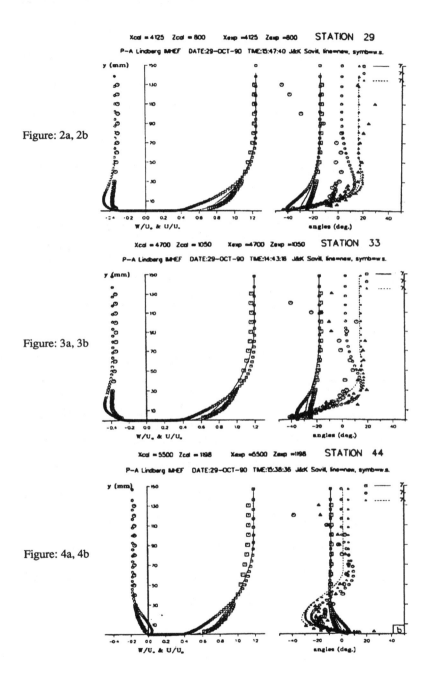

Figure: 2a, 2b

Figure: 3a, 3b

Figure: 4a, 4b

COMPUTATION OF THE 3D TURBULENT BOUNDARY LAYER IN AN
S-SHAPED CHANNEL

J. Wu [1], U.R. Müller[2], E. Krause

AERODYNAMISCHES INSTITUT, RHEINISCH-WESTFÄLISCHE TECHNISCHE
HOCHSCHULE AACHEN, F.R. GERMANY

1 SUMMARY

This report summarizes the contribution presented at the
Workshop on Numerical Simulation of Unsteady Flows, Transition to Turbulence and Combustion, Lausanne, March 26-28,
1990. The "Boundary layer in an 'S' shaped channel" (Problem T1) was computed with an efficient implicit finite-difference algorithm employing a low-Reynolds- number k-ε -
turbulence model. The results are compared with the experimental data provided by EPF- Lausanne after the workshop.

2 INTRODUCTION

Investigations of three- dimensional boundary layers form
part of the turbulence research at the Aerodynamisches
Institut and in the aeronautical industry as well. A recent
cooperation resulted in the work of Wu (1989), who developed an efficient finite- difference algorithm, which simultaneously solves all governing equations. The options for
turbulence modeling include algebraic, low- Reynolds- number k-ε , and three- equation Reynolds stress closure assumptions, as well as anisotropy of eddy viscosities. The
method was validated in comparison with various experiments
in flows with moderate to strong pressure gradients, and
then used to predict the boundary layer flow in question.

3 FINITE- DIFFERENCE METHOD

3.1 Coordinates
The Cartesian coordinates x-y-z are transformed into curvilinear, non-orthogonal ξ- η- ζ ones. The in- plane coordinates ζ and ξ are defined as follows: ζ is parallel to z,ξ
is defined by means of the limiting streamlines defining
the lateral extent of the Raetz domain of influence.

---

1   ZARM, University Bremen, D-2800 Bremen
2   Deutsche Airbus, D-2800 Bremen

The direction of ξ on the lateral boundaries is given by the directions of the limiting streamlines, at interior grid points the direction is linearly interpolated in ζ - direction between the boundary values. The y-axis is perpendicular to the other coordinates. The metric coefficients are constant in the normal direction. The η- coordinate is obtained by stretching the y- coordinate by $\sqrt{Re}$. Secondly, it is scaled by a prescribed, estimated development of the boundary- layer thickness, thereby virtually eliminating the need to increase the number of grid points due to boundary- layer thickening.

3.2 Algorithm
The conservation equations for mass and momentum as well as the turbulence transport equations for kinetic energy and its dissipation rate were discretised by a three- dimensional version of the Laasonen scheme and yielded the numerical accuracy of $O(\Delta\xi, \Delta\eta^2, \Delta\zeta)$. At every grid point, the CFL condition was accounted for by upwind differencing. The difference equations are Newton- linearised and solved simultaneously by a method for inverting block tridiagonal matrices. As shown by Wu (1989), this coupled solution procedure reduces the computer time required by a decoupled method to about one third.

The low- Reynolds- number k-ε model of Chien (1982) was extended to a form suitable for three- dimensional flows; for the present computation, isotropic eddy viscosities were used.

3.3 Initial and Boundary Conditions
Given the external pressure distribution, the inviscid external flow is computed by the Euler equations for two-dimensional flow. With $\partial U/\partial y$ and $\partial W/\partial y$ vanishing at the outer edge of the boundary layer, the equations are discretised explicitly and yield algebraic equations for the two unknown velocity components. The computation is advantageously started at x= 0.695, with the measured pressure coefficient being $c_p \approx -0.16 =$ constant and the one- dimensional velocity $U_e \approx 1.08$ as obtained from Bernoulli's equation. Downstream, the pressure coefficients were linearly interpolated between the experimental data.

Using the external velocities as a boundary condition, a two-dimensional boundary layer computation matching the given experimental data was run towards x=1.945. Then this flow was marched downstream in a three- dimensional manner towards the position x=6.0.

3.4 Grid and Accuracy

The computation employed the standard step sizes $\Delta\xi$ =0.02 and $\Delta\zeta$ =0.03; the reference length used for nondimensionalising is L=1m. Keeping the number of grid points constant in crosswise direction automatically decreased the step size with shrinking lateral extent of the Raetz domain. In regions of large skewing, the step size was limited by the CFL condition. A nonuniform grid was prescribed for the normal direction. For resolving the viscous sublayer, a step size of the order of Δy^+=1 was considered adequate. Stretching the grid towards the outer edge yielded a total of about 80 nodes across the boundary layer.

In test calculations, the streamwise step size was varied by a factor of four down to $\Delta\xi$ =0.005 and combined with a reduced convergence criterion. Halving the step size from 0.01 to 0.005 neither effected the integral lengths nor the skin friction, but nearly doubled the CPU time. Corresponding to previous experiences, we demonstrated that the use of the k-ε model was most efficient when lagging the computation of the eddy viscosity by one step size. Compared to an iterative inclusion of its calculation, this saved more than 50 percent of the CPU time at the expense of an uncertainty of 2 percent in c_f.

As a further test, we evaluated the integral balances of momentum and kinetic energy in order to check both the downstream development of the contributions to the balances and the accuracy as well.

4 RESULTS AND DISCUSSION

The flow develops from an accelerated, nearly two- dimensional boundary layer towards a fully three- dimensional one. Lateral pressure gradients (positive in the prescribed left- hand Cartesian frame of reference) drive the near-wall flow into the negative z- direction, then the flow turns back, thereby producing cross- over W- profiles at about x=5.255. Finally the inner layer also skews into the opposite direction. Throughout the flow field, the measured and computed mean velocities and Reynolds shear stresses, an example is given in Figs. 1 and 3, show good overall agreement, at least a much better one than achieved in any previous comparison calculation.

Some deviations of the outer edge velocities may be traced back to inconsistencies between the measured velocities and pressures. The reasons for the close agreement between experiment and calculation are that the computed boundary layer proved to be an equilibrium one due to moderate pressure gradients, that moderate skewing was below 10°, and that the "infinite swept wing" like behaviour showed little variation in the lateral direction. Though the shear stress profiles in Fig. 3 reveal considerable scatter, they seem to approach the inner- layer asymptotes $\tau_i = \tau_{wi} + \partial p\ \partial x_i * y$. In general, however, the measured skin friction data (reproduced by assuming a constant freestream velocity of 18m/s) do not fit well but seem to be overestimated by about 50 percent. The computed flow field obviously is an isotropic one with coinciding directions of the shear stress vector and the velocity gradient vector, Fig. 2. Disregarding scatter within a near- wall error band of ± 25° , both directions seem to coincide and point towards isotropic eddy viscosities.

5. CONCLUDING REMARKS

The three- dimensional turbulent boundary layer in an S-shaped duct was computed by means of an implicit finite – difference method and a low- Reynolds- number k- ε model. Close agreement between experimental and computational results was achieved throughout the flow field, even for the two Reynolds shear stresses. Future experimental test series are recommended to strive towards enhanced skewing of the flow, noticable departure from local equilibrium, and generation of non-isotropic flow features.

6. REFERENCES

J. Wu (1989) Berechnung zwei- und dreidimensionaler turbulenter Grenzschichten. Dissertation, Rheinisch- Westfälische Technische Hochschule, Aachen, FRG.

K.-Y. Chien (1982) Predictions of channel and boundary layer flows with a low- Reynolds- number turbulence model. AIAA J. 20, no. 1, 33.

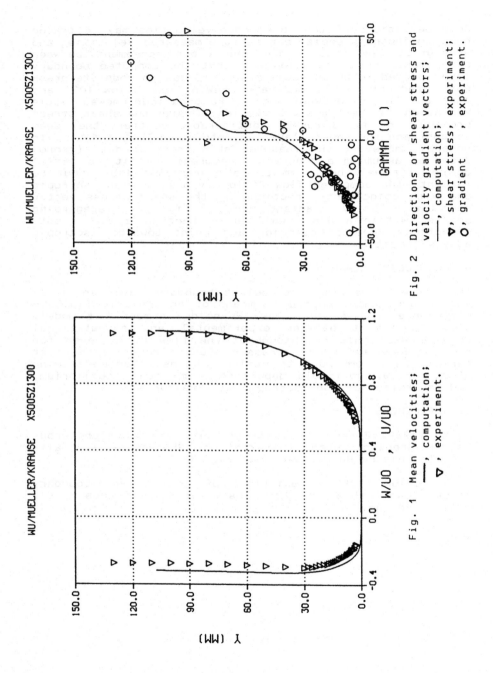

Fig. 2 Directions of shear stress and
velocity gradient vectors;
—— , computation;
▷ , shear stress, experiment;
○ , gradient , experiment.

Fig. 1 Mean velocities;
—— , computation;
▷ , experiment.

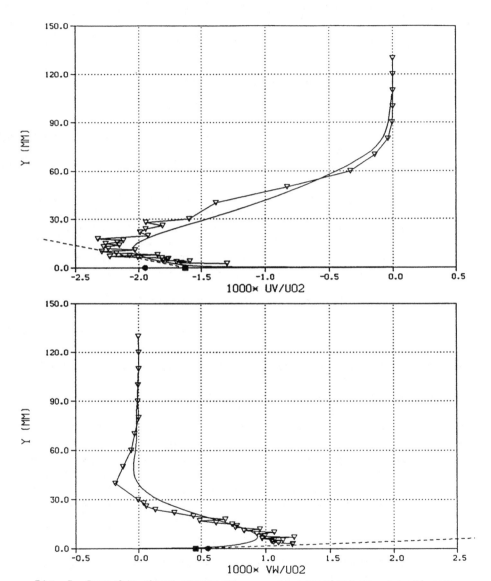

WU/MUELLER/KRAUSE   X5005Z1300

Fig. 3   Reynolds shear stresses; ———, computation; ▽ , experiment.
■ / ● , computed/ experimental wall shear stress component;
---, $\tau_i = \tau_{wi} + \partial p / \partial x_i * y$.

A New Statistical Approach to Predict Turbulence

A. NAKKASYAN

IMHEF/Ecole Polytechnique Fédérale de Lausanne , 1015 Lausanne, Switzerland

1 INTRODUCTION

All TBL calculations, regardless their degree of complexity, are accompanied by a number of hypotheses which, individually, may or may not be tested experimentally. As a general rule, the algebraic expressions for $(-\overline{u'v'}, -\overline{w'v'})$ in simple differential methods can be tested for self-consistency with actual data and eventually be discarded or improved safely. On the other hand, a number of hypotheses which are incorporated in complex differential expressions, such as the transport equations for $(\overline{u_i'u_j'})$, will probably not find direct experimental support for a long time. Nevertheless, these non-algebraic methods have conserved the unique advantage of predicting a greater number of TBL properties; typically all six $\overline{u_i'u_j'}$ [as compared to $(-\overline{u'v'}, -\overline{w'v'})$, only].

In this paper, a new approach is used, to predict the measured ($\overline{u'^2}$, $\overline{w'^2}$, $\overline{v'^2}$, $\overline{u'w'}$)- profiles for the T1-test case with the help of the previously calculated profiles[0] of $(-\overline{u'v'}|_{calc} ; -\overline{w'v'}|_{calc})$. The *basic hypothesis* and *new simplifying relationships* are obtained from the (re)analysis of two, practically identical, "infinite" swept-wing experiments[1][2]. The justification for our hypothesis is derived from the success of a *self-consistency test*. The test answers to the question "<u>can we reproduce</u> with reasonable precision the <u>measurements of the 6 double-correlations</u> $(\overline{u_i'u_j'})$, with the knowledge that, the <u>distribution law</u> $P_{\hat{A}}(\vec{u}_1' , \vec{u}_2' , \vec{u}_3')$ of the components of the fluctuation vector $\vec{q'}(t)$ in the particular coordinate system of the eigenvectors $(\hat{a}_1, \hat{a}_2, \hat{a}_3)$ should be <u>simpler than</u> the corresponding <u>joint-probability distribution</u> $P_{\hat{X}}(\vec{u}_1', \vec{u}_2', \vec{u}_3')$ in the measurement system $(\hat{x}_1, \hat{x}_2, \hat{x}_3)$; <u>because</u> the correlation matrix of the eigenvalues $(\overline{u_i'u_j'})$ satisfies by definition $[\overline{u_i'u_j'} \equiv 0;$ if $i \neq j]$, which is the additional condition for statistically-independent variables ?".

Our *basic (trial) hypothesis* is inspired from the fundamental approximations of statistical physics[3]. Accordingly, the components of the fluctuation-vector $\vec{q'}(t)$ in $(\hat{a}_1, \hat{a}_2, \hat{a}_3)$ are assumed to be statistically independent variables and to obey three centered-Gaussian laws:
$P(\vec{u}_1', \vec{u}_2', \vec{u}_3') = P_1(\vec{u}_1') \cdot P_2(\vec{u}_2') \cdot P_3(\vec{u}_3')$; with $P_i = (2\pi\overline{u_i'^2})^{-1/2} \exp(-u_i'^2/(2\overline{u_i'^2}))$.

172

Matrix-diagonalizations and Monte Carlo (MC-) calculations[4] (the latter allow the prediction of the double-correlations by simulation on computer) are used as natural tools for the exploration of the unknown distribution-laws. Since a good agreement is found between our MC-predictions $<\vec{u'}_{i\text{-MC}}\vec{u'}_{j\text{-MC}}>$ and the measurements $(\overline{u'_i u'_j})$ for the "infinite" swept-wing experiments, the "three centered-Gaussian hypothesis" for the vectors $(\vec{u}'_1, \vec{u}'_2, \vec{u}'_3)$ is kept as a *hypothesis* for the modelling of the double-correlations (only!) of the S-duct experiment. Subsequently, *to test the validity* of our working hypothesis, we have to proceed, as for the "infinite" wing case, in the following order:

(a) the locally measured $(\overline{u'_i u'_j})$-matrices are diagonalized to yield the "experimental" eigenvalues $\overline{u'_i{}^2}$, eigenvectors $\hat{d}_i$ and the transformation matrix $\mathcal{T} \equiv (d_{ij})$. Note, the j-th column of (d_{ij}) contains the coordinates (d_{1j}, d_{2j}, d_{3j}) of the eigenvector $\hat{d}_j$ in $(\hat{x}_1, \hat{x}_2, \hat{x}_3)$. Therefore, coordinate changes between the local measurement-system and the eigenvector-system $(\vec{u}'_1, \vec{u}'_2, \vec{u}'_3) \overset{\mathcal{T}}{\leftrightarrow} (\vec{u}'_1, \vec{u}'_2, \vec{u}'_3)$ can be made with $\mathcal{T}$ or its inverse.

(b) our MC-calculations start by simulating "events" (MC-events) in the coordinate system of the eigenvectors $(\hat{d}_i)$ (see APPENDIX). As a general rule, these computer generated events are submitted to any well defined hypothesis (ambiguity is avoided by testing one hypothesis at a time). Mathematically speaking, the "event-n" is *defined* as a *hypothetical* vector $\vec{q}'(n)_{\text{-MC}} = \vec{u}'(n)_{\text{-MC}} + \vec{w}'(n)_{\text{-MC}} + \vec{v}'(n)_{\text{-MC}}$ whose three components obey prescribed density-probability laws (distribution-laws). For example, the required law for $(\vec{u}'_i(n)_{\text{-MC}}; i=1,2,3; n=1,...,\mathcal{N})$ may be three, centered-Gaussians whose corresponding (r.m.s.)-values $[\overline{u'_i{}^2}]^{1/2}$ are given experimentally. Note: if $\mathcal{N} \to \infty$, statistical-errors $\to 0$. If, in addition, one or more centered-Gaussians are shifted by a small fraction of their (r.m.s.)-values along their respective eigenvectors, the distribution-laws are referred as shifted-Gaussians (or, more generally speaking, as *quasi*-Gaussians).

(c) as soon as generated, each MC-event(n) is transferred from the eigenvectors-system to the measurement-system $(\vec{u}'_i(n)_{\text{-MC}} \overset{\mathcal{T}}{\leftrightarrow} \vec{u}'_{i\text{-MC}})$ where the predictions are finally made using the scalar-product rule for vectors: $(1/\mathcal{N})\sum (\vec{u}'_{i\text{-MC}} \cdot \vec{u}'_{j\text{-MC}})_n = <\vec{u}'_{i\text{-MC}} \cdot \vec{u}'_{j\text{-MC}}>_{\mathcal{N}}$ where summation is performed from $n=1$ to $n=\mathcal{N}$. Since the comparisons between these MC-predictions and corresponding measured quantities were good within the errors (i.e.; {the measurement errors $(\pm\Delta f)_m/f$ + the statistical-errors $(\pm \sqrt{\mathcal{N}})/\mathcal{N}$ + systematic errors}), we conclude that the *basic hypothesis* is applicable to the double-correlations of experiments "similar" to the above mentioned "infinite" swept-wing experiments. Because correlations of higher order than two, such as $(\overline{u'_i u'_j u'_k})$, can detect smaller differences between *quasi*-Gaussian laws and ideal Gaussian laws, they require better knowledge (approximations) of the actual probability-laws (see section 5) .

2 THE EIGENVALUES AND THE EIGENVECTORS OF $(\overline{u_i'u_j'})$

Even if the MC-predicted double-correlations and their corresponding measurements agree at all measurement-stations, there remains the problem of the determination of 12 unknowns at a given calculation-station, that is, the 9 elements of the $T \equiv (\text{dij})$ matrix, and the 3 eigenvalues $\overline{u_i^2}$. Nevertheless, *we have been able to reduce the unknown variables from 12 to 2 with the help of semi-empirical, simplifying relationships*, which were derived from our analysis of ref[1]. Subsequently, the predictions of the $(\overline{u'^2}, \overline{w'^2}, \overline{v'^2},$ $\overline{u'w'})$ were made possible by evaluating the remaining two unknowns with the help of the algebraic model[0]:

(a)   A non-trivial approximation was found while visually analysing a small figure, made of 3x3 sub-figures disposed in a way to represent the ij-th elements of all (dij)-matrices. The quality of the approximation was later tested mathematically at the measurement stations and found to be good. It consisted of replacing the coordinate-transformation matrix (dij) with the product of two axial rotations: one of them about the horizontal eigenvector $\hat{d}_2$ with an angle θ, the other about the vertical axis $\hat{x}_3$ with an angle φ.

$$(\text{dij}) \approx \mathbf{R}(-\varphi)\mathbf{R}(-\theta) = \begin{pmatrix} \cos\varphi & -\sin\varphi & 0 \\ \sin\varphi & \cos\varphi & 0 \\ 0 & 0 & 1 \end{pmatrix} \begin{pmatrix} \cos\theta & 0 & -\sin\theta \\ 0 & 1 & 0 \\ \sin\theta & 0 & \cos\theta \end{pmatrix} \tag{1}$$

Thus, the angles $[\varphi ; (\pi/2)-\theta]$ are the *spherical angles of* $\hat{d}_1$ in system $(\hat{x}_1, \hat{x}_2, \hat{x}_3)$; eigenvector $\hat{d}_2$ is horizontal (parallel to the wall); the angle between $\hat{x}_3$ and $\hat{d}_3$ is θ.

When the general properties of *statistically independent* variables are taken into account, a *mnemonic* relationship *for* evaluating the *2nd*-order correlations is obtained [*!! Note: the relationship is not valid as it is :* $\overline{u^i} = 0$, $[\overline{u'^2}]^{1/2} \neq 0$, etc.]   :

$$\begin{matrix} \overline{u^i} \\ \overline{w^i} \\ \overline{v^i} \end{matrix} \overset{!!}{\cong} \begin{pmatrix} \cos\varphi\cos\theta & -\sin\varphi & -\cos\varphi\sin\theta \\ \sin\varphi\cos\theta & \cos\varphi & -\sin\varphi\sin\theta \\ \sin\theta & 0 & \cos\theta \end{pmatrix} \begin{matrix} [\overline{u'^2}]^{1/2} \\ [\overline{w'^2}]^{1/2} \\ [\overline{v'^2}]^{1/2} \end{matrix} \tag{2}$$

[for statistically-independent (α,β): $\overline{F(\alpha,\beta)+G(\alpha,\beta)} = \overline{F} + \overline{G}$; $\overline{f(\alpha).g(\beta)} = \overline{f(\alpha)}.\overline{g(\beta)}$].

By following the next two examples of utilisation of (2) :

$$\overline{u^i}\overline{v^i} \Rightarrow \overline{u'v'} = \sin\theta\cos\theta\cos\varphi\,\overline{u'^2} - \sin\theta\cos\theta\cos\varphi\,\overline{v'^2}$$
$$\overline{w^i}\overline{v^i} \Rightarrow \overline{w'v'} = \sin\theta\cos\theta\sin\varphi\,\overline{u'^2} - \sin\theta\cos\theta\sin\varphi\,\overline{v'^2} \tag{3}$$

one can readily extend the procedure to the remaining $(\overline{u'^2}, \overline{w'^2}, \overline{v'^2}, \overline{u'w'})$-correlations. Note also; both $\overline{u'v'}$ and $\overline{w'v'}$ are proportional to $(\overline{u'^2} - \overline{v'^2})$.

(b) In addition, the following *new simplifying relationships* are assumed valid for experiments "similar" to the "infinite" swept-wing experiments:

$$|\theta\,(C_p(X,Z))|\approx|\theta_0|\approx22.5°\pm3.5°\;;\quad [\overline{\mathcal{U}'^2}/\overline{q'^2}]\cong2[\overline{\mathcal{W}'^2}/\overline{q'^2}]\cong4[\overline{\mathcal{V}'^2}/\overline{q'^2}]\cong4/7 \qquad (4)$$

Note, we readily obtain "Bradshaw's constant for a 2D-flow" by setting $\varphi=0$ in (2) - or in (3)- and by using (4) : $a=\overline{(u'v')}/\overline{(q'^2)}=(0.3535)\cdot(4/7 - 1/7)\cong0.15$.

(c) Since the last two unknowns $\overline{q'^2}=\overline{\mathcal{U}'^2}+\overline{\mathcal{W}'^2}+\overline{\mathcal{V}'^2}$ and $\varphi=\varphi(C_p(X,y,Z))$ are present in equations (3), we replace their left hand-terms with $-\overline{u'v'}|_{calc}$ & $-\overline{w'v'}|_{calc}$ which yields

$$\varphi(y) = \mathrm{ATAN}((-\overline{w'v'}|_{calc})/(-\overline{u'v'}|_{calc}))\quad\text{and}\quad \overline{q^2}(y) = \frac{[\,(\overline{u'v'})^2|_{calc}+(\overline{w'v'})^2|_{calc}\,]^{1/2}}{(4/7-2/7)((\sin2\theta_0)/2)} \qquad (5).$$

3   GENERATION OF MONTE CARLO EVENTS

We have shown in APPENDIX how, exactly, we have generated on the computer the centered-Gaussian distribution of a variable X.

By extension, if the probability density $P(X1,X2)$ of two, *statistically independent* variables $(X1,X2)$ is the product of two distinct centered-Gaussians $P1(X1;\sigma_1)\cdot P2(X2;\sigma_2)$ along the coordinate axis, it generates a hill-shaped surface with two planes of symmetry, each containing the vertical axis P. The intersection of any horizontal plane at $P=Pn$ with this hill-shaped surface gives an elliptical contour whose principal axis $(Xn1,Xn2)$ are aligned with the coordinate axis. All pairs of variables $(X1,X2)$ which satisfy the condition $(|X1|<Xn1;\ |X2|<Xn2)$ are contained inside this "ellipse". The condition $(|Xn1|<3\sigma_1;\ |Xn2|<3\sigma_2)$ corresponds to 99.5 % of all possible pairs $(X1,X2)$.

If, on the other hand, the coordinates $(X1,X2)$ are rotated with an angle α (with $\alpha\neq k(\pi/2)$; $k=1,2,3,4$) around the vertical axes P to become a new system $(X\alpha1,X\alpha2)$, the new variables are not statistically-independent. The sign of the correlations $\overline{X\alpha1\cdot X\alpha2}$ is determined by the over-population in the two diagonally disposed quadrants: {I & III $\leftrightarrow$ $\overline{X\alpha1\cdot X\alpha2}>0$; if $(0<\alpha<\pi/2)$} or {II & IV$\leftrightarrow$ $\overline{X\alpha1\cdot X\alpha2}<0$; if $(\pi/2<\alpha<\pi)$}. Similarly, in a non-privileged measurement system the sign of $(\overline{u_i'u_j'})$ is plus or minus.

When the number of the variables is increased from 2 to 3 and corresponding distribution-laws are 3 centered-Gaussians along the axis $(X1,X2,X3)$; the "ellipses" become "ellipsoids" with 3 principal-axis along $(X1,X2,X3)$. This means $\overline{X1\cdot X2\cdot X3}=0$.

4   PREDICTIONS OF $(\overline{u'^2}, \overline{w'^2}, \overline{v'^2}, \overline{u'w'})$ FOR THE T1-TEST CASE

In Fig 1, the predicted $(\overline{u'^2}, \overline{w'^2}, \overline{v'^2}, \overline{u'w})$-profiles are shown for the stations X4700Z0730 (top row) and X5500Z1198 (bottom row). Since the negative contributions of the badly calculated[0] $(\overline{u'v'})$-profile of X4700Z0730 can be identified on the presently predicted $(\overline{u'^2}, \overline{w'^2}, \overline{v'^2})$-profiles, we conclude that the quality of the latter is reasonably good.

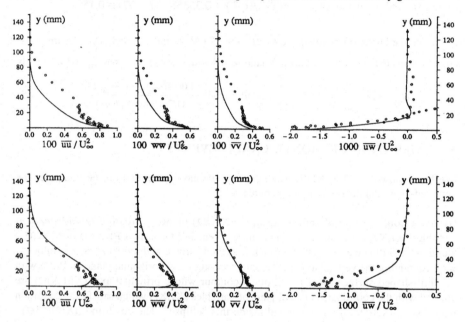

5   POST ERCOFTAC REMARKS

Our recent intrinsic analysis of the T1-test case[5] shows that *(a)* the spreads of the experimental $\{|\,\theta\,|\,;\ (\overline{\mathcal{V}_i'^2}/\overline{q'^2})\}$ about the idealized values $\{22.5°;\ (4/7, 2/7, 1/7;\ i=1,2,3)\}$ are slightly larger than previously[1][2]; *(b)* the calculated residuals for the approximation $\mathcal{T}(\mathrm{d}ij) \approx \mathbf{R}(-\varphi)\mathbf{R}(-\theta)$ are moderately good. Nevertheless, our MC-predictions *for* $(\overline{u_i'u_j'})$ with experimental-(dij), at all 24 measurement-stations, are excellent. Therefore, it is still reasonable to assume "3 *centered*-Gaussian laws" along the eigenvectors of $(\overline{u_i'u_j'})$, for the predictions of the double-correlations. Our conclusion is further supported by the fact that, all (24x16) locally measured $(\overline{u_i'u_j'})$- <u>and</u> $(\overline{u_i'u_j'u_k'})$-profiles could be MC-calculated given 3 *shifted*-Gaussians along the local eigenvectors of $(\overline{u_i'u_j'})$. The quality of these MC-predictions was very good for $(\overline{u_i'u_j'})$ and average/good for $(\overline{u_i'u_j'u_k'})$, when the particular choice $\{\ \mathcal{V}_i'\,|_{\text{shf.G.}} = \mathcal{V}_i'\,|_{\text{c.G.}} + \Delta\mathcal{V}_i\,;\ \Delta\mathcal{U}' = -.5,\ \Delta\mathcal{W}' = -.5,\ \Delta\mathcal{V}' = 0\}$ was applied. Note, the accumulation of the shifts $\Delta\mathcal{U}'=-.5$ for each MC-event corresponds globally to $\overline{\mathcal{U}'}|_{shf.G.}\approx -0.1[\overline{\mathcal{U}_i'^2}]^{1/2}$.

6 REFERENCES

[0] NAKKASYAN A., PRAHLAD T.S., GILQUIN P., "An Algebraic Model based on the Velocity-Profile Skewing". *This Proceeding.*

[1] BERG B.van den, ELSENAAR A., LINDHOUT J.P.F. & WESSELING P., (1975) Measurements in an incompressible three-dimensional turbulent boundary layer, under infinite swept-wing conditions, and comparison with theory. *J.F M.* 70.

[2] PONTIKOS N.S., (1982) "The Structure of three-dimensional Turbulent Boundary Layers", Dept.of Aeron., Imperial College, University of London, *Ph.D. Thesis*

[3] REIF F., (1965) "Fundamentals of statistical and thermal physics" McGraw-Hill, Inc.,

[4] METROPOLIS N. and ULAM S., (1949) "The Monte Carlo Method", ⌊ pg. 265, 581. *J. Amer. Statis. Assoc.*,44.

[5] NAKKASYAN A., "A new statistical approach to predict simultaneously the double- and triple-correlations of the velocity fluctuations " -*(paper in progress)-* .

APPENDIX

```
C  EXAMPLE:  Generation of a 1-dimensional ( i≡1) MC-event :
C  FAC≡(1/√(2*π)) ; X≡𝒰'ᵢ with [-a*SIGU<X<a*SIGU] , a=3 (tail-less) ; SIGU=[𝒰'ᵢ²]¹ᐟ² -(mesur.)- ;
C !!  the name RNUNF( ) for a random-number-generator function  is computer-dependent
    101  CONTINUE
            X0=RNUNF( )              ! we get a new random-number  0≤ X0≤1
            P0=RNUNF( )              ! we get a new random-number  0≤ P0≤1
            X= -3.*SIGU+6.*SIGU*X0   ! scale change on axis - 3*SIGU ≤X(X0)≤+3*SIGU
            P=FAC*P0                 !! P=P(X) is the probability of occurrence of variable X
            B1=((X/SIGU)**2)/2.      -see drawing below-
            Plaw=FAC*EXP(-B1)        !if (P(X)<Plaw(X)), P is kept (•); if not, P is rejected (x)
            IF(P.GT.Plaw)  GO TO 101 ! these candidates are rejected
            ICKEEPU=ICKEEPU+1        ! these are new MC-events (𝒰ᵢ'(n)₋MC ; i=1 ),
    102  CONTINUE                    ! ICKEEPU≡n;  (1≤(n)≤𝒩 ; (𝒩≳5000)
```

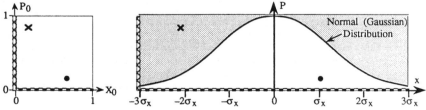

Problem T1: Predictions Of The Boundary Layer Flow In An S-shaped Duct.

D.S.Clarke

CFD Department,
Harwell Laboratory,
Oxon, OX11 0RA, UK

1 NUMERICAL METHOD

The calculations have been made using the HARWELL-FLOW3D fluid-flow computer program. FLOW3D is a finite volume code for the prediction of laminar and turbulent flow, and heat transfer (Burns and Wilkes (1987)). The code uses body-fitted coordinates enabling the calculation of flows through complex geometries. The SIMPLEC algorithm (Van Doormal and Raithby (1984)) is used to solve the coupled system of non-linear equations. In this algorithm the equations are linearised and each linear transport equation is solved using an iterative method. The treatment of the velocity-pressure coupling gives rise to an equation for a correction to the pressure.

FLOW3D uses the $k-\varepsilon$ turbulence model as the default turbulence model but has the options of more advanced Reynolds stress turbulence models, namely algebraic and differential stress models (Clarke and Wilkes (1988), (1989)). For the problems described here calculations have been made using the $k-\varepsilon$ and differential stress models (DSM).

The momentum equations are solved for velocity components in the fixed Cartesian directions on a non-staggered grid. The problem of chequerboard oscillations appearing in the pressure and velocity variables when using such a grid is well known. To avoid these oscillations FLOW3D uses the Rhie-Chow algorithm.

The DSM eliminates these oscillations by modifying the Rhie-Chow corrections to the computation of the normal velocity components on control volume faces prior to the computation of the convection coefficients. The error term introduced into the continuity equation by these velocity corrections is such that it ensures that the pressure field remains smooth.

In the DSM, there is no diffusion term in the momentum equations involving velocity at adjacent points, so that the elimination of chequerboard oscillations in the pressure does not necessarily imply the elimination of such oscillations in the velocity. For numerical stability a diffusion term is added to both sides of the momentum equations and oscillations in the velocity are eliminated by a procedure analagous to the Rhie-Chow interpolation. The overall error introduced by this second interpolation formula can be shown to involve the fourth derivative of the velocity, thus

178

ensuring a smooth velocity field. Near walls however, this fourth derivative may be large and special care has to be taken.

Results have been obtained using the k–ε and DSM turbulence models, with the upwind and higher upwind differencing schemes available in FLOW3D.

2 CALCULATIONS

2.1 Grids

For this problem two sizes of non-orthogonal grid were used: 39×12×18 (coarse) and 67×21×32 (fine). Both grids are body-fitted avoiding the use of steps in the walls of the 'S' bend which would have introduced unneccessary errors into the boundary layer calculations. Figures 1(a) and 1(b) show the coarse grid in the x–z and y–z planes respectively.

2.2 Turbulence Models and Differencing Schemes

On the coarse grid, calculations were made using the k–ε turbulence model with both upwind and higher upwind differencing schemes. It was not possible to use the DSM on this grid as there are too few control volumes between the external and internal walls on the 'S' bend. To ensure a smooth pressure field the calculation of the pressure gradient at a wall requires four points to lie inside the flow away from the wall.

On the fine grid calculations were made using the k–ε model with upwind and higher upwind differencing, and the DSM with upwind differencing.

2.3 Initial and Boundary Conditions

At all walls logarithmic boundary conditions are used for the velocities. The values of the Reynolds stresses at the wall are calculated by linear extrapolation from the values in control volumes interior to the flow.

The internal walls which lie parallel to the wall of the 'S' bend were modelled as infinitely thin surfaces, allowing fluid to flow between the internal and external walls. The flow above the test plate was not modelled.

The data given at each of the 21 inlet stations were restricted to the region $0.002 \leq y \leq 0.07$, $-0.940 \leq z \leq -0.06$m. This data included profiles of velocities, Reynolds stresses and flow angles γ, γ_τ, and γ_g. Calculations were started from the point $x = 1.18$m. The profiles at this point were interpolated from those given at the neighbouring stations. For the remaining interior points, where data was not specified, a plug profile was used for the velocity and Reynolds stresses. Towards the walls the data was interpolated logarithmically. The turbulent kinetic energy k and dissipation rate ε were taken to be:–

$$k = \frac{1}{2}\left(\overline{uu} + \overline{vv} + \overline{ww}\right),$$

$$\varepsilon = \frac{k^2 C_\mu \dfrac{\partial U}{\partial y}}{\overline{uv}} \text{ for } y \leq 0.07,$$

$$\frac{k^{\frac{3}{2}}}{0.3 \; 0.155} \quad y > 0.07$$

where C_μ is a turbulence model constant. The profile for ε taken as above, was reasonably smooth at the point $y = 0.07$.

Neumann boundary conditions were imposed at the exit of the channel.

3 RESULTS

3.1 Comparison of Turbulence Models, Differencing Schemes and Grids

Figures 2 and 3 show comparisons between the different turbulence models on different sizes of grid and using upwind and higher upwind differencing, at the station $x = 4125$mm, $z = -646$mm. Figures 2(a) and (b) show profiles of the non-dimensionalised velocities $\dfrac{U}{U_\infty}$ and $\dfrac{W}{U_\infty}$ respectively,

as functions of y(mm). Similar profiles of the non-dimensionalised Reynolds stresses $\dfrac{\overline{uv}}{U_\infty^2}, \dfrac{\overline{uu}}{U_\infty^2}$, are

shown in Figures 3(a)–(b) respectively. All results seem to show reasonable agreement. On each grid, results using the $k-\varepsilon$ model with different differencing schemes are in close agreement. The results using the $k-\varepsilon$ model on the fine grid lie closest to those using the DSM. The differing levels of the normal stresses represent the anisotropy of the turbulence. Some differences can be seen between the predictions for the normal and shear stresses from the two models.

3.2 Comparisons with Experimental Data

Comparisons between the predictions and experimental data are shown in Figures 4(a)–(b) at the station $x = 4700$mm, $z = -1050$mm. The predictions are those on the fine grid, using the $k-\varepsilon$ model with higher upwind differencing and the DSM with upwind differencing. Figures 4(a)–(b) show profiles for the velocities $\dfrac{U}{U_\infty}, \dfrac{W}{U_\infty}$ respectively. Comparisons of the Reynolds stresses with

experimental data showed that predictions using the $k-\varepsilon$ model of $\overline{uu}$ were not in as good agreement with experimental data as those of $\overline{vv}$ and $\overline{ww}$ Results for the normal stresses using the DSM all compared well with experiment. At the wall both turbulence models over-predicted the value of the shear stress $\overline{uv}$, particularly the $k-\varepsilon$ model. This behaviour could be improved by using a finer grid with a greater number of grid points near the wall.

3.3 Cpu Time

Table 1 gives the running times for each calculation on a Cray 2. It should be noted that no attempts were made to optimize these times.

Model	Grid	Differencing	CPU (secs)
$k-\varepsilon$	Coarse	UW	184
$k-\varepsilon$	Coarse	HUW	+145
$k-\varepsilon$	Fine	UW	1500
$k-\varepsilon$	Fine	HUW	+1400
DSM	Coarse	UW	1930

Table 1. Running times on a Cray 2

4 REFERENCES

Burns, A.D. and Wilkes, N.S. A finite difference method for the computation of fluid flows in complex three dimensional geometries. AERE-R 12342 (1987).

Clarke, D.S. and Wilkes, N.S. The calculation of turbulent flows in complex geometries using an algebraic stress model. AERE-R 13251 (1988).

Clarke, D.S. and Wilkes, N.S. The calculation of turbulent flows in complex geometries using a differential stress model. AERE-R 13428 (1989).

Van Doormal, J.P. and Raithby, G.D. Enhancements of the SIMPLE method for predicting incompressible fluid flows, Numer. Heat Transfer, 7 pp 147-163 (1984).

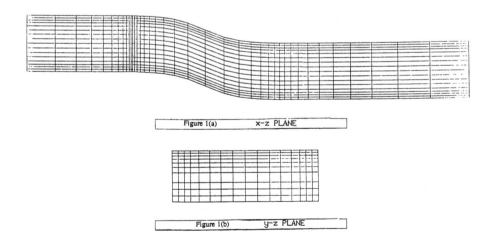

Figure 1(a)   x-z PLANE

Figure 1(b)   y-z PLANE

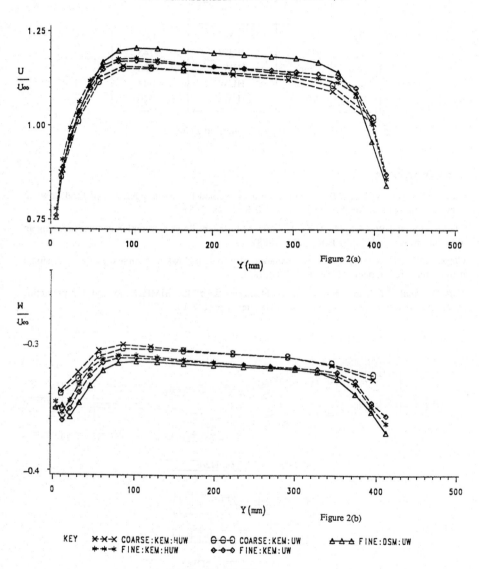

PROFILES AT X=4.125M , Z=-0.646M.
Non dimensionalised with U∞ = 18.0 m/s.

Figure 2(a)

Figure 2(b)

KEY   ✕✕✕ COARSE:KEM:HUW   ⊖⊖⊖ COARSE:KEM:UW   △△△ FINE:DSM:UW
      ✳✳✳ FINE:KEM:HUW   ◇◇◇ FINE:KEM:UW

COMPARING TURBULENCE MODELS AND DIFFERENCING SCHEMES.

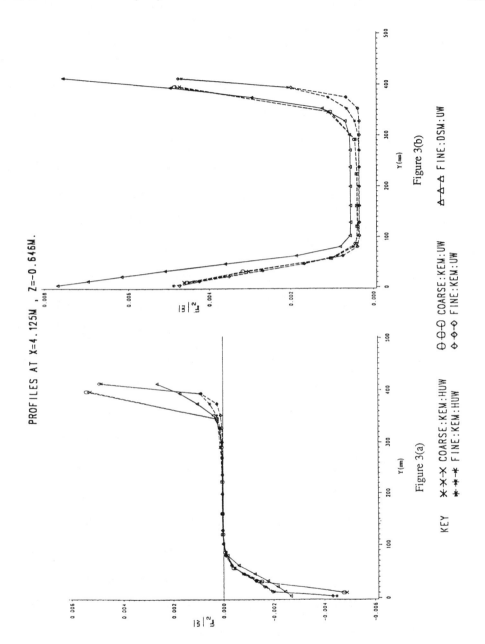

Figure 3(a)

Figure 3(b)

KEY   ✳–✳–✳ COARSE:KEM:HUW   ⊖–⊖–⊖ COARSE:KEM:UW   ▵–▵–▵ FINE:DSM:UW
       ✳–✳–✳ FINE:KEM:HUW   ◇–◇–◇ FINE:KEM:UW

COMPARISONS OF FLOW3D AND EXPERIMENTAL DATA
X(MM) = 4700
Z(MM)= −1050

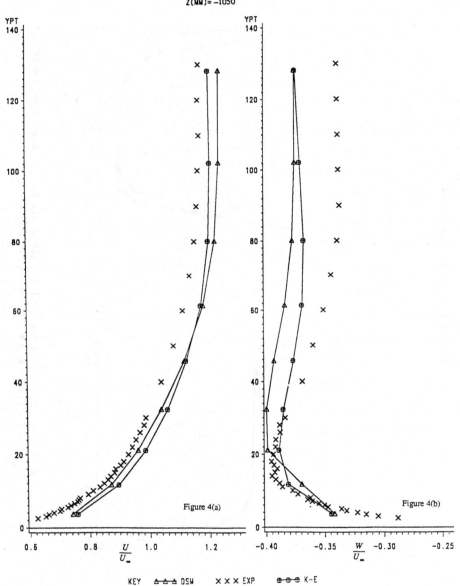

Figure 4(a)

Figure 4(b)

KEY   △-△-△ DSM   × × × EXP   ⊕-⊕-⊕ K-E

A 3D COMPRESSIBLE NAVIER-STOKES CODE
FOR A BOUNDARY LAYER IN A S-SHAPED CHANNEL CALCULATION

D . DUTOYA   (ONERA)
Ph. PONCELIN DE RAUCOURT (SNECMA)

1 - INTRODUCTION

This paper presents the method used for the calculation of the flow in the "S-shaped" channel. Before presenting the method in detail, however, a few remarks are in order to explain our motivations :

- The code used, referred to as "MATHILDA", is a time-marching 3-D Navier-Stokes solver, designed for computations in complex geometries, and covering a wide range of Mach numbers conditions. Therefore the pressure field is, in our case, a result of the calculation. In fact the flow through the whole geometry, including the main channel and both the secondary side flow traps, was computed in one piece.

- The storage requirements of MATHILDA and the memory size of the CRAY XMP computer, limited the calculations to a "coarse" 100000 cells grid to cover the isentropic core, boundary layers and side channels flows. Thus, one should not expect more than a crude evaluation of the effects of the boundary layer structure and development on the main flow.

  - The turbulence model used in the calculations was a simple mixing length model.

Nevertheless, our aim in this workshop is to evaluate a turbulence model in a truly predictive situation. Therefore

185

the emphasis is on a comparison between computed and measured head losses and pressure fields. In these conditions, we try to evaluate the validity of a simple equilibrium turbulence model.

2 - NUMERICAL METHOD

2.1 Equations

The MATHILDA code solves the full unsteady compressible Navier-Stokes equations. The fluid is a perfect gas. The turbulent fluxes are expressed in terms of the flow field gradients through the Boussinesq approximation. The turbulent eddy viscosity and diffusion coefficient are computed with a mixing length model.

2.2 Space grid and geometry

A curvilinear, non orthogonal, boundary fitted grid is used for space discretisation. The nodes, whose cartesian coordinates are the input data, define a set of adjacent hexahedral cells. Each of the six facets of a cell is either shared with an adjacent cell or with a boundary of the calculation domain (wall, symetry plane, flow inlet or outlet,...). The volume (Ω) of each cell, and the cartesian components of the average surface vectors associated with its facets ($\underline{\sigma} = \sigma.\underline{n}$) , are calculated through a geometricaly conservative procedure, and then stored.

An (i,j,k) numbering system is used for the grid. A cell (i,j,k) may be either a flow cell or a virtual cell located outside the domain. The grid points are identified by the indices (i+1/2,j+1/2,k+1/2) and the facets by (i+1/2,j,k), (i,j+1/2,k), (i,j,k+1/2).

2.3 Space discretisation

The primary flow quantities are the average specific mass, momentum and energy in each cell : $q = (\rho, \rho \underline{v}, \rho E)$. Space integration of the conservation equations over the cells leads to a set of ordinary differential equations :

$$\frac{d}{dt} (\Omega q) = \Sigma_\alpha \sigma_\alpha \Phi_\alpha \qquad (1)$$

The fluxes Φ_α through the facets α are then expressed in terms of the values of q in surrounding cells.

The inviscid part of the flux (Euler) is calculated by using a non-centered scheme which takes into account the different wave propagation mechanisms. Different methods have been implemented in the code: Steger and Warming flux splitting, and various approximate Riemann solvers. The scheme used for the present calculations, to similar Roe's, is briefly described below.

On each surface element, two second-order approximations of the flow state are calculated. For instance :

$$q_l(i+1/2,j,k) = -\beta/2 \; q(i-1,j,k) + (1+\beta/2) \; q(i,j,k)$$

$$+ (1-\beta)/2 \quad q(i+1,j,k) \qquad (2)$$

$$q_r(i+1/2,j,k) = -\beta/2 \; q(i+2,j,k) + (1+\beta/2) \; q(i+1,j,k)$$

$$+ (1-\beta)/2 \; q(i,j,k) \qquad (3)$$

with $0 < \beta < 1$.

The corresponding left and right values of the fluxes are then deduced as :

$$f_1 = f(q_1) \qquad\qquad\qquad\qquad (4)$$
$$f_r = f(q_r)$$

The flux through a given facet is expressed as :

$$f = (f_1+f_r)/2 + S(J) \quad (f_1-f_r)/2 \qquad (5)$$

$S(J)$ is the "sign jacobian operator". The eigenvectors of the matrix $S(J)$ are those of the jacobian matrix $J=\partial f/\partial q$. The eigenvalues of $S(J)$ are the sign (+1 ,or -1) of the eigenvalues of J. The matrix S is calculated at an intermediate state $q=(q_1+q_r)/2$.

This scheme is stable throughout the transonic range, and allows shock capturing. According to Roe and Van Leer (1), however , numerical diffusion in the vicinity of contact discontinuities is not as strong as for more standard flux splitting methods. This property is essential for shear layer and boundary layer calculations. Coarser grids can be used to capture these mixing layers.

Turbulent shear-stress and heat flux are expressed in terms of the flow field gradients $\nabla \underline{V}$ and ∇ T . A Green integral formula is used to calculate the average cell values of $\nabla \underline{V}$ and ∇ T. The values of the stress tensor and heat flux vector on each facet are deduced, and then projected onto the surface vectors to get the shear stress and heat flux.

The flow near the wall is assumed to be linear or logarithmic depending on the value of y^+. A wall function based on Spalding's formula is used on solid boundaries. Note that the no-slip condition and the given wall temperature are imposed on the flow system only through these wall functions.

2.4 Time integration

The ODE system is integrated by using a 3 steps implicit ADI procedure. This method allows large time steps, scaled on the physical characteristic times of the problem rather than grid dependent CFL time steps. CPU time and storage requirements remain reasonable. Vectorisation is possible since the block tridiagonal linear systems are uncoupled.

The steady state obtained is independant of the time step. The non-centered scheme ensures linear stability for all values of the time step.

3 - CALCULATIONS IN THE S-SHAPED CHANNEL

3.1 Mesh

The entire geometry, from the inlet plane X=0 to the exit section X=9000 mm, including the boudary layer traps flows, is considered, from the upper wall Y=0 to the mid Y plane Y=210 mm. The walls which separate the main flow and the boundary layers trap flows are assumed to be infinitely thin.

The grid is made of 80 x 60 x 20 cells, decomposed as follows :
- in the X direction 20 grid cells for 0 < X < 3000
                       40 grid cells for S shaped region
                       20 grid cells for 6000 < X < 9000

- in the Y direction there are 20 grid cells, the grid being refined near the wall

- in the Z direction there are 40 grid cells across the main channel, 10 across each secondary flow.

3.2 Boundary conditions

At the inlet plane X=0 , total pressure P_a and total temperature T_a are imposed, along with the direction of the flow :

$$P_a = 100250 \ Pa$$
$$T_a = 300 \ K$$

At the exit plane, the static pressure p_a is fixed :

$$p_a = 100000 \ Pa$$

In 3-D calculations, these data lead to an inlet (X=0) average velocity $U_\infty=18$ m/s and a density $\rho_\infty=1.168$ Kg/m³

On the wall, mass flux, momentum and energy fluxes are set to zero. The wall shear and heat flux are computed assuming zero wall velocity and a given wall temperature $T_w = 300$ K. The symmetry plane Y=210 mm acts as a wall with zero shear and heat transfer.

3.3 Result of the calculations

Preliminary 2-D calculations in the X-Z plane were performed, giving a pressure field similar to the 3-D flow pressure field. The head loss was much lower, because of the absence of boundary layer on the floor. The effect of the grid was also tested, showing that a 80 x 40 grid in the main channel is sufficient to obtain a grid-independant solution.

An Euler calculation, assuming a zero viscosity and zero wall shear led to very weak total head loss : the numerical dissipation at steady state was found to be acceptable.

As noted above, the 3-D calculation were limited to 20 cells across the boundary layer. Different grid refinements near the wall were tested, but no significant influence on the pressure and velocity fields was observed.

Note that the reference pressure   and velocity used for non-dimensionalization are the inlet (X=0) average values of p and U.

The comparison between calculated and measured pressure fields shows that the numerical simulation underestimates the head loss, as well as the difference between maximum and minimum static pressure. Fitting two sets of data would require a lower value(20%) of the dynamic pressure   used to compute the pressure coefficient. The origin of this overall defect should become clearer after comparison of calculated and measured boundary layer profiles.

REFERENCE

B VAN LEER JL THOMAS PL ROE RW NEWSOME - 1987
A comparison of Numerical Flux Formulas for the Euler and Navier-Stokes Equations - AIAA 87-1104

Figure 1 to 24 : profiles of mean velocity and angle

Problem T1 : Boundary Layer in a "S"- Shaped Channel.

M. GABILLARD, G. MARTY, P.L.VIOLLET

Electricité de France
Laboratoire National d'Hydraulique
6, quai Watier - 78401 CHATOU, France.

Date : 15/09/89

Characteristics : 3D NS isothermal turbulent flow

Code : ESTET - Version 3.0

1 - CODE DESCRIPTION

The 3 D code ESTET 'Baron (89), Mattei (88)' of Laboratoire National d'Hydraulique is based on finite difference and volume discretization and an incremental version of the original fractional step method described in 'Viollet (87)'. In principle, the fractional step method splits transport equations into elementary partial differential equations and leads to the solution of operators of the type :
   -advection solved with the three-dimensional method of characteristics
   -parabolic, including source terms, solved with a direct semi-implicit method (Gauss elimination) after splitting into three orthogonal directions
   -elliptic, for the pressure, solved with an iterative method (Gauss-Seidel or conjugate residuals).

Consider the simple case of the transport equation for a variable G :

$$\rho \frac{\partial}{\partial t} G + \rho \, u_j \frac{\partial}{\partial x_j} G = \frac{\partial}{\partial x_j}\left[\rho \, K \frac{\partial}{\partial x_j} G \right] + S_x + S_i \, G \qquad (1)$$

and introduce the increment of variable G over time step dt :

$$\delta G = G^{n+1} - G^n \quad \text{with} \quad G^n = G(t) \qquad G^{n+1} = G(t+\delta t) \ .$$

192

Subsequently, discretization in time of the transport equation (1) gives the following equation on the increment δG, solved with the fractional step method described above :

$$\rho^n \frac{\delta G}{\delta t} - \theta_1 \frac{\partial}{\partial x_j}\left[\rho^n K^n \frac{\partial}{\partial x_j}\delta G\right] - \theta_2 S_i^n \delta G = \rho^n \frac{\widehat{\delta G}}{\delta t} \quad (2)$$

with θ_1 and θ_2 the time-implicitation coefficients $(0 < \theta_i < 1)$.

The right hand side of eqn. (2) contains the fully explicit solution of equation (1) :

$$\rho^n \frac{\widehat{\delta G}}{\delta t} = \rho^n \frac{\widehat{G} - G^n}{\delta t} + \frac{\partial}{\partial x_j}\left[\rho^n K^n \frac{\partial}{\partial x_j}G^n\right] + S_x^n + S_i^n G^n \quad (3)$$

with $\widehat{G}$ the solution of the advection step solved by the method of characteristics :

$$\widehat{G}_{x_j} = G_{x_j}^n - \int_{\delta t} u_j^n \frac{\partial G}{\partial x_j}\,dt = G_{x_{cj}}^n \quad \text{and} \quad x_{cj} = x_j - \int_{\delta t} u_j^n\,dt$$

For the momentum equation, explicit pressure gradient appears in eqn. (3), and mass conservation is achieved by solving Poisson equation for the pressure increment. The parabolic step for k and epsilon includes linear coupling of their increments.

Due to implicitation of substeps in the fractional step method, the relative error on the increment is of first order in time. Therefore, in steady flow computations, as the increment tends towards zero with convergence in time, the absolute error on the variable itself is minimized.

A "space variable time step" technique allows a great saving of computing time for problems having important variations of the Courant number.

2 - TURBULENCE MODEL

The standard k-epsilon turbulence model is used with the fixed empirical constants of 'Launder, Spalding (74)' :

$$c_\mu = 0.09 \; ; \; c_{1\varepsilon} = 1.44 \; ; \; c_{2\varepsilon} = 1.92 \; ; \; \sigma_k = 1.0 \text{ and } \sigma_\varepsilon = 1.3.$$

3 - TEST CASE DESCRIPTION.

3.1 Geometry

The geometry of the test case is described on figure 1. Only half of the experimental channel has been simulated using a symmetry condition for the middle plane.

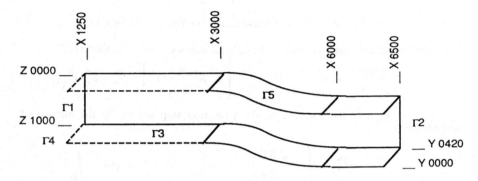

Figure 1 : Geometry of the simulated channel

3.2 Boundary conditions

Let Ω be the computational domain, where the flow is to be simulated. The boundary Γ of Ω can be decomposed into : inflow, outflow, slip boundaries, symmetry plane and solid boundaries.

Γ_1 :   entrance of the flow
           u,v,w, k and epsilon are given

Γ_2 :   exit of the flow
           u,v,w,k and epsilon are computed through advection
           u,v,w are corrected for global continuity

Γ_3 :   Slip boundary

Γ_4 :   symmetry plane

Γ_5 :   solid boundaries
           law of the wall for u,v,w,and imposed value for k and epsilon function of the
           shear velocity.

3.3 Physic characteristics

. The dimensions are given on the figure 1

. Experimentals profiles are used for the inlet conditions of velocity.

.Reynolds number of the flow : Re = 1.26 E+ 6
(Ub = 18 m/s ,Tb = 25 °C, channel width L = 1m)

3.4   References

Baron F., Gabillard M., Lacroix C.(1989), "Experimental study and 3D numerical prediction of recirculating and stratified flows in PWR.", to be presented at the 4th International Topical Meeting on Nuclear Reactor Thermal-Hydraulics (NURETH 4), Karlsruhe October 10-13,1989.
Mattei J.D., Laurence D.(1988), "Application of the ESTET code to vehicle aerodynamics.", presented at the International Conference on Supercomputing in the Automotive Industry, Séville Octobre 1988.
Viollet P.L.(1987), "On the numerical modelling of stratified flows", Physical processes in estuaries (J. Dronkers and W. van leussen Eds), pp. 257-277, Springer-Verlag.
Launder, B.E., D.B. Spalding (1974), "The numerical computation of turbulent flows", Comp. Meth. in Appl. Mech. and Eng., 3, p. 269.

4 COMPUTATION DESCRIPTION

3D turbulent flow

standard k-epsilon model

Space variable time step technique

Numerical parameters :
Diffusion Step : time implicitation coefficients $\theta_i = 1$
Pressure-continuity Step : (conjugate residuals) relative mass residuals : 1E-3

Mesh (figure 2) :

The mesh processed with the code CEZANNE, is composed of 39294 nodes in cartesian coordinates (37*18*59).

Stationnary state

Chronological account of convergence is given on figure 3 at three different locations in the flow.
The stationnary state is obtained after about 450 time steps : 2 hours 11 minutes of CPU time on CRAY XMP28 CFT 1.15
CPU time /time step/1000 nodes : 0.44 second.

5 - COMPUTATION RESULTS

On the figure 4, the computed velocity field is given in the symmetry plane, and the computed pressure coefficient field near the roof of the channel is compared to experimental data. The different profiles (U/U_b, W/U_b, $100k/U_b^2$, γ, γ_g) are available at the different experimental locations.

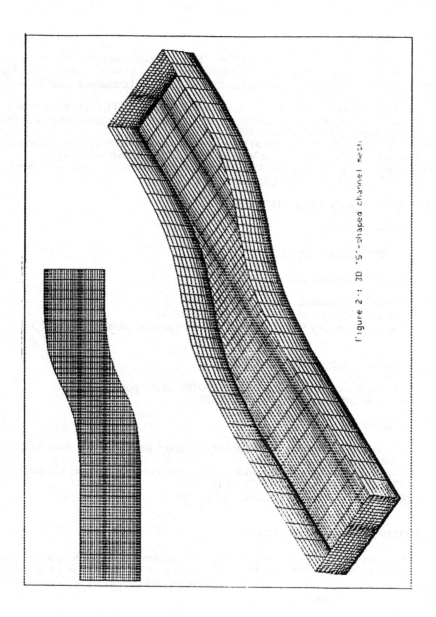

Figure 2 : 3D "S"-shaped channel mesh

Summary and Conclusions for the Test Case T1

I.L. Ryhming, T.V. Truong and P.A. Lindberg

Fluid Mechanics Laboratory
Swiss Federal Institute of Technology, Lausanne, Switzerland

1. INTRODUCTION

The boundary layer flow in an S-shaped duct has been proposed as a test case since it covers a number of features of practical as well as fundamental interest. Some of these are:

i)   The overall (viscous) flow in the S-duct and the wall bounded 3-D boundary layer on the test plate, both represent challenging computational problems involving complex flow phenomena. Their solution requires sophisticated modelling as well as advanced numerical techniques.

ii) From the fundamental point of view, the S-duct geometry allows us to study the behaviour and the development of the mean-flow and of the Reynolds stresses over long distances (30-40 boundary layer thicknesses). In particular the S-duct geometry will induce cross-over profiles in the boundary layer far downstream, a feature that so far has not been investigated in detail.

iii) The S-shaped geometry induces a pressure gradient/shear driven 3D boundary layer flow on the walls of the duct which in turn generates the secondary flow motion. Such flows are of considerable industrial interest in view of the numerous technical applications. Furthermore, for the validation of codes in which Reynolds stress modelling at high Re is essential, there are relatively few relevant test cases available which are sufficiently well documented. For this purpose, the S-duct geometry is particularly well suited as it is of interest for the prediction of viscous internal flows in ducts as well as for the general wall-bounded 3D turbulent boundary layer flows.

1.1 Some remarks concerning the physics of the flow in the tunnel

The flow in the tunnel can be described in different ways. Information concerning the anisotropic turbulence behaviour is obtained by observing the changes in the flow direction as revealed by the profiles of the angles γ, γ_τ and γ_g defined previously in the description of the test case T1 and obtained by following the streamline paths determined from the measured pressure distribution on the plate. Two such paths will be used to illustrate the measured 3D boundary layer flow.

197

We select first a calculated streamline path adjacent to the right hand sidewall of the tunnel. This path is marked "Right" in Fig. 1 (*same as Fig. 2 in the Specification document*) and it comprises a number of measurement stations:

X	2945	3320	4125	4700	5005	5255	5500
Z	660	660	800	1050	1150	1150	1198

The development of the angles γ, γ_τ and γ_g, as functions of the distance normal to the wall (test plate) at these stations and illustrated in Fig. 2, shows that at the first station the pressure gradients in the streamwise and crosswise directions have already started to induce a small cross-flow giving rise to a change in γ_g from negative to positive values close to the wall. This development is increased at the second station where γ_τ lags behind γ_g by some 10° over essentially the entire height of the boundary layer. At the third station which is situated well into the S-bend part of the duct, the γ_g behaviour is very pronounced with a variation ranging from -25° close to the plate to +20° at a height corresponding to about a third of the boundary layer thickness. At this position, γ_τ lags behind by as much as 25° maximum. At the 4th station, the γ_g variation extends from -37° to +16° with the maximum value higher up in the boundary layer. For the most part, γ_τ still lags behind over the boundary layer height with as much as 10°. At the next station, which is close to the cross-over point, γ_τ has overtaken γ_g except very close to the wall, where γ_τ is always behind γ_g. At the following two stations, which are situated downstream of the cross-over point, γ_τ exceeds γ_g with as much as 10°. The γ_g profile in these two stations peaks close to the wall indicating that a profound change in the mean-flow behaviour has occurred.

The wall streamline pattern on the test plate, see Fig. 3, indicates that the boundary layer fluid moves towards the right hand sidewall already at the X-location of the first station considered above. This behaviour is maintained through the S-bend until $X \approx 5255$ where a first streamline leaves the right sidewall and moves on a curved path across the test plate surface. This streamline pattern indicates that the boundary layer fluid on the test plate feeds into the boundary layer on the vertical wall thus causing the horizontal streamlines on the right hand sidewall to converge initially and then diverge after the X-location mentioned above has been passed. This boundary layer flow, which is present on the test plate as well as on the tunnel floor, is responsible for the onset and the subsequent decrease of the secondary flow motion in the tunnel.

In view of the observation that γ_τ lags behind γ_g initially and eventually catches up with and overtakes γ_g, it is to be expected that an anisotropic behaviour of the Reynolds stresses will be found in the boundary layer in this part of the flow. Hence, it can be expected that turbulence models based on an isotropic eddy viscosity in general will not represent the details of the flow as well as non-isotropic models.

A much different boundary layer flow field is found along the path marked "Left" in Fig. 1. This path is close to the left sidewall in the tunnel and it can, therefore, be expected that the last three stations would, from the computational point of view, be situated in a region determined by the initial conditions along the initial data line as well as the inflow boundary conditions along some streamwise surface defined in the upstream and left region of flow in the S-bend. Indications of such behaviour can be seen from Fig. 4 where again γ_τ, γ_g and γ are shown as a function of y at the three stations:

X	4700	5005	5225
Z	730	850	850

At the last of these three stations, considerable differences between γ_g and γ_τ develop such that close to the plate the magnitude (negative) of γ_g exceeds γ_τ by 15-20° whereas at ~ $\delta/3$ γ_τ (positive) exceeds γ_g by about the same amount. The development of the angles γ_τ in the flow is of course reflecting the bumpy character of, in particular, the $\overline{uv}$ profiles at the XZ stations of the "Left" path shown in Fig. 1. To illustrate, Fig. 5 shows one pair of profiles of $\overline{uv}$ and $\overline{vw}$ at the last "Left"-station, and it is evident that the "Left" path way data is influenced to some extent by the complex flow conditions along the left hand S-bend corner region formed by the horizontal test plate and the left vertical sidewall. We shall return to this point later on.

2. DESCRIPTION OF THE METHOD USED IN THE CONTRIBUTIONS

Nineteen contributions were received for the T1 test-case, thirteen of those were based on the 3D B.L. equations and six were based on the 3D N.S. equations. In the latter group some general purpose codes were used, e.g. FLOW 3D (Harwell), ESTET-3D (Laboratoire National d'Hydrologique), MATHILDA (ONERA-SNECMA). A brief overview of the different contributions is shown in Table I where code basis, turbulence model and some numerical features are presented. For a complete description of any one method we refer to the individual papers in these proceedings.

In the B.L. methods all contributions except one [Krogstad & Skåre (1)], have used the measured and smoothed pressure distribution as an outer boundary condition. In their contribution (1), Krogstad & Skåre used the inviscid pressure distribution obtained by solving the Euler equations. However, the inviscid pressure distribution deviates somewhat from the measured distribution, and this causes additional discrepancies in the flow characterisation of the boundary layer. For this reason we will not consider this contribution any further in the comparison with measurements. The smoothed measured pressure distribution has also been used to calculate the external streamlines by all contributors. In the B.L. methods, finite-difference techniques have been used in all cases except one [Gleyzes et al. (1)], where an integral method is employed. The S-duct geometry will cause cross-over profiles to occur and it is difficult to take this feature into consideration in an integral method. No useful results for the intended purposes were obtained by the integral method in the present case, and we refer to the detailed account of the problems encountered in the paper by Gleyzes et al. Consequently, we do not consider this contribution any further in the evaluation. In the finite-difference methods employed, it is important to take into account properly the hyperbolic character of the equations, which feature transpires if the sub-set of the B.L. equations containing only first order partial derivatives is considered. The existence of a domain of dependence must be observed when choosing the finite-difference molecule. This requirement has been observed in all but two contributions [Krogstad & Skåre (1), (2)], which consider the small cross-flow approximation to the equations based on the observation that the current S-duct configuration (for this test case) produces only a mild cross-flow (~ 10-15°). In the small cross-flow approximation the lateral influence is suppressed so that the lateral and streamwise momentum components are decoupled in a coordinate system using the external streamlines and their orthogonals as reference lines.

Whereas the numerical method is practically the same or very similar for all the nine B.L. contributions, the N.S. contributions reflects the great variety of methods currently in use to solve the N.S. equations. Hence, the Harwell-Flow 3D is a finite-volume code in which a modified SIMPLE algorithm is used to couple the pressure and velocity fields, whereas the fractional step method used in the ESTET code splits the transport equations into an advection part, a parabolic part including source terms, and an elliptic part for the pressure. The ESTET code is based on finite-difference/volume discretization. The MATHILDA code is based on time-marching of the unsteady equations and different flux splitting methods are employed for the space discretization. The code developed by the Rodi group is designed to calculate incompressible elliptic flows in complex geometries by a finite-volume procedure. Pressure-velocity coupling is obtained via the SIMPLE algorithm. In the calculation from Hecht et al., only the pressure results in different sections are displayed in graphical form not useful to compare with the experimental data; for this reason we will not consider this contribution any further in the comparison with measurements.

Some of the N.S. methods suffer from insufficient resolution close to the wall due to computer storage problems. Hence, 10 points or so used in the N.S. calculations across the boundary layer are not sufficient to capture the details of the flow close to the wall. By comparison, the B.L. methods employ typically 100 to 150 points to obtain a good resolution. Apparently multiblock approaches are not yet fully operational, which if they were, could alleviate the resolution problems with the N.S. methods.

Turbulence modelling is needed in order to calculate the high Re number flow in the S-duct. The contributors have considered simple mixing length formulations like those of Michel or Cebeci-Smith, suitably modified to cope with a 3D boundary layer, or more complex formulations involving transport equations. The simple mixing length formulations have been used in seven contributions. The standard low Re form of the k-ε model has been used in six contributions. All these models are isotropic and, thus, do not distinguish between the angles γ_g and γ_τ. A simple algebraic non-isotropic model is used by Nakkasyan et al., which is to some extent inspired by the Rotta T-model. The Johnson-King (J-K) model extended in different ways to 3D flows have been used in two contributions in which also a transport equation for the anisotropy factor T has been included. The more sophisticated models using several transport and auxiliary equations have been used by Lindberg (1) and Clarke et al. (2).

2.1 Comparisons between calculated and measured data

The contributions to this test case can be classified in different ways e.g. with respect to the numerical methods chosen, the turbulence model used etc... In order to have as good a basis as possible for the comparisons, we have chosen to group contributions together which are based on a similar numerical method or on the same or a similar turbulence model. This means that all contributions based on the B.L. equations form one group within which subgroups are formed with respect to the turbulence model used. Another group is formed by contributions using the N.S. equations within which again subgroups are formed according to the turbulence model used.

With respect to the B.L. methods, almost all participants have used the measured pressure distribution suitably smoothed to determine the outer streamline pattern. It must be observed, however, that the prediction of the external mean-flow streamlines from the given pressure data is not a unique procedure and therefore differences in calculated shape of the streamline will, as we shall see, have an influence on the calculated flow downstream of the given initial data line. This follows because the streamlines are used as characteristics in many of the boundary-layer integration methods presented in the workshop. Hence, the shape of the domain of dependence downstream of the initial data-line is found from the mean-flow streamlines which diverge the most from the external streamline, and the domain of influence thus determined has a direct influence on the predicted flow behaviour. This has led to differences in the predictions as indicated previously; we will come back later to a sensitivity analysis of the effect thereof.

In order to distinguish the effect of the turbulence modelling from other influences on the calculated results, it would be preferable to use one single numerical method that can accommodate different turbulence models. Such a calculation has been performed and will be presented subsequently. However, the purpose of this test case is not only to examine turbulence modelling. It is also a question of predicting as closely as possible the measured data in a duct with any one method one chooses to use and the differences found in the predictions must be judged according to different criteria.

In what follows, we compare the different contributions at a few measurement stations which we have selected because of their significance with respect to important flow phenomena. The contributions have been grouped together as indicated and five groups have been chosen as shown in Table II where the measurement stations chosen as representative are given as well. The first three groups are the B.L. contributors who all have calculated the requested variables at the three stations chosen which, for almost all contributions, fall within a domain of dependence containing the measured data along the initial data lines. The three stations are also within the "Right" pathway, discussed previously along which important 3D changes in mean-flow as well as in the Reynolds stresses occur. Although the measured data along the three measurement axes X = 1130, 1193 and 1255 indicate that the flow in these sections of the tunnel is not precisely parallel, an upstream flow influence along the left hand side of the tunnel makes the flow not perfectly parallel to the vertical side walls, almost all contributors have ignored this feature. The influence on the results of this slightly non-parallel flow will be discussed subsequently.

The fourth group in Table II is a pure N.S. group. Since the N.S. calculations solves for the flow in the entire tunnel, we compare the results at one station common with the B.L. groups, i.e. the station X4700/Z1050 and also two other stations close to the left and right hand side walls. In the fifth group we compare three calculations based on the B.L. and the N.S. equations either with very simple turbulence modelling (Dutoya & Poncelin) or with a complex modelling involving transport equations (Lindberg and Clarke) at the common station X4700/Z1050 and at two other stations close to the left hand side wall.

The comparisons are shown in Figs. 6 - 10. The contributors were also asked to furnish calculated results for external flow parameters, boundary layer profile and wall parameters. These results, compared with measured data, are contained in Tables III and IV and will be discussed subsequently.

2.2   Description of the results of the B.L. calculations, Group I

The contributors in this group, Bettelini et al., Gleyzes et al., Krogstad et al. (version 2), Krogstad et al. (version 3) have all used essentially the same turbulence model, i.e. the eddy viscosity-mixing length formulation of Cebeci-Smith-Michel suitably extended to 3D flows. Differences are to be found in the numerical schemes, see Table I. At the first station X4125/Z0800, see Fig. 6a, the U/U_o profiles differ very little, whereas the W/U_o results of Bettelini et al. show a greater overprediction than the other results do. The turbulence model is isotropic, so there is no difference between γ_g and γ_τ in these calculations. With respect to the shear stresses, the calculated $\overline{uv}/U_o^2$ profiles agree better with the measured data than does the $\overline{vw}/U_0^2$ profiles. This feature which is related to the simple turbulence model has been observed earlier in other test cases. In particular, the calculated $\overline{vw}/U_0^2$ profile has a zero point where the calculated W/U_0 has its maximum. The measured $\overline{vw}/U_0^2$ profile, however, does not show a region of negative values. No further Reynolds stress data can be obtained from the simple model.

In the next station (X4700/Z1050), see Fig. 6b, the predictions of the U/U_0 and W/U_0 differ more between each other and the data than in the previous station. Similarly, the calculated shear stresses show more spread between each other and the data. In particular, with a good agreement with the $\overline{uv}/U_0^2$, data goes a less satisfactory agreement with the $\overline{vw}/U_0^2$ close to the wall. This station is situated upstream of the crossover point which is seen in the $\gamma_g - \gamma_\tau > 0$ data. In the following station (X5500/Z1198), see Fig. 6c, $\gamma_g - \gamma_\tau < 0$ and here the calculations have to reproduce the crossover features. In order to appreciate these features, it is also useful to compare the large differences in the W/U_0 distributions in the two consecutive stations represented in Figs. 6b and 6c. As can be seen, there is considerable spread in calculated U/U_0 and W/U_0 distributions, the best agreement being achieved in the Gleyzes et al. (version 2) calculation. This spread is apparent also in the calculated $\overline{uv}/U_0^2$ and $\overline{vw}/U_0^2$ distributions where there is a clear tendency of overestimation of $\overline{vw}/U_0^2$ away from the wall region in all the results. Since the turbulence model is the same in all these results, the differences are due to the different numerical treatment of the problem. The differences become more pronounced with increasing distance downstream of the initial data line. A probable cause is that the shape of the external streamlines differs, except in the calculations of Krogstad et al., versions 2 and 3, where the same streamline pattern has been used. In fact, these two calculations represent two extremes in the data where paradoxically the potentially more accurate method gives nearly everywhere the less satisfactory result.

2.3   Description of the results of B.L. calculations, group II

The contributors in this group, Nakkasyan et al., Lindberg version 2, and Savill et al. have all used different anisotropic turbulence models. The model used by Nakkasyan is a 3D mixing length model corrected for crossflow, whereas the models used by Lindberg and Savill et al. represent different 3D extensions of the Johnson-King model. In the first station (X4125/Z0800), see Figs. 7a and 7b, the Nakkasyan calculation reproduces the data best with, in particular, a reasonable agreement with the shear stress profiles. The J - K models are seen to be less accurate, which depends apparently on problems encountered in the computation of the maximum shear stress and related values, see the papers by Lindberg and by Savill et al. in these proceedings. In particular, the W/U_0 profiles are strongly overpredicted and the U/U_0 profiles are underpredicted in both calculations in the wall region. The results reported at station (X4700/Z1050), see Figs. 7c and 7d, show similar trends although differences between measured and calculated data become more pronounced. At station (X5500/Z1198), see Figs. 7e and 7f, downstream of cross-over, the results show considerable spread with important differences in different parts of the flow. In comparison with the results from the previous group, it is seen that the J-K models do not improve the results over the simple mixing length models, but that the model of anisotropic behaviour deduced from the mean-flow gradients and introduced by Nakkasyan et al. represents a clear improvement for this particular case.

2.4 Description of the results of B.L. calculations, Group III

In this group we compare results of two low Re k-ε model calculations of Lindberg version 3, and Wu et al. with the results of a mixing length calculation (Cebeci-Smith) of Lindhout et al. These are all calculations using an isotropic turbulence model. Hence, there is no difference between γ_g and γ_τ. The k-ε model is seen to predict reasonably well the $\overline{uv}/U_0^2$ shear stress distributions at all three stations considered; the $\overline{vw}/U_0^2$ distributions are less well captured, see Figs. 8a - 8f. Differences in results between the k-ε model and with the simpler mixing length model used by Lindhout et al. are primarily found in the outer layers of the flow. These differences are, however, rather small and indicates that the transport capability offered by the k-ε model over the mixing length formulation gives only marginal improvements along a pathway running close to the right hand border of the duct. In this series of calculations Lindberg has computed the $\overline{uw}/U_0^2$ shear stress as well as the three normal components of the Reynolds stress tensor. In general, it is seen that the $\overline{vv}/U_0^2$ and the $\overline{ww}/U_0^2$ distributions are captured rather well, whereas considerable differences are found between calculated and measured data for the $\overline{uw}/U_0^2$ and the $\overline{uu}/U_0^2$ distributions.

2.5 Description of the result of N.S. calculations, Group IV

The contributors in this group have solved the N.S. equations by different numerical techniques but they have all used the standard form of the low Re k-ε turbulence model. At the first station selected for comparison of the results, i.e. X4700/Z0730, see Figs. 9a and 9b, close to the left hand boundary, the mean-flow profiles according to the Clarke and the Rodi calculations are very close, whereas the same profiles obtained by Gabillard et al. show considerable differences in particular in the U/U_0-profile. In comparison with the measured data, differences are found in the inner portions of the boundary layer. The $\overline{uv}/U_0^2$ and $\overline{vw}/U_0^2$ profiles of Clarke and Rodi are also very close, although some resolution problems are evident in the Clarke calculation due to too few points in the boundary layer. In comparison with the measured data, there are the same tendencies in the shape of these profiles, however, over the inner half of the boundary layer thickness the magnitudes are not so well predicted, which explains the differences between data and calculations in the U/U_0 profile. The $\overline{uw}/U_0^2$ - profile is poorly predicted, whereas in the normal stress predictions good agreement with the data is found for the $\overline{vv}/U_0^2$ and

$\overline{ww}/U_0^2$ calculations. However, substantial differences between prediction and measured data are again found in the $\overline{uu}/U_0^2$ distribution. Gabillard et al. have not submitted any Reynolds stress results.

In the next station chosen, i.e. X4700/Z1050, see Figs. 9c and 9d, important differences in the calculated U/U_0-distribution are found, whereas the W/U_0-distribution show much less differences. The corresponding $\overline{uv}/U_0^2$ and $\overline{vw}/U_0^2$ profiles, in particular in the Clarke calculations, agree very well with the measured data, except very close to the wall again due to insufficient resolution. The $\overline{uw}/U_0^2$ distribution is not captured at all and in the normal stresses the $\overline{uu}/U_0^2$ profile is less well predicted. At the outer edge of the boundary layer the turbulent kinetic energy does not tend to zero in either of these two calculations as is clearly noticeable from the three normal stresses calculated.

In the last point chosen for this group, i.e. X5500/Z1328, see Figs. 9e and 9f, (after cross-over and close to the right hand sidewall) the tendencies found at the upstream stations are accentuated. In particular important discrepancies are found between predictions and data, e.g. in the kinetic energy at the outer edge of the boundary layer. The results of Rodi et al, however, at this station show better agreement than those of Clarke, whereas the situation was the reversed at one of the upstream stations considered.

2.6 Description of the results of N.S. and B.L. calculations, Group V

In this group we compare results of two N.S. calculations with the results of one B.L. calculation. In the N.S. calculation, Dutoya et al have employed a simple mixing length model, whereas Clarke has used the differential stress model. In the B.L. calculation, Lindberg has used a new anisotropic turbulence model based on the transport equations.

At the first station chosen for the comparisons, i.e. X4700/Z0730, see Figs. 10a and 10b, the N.S. predictions of Clarke, despite resolution problems close to the wall, are in general in very good agreement with measured data. Some problems in predicting the $\gamma_g - \gamma_t$ lag angle in the middle of the boundary layer show up only in the $\overline{uw}/U_0^2$ and $\overline{vw}/U_0^2$ profiles, which in turn has only a minor influence on the U/U_0 profile. The B.L. predictions are also in good agreement with the data. However, to predict the flow behaviour well at this station in a B.L. calculation, inflow boundary conditions are needed. In the Lindberg calculation all crossflow derivatives required upstream and along a left hand boundary were set equal to zero since no information on the inflow was provided in the test case specifications. The effect of neglecting inflow boundary data is discussed in a subsequent section of this report.

At the second station chosen for the comparisons, i.e. X4700/Z1050, see Figs. 10c and 10d, a station common with all the other groups, the results of the more complex turbulence models are hardly better than the simple mixing length model results. Again effects of the turbulent transport terms are apparently not significant or cancel out. The complex models allow, however, for the determination of all the Reynolds stresses, and it is seen that in the Clarke calculation, the $\overline{vw}/U_o^2$ and $\overline{uw}/U_o^2$ distributions are closer to the data than the $\overline{uv}/U_o^2$ results. In the normal stresses there is a tendency towards overprediction. The B.L. results of Lindberg are of similar quality as at the first station considered. However, the B.L. results at this station are dependent only on the initial data since no inflow data are needed at this station.

At the last station chosen for the comparisons, i.e. X5255/Z0850, see Figs. 10e and 10f, the experimental data show a bumpy character of, in particular, the $\overline{uv}/U_o^2$, $\overline{uu}/U_o^2$ and $\overline{vv}/U_o^2$ distributions. Evidently, the flow behaviour at this station is difficult to predict and even though, the trends appear right; the N.S. calculation suffers because of insufficient resolution and, for the B.L. calculation, inflow boundary conditions are missing.

2.7 Comparison of external flow & boundary-layer-profile/wall parameters

The contributors to the test case T1 have been requested to furnish external flow and boundary-layer-profile and wall parameters as a result of their calculations. These parameters are Q_e and α_e of the external flow, the boundary-layer-profile parameters δ_{995} and H and the wall parameters τ_{wall}, u_τ, C_f and γ_{wall}. These parameters from all contributions are shown in Table III at the stations chosen for the comparisons of measured and computed data. As can be seen from Table III, there are great variations between calculated and measured properties related to the various methods used as well as with respect to the location where the data are compared. We notice, in particular, large deviations in the calculated values of α_{ext}, Q_{ext}, γ_w and u_τ, see Table IV, of up to $\pm 3^o$ in α_{ext}, $\pm 14\%$ in Q_{ext}, $\pm 15^o$ in γ_w and $\pm 30\%$ in u_τ.

3.   SENSITIVITY STUDY

3.1   Influence of the shape of the external streamlines

As remarked previously, the shape of the streamlines outside of the 3D boundary layer influences the calculated properties within the boundary layer. The external streamlines are obtained by integrating the Euler equations on the basis of the measured pressure distribution. As initial conditions for the integration the measured free stream velocity vector (magnitude and direction) should be used. In a first case, the magnitude of the velocity is available at the entrance to the test section through the reference Pitot reading and the C_p-values at the station X0195. In a second case further downstream, i.e. at the stations X1130, 1193 or 1255, detailed measurements are also available. However, the flow in the tunnel at those stations is not perfectly well aligned with the parallel side walls.

Furthermore, the flow development downstream of these station until X2445 has not been measured in sufficient detail to allow for an accurate integration of the Euler equations. Hence, if the integration is started from any one of the three X-stations mentioned above, sparsely known C_p-values are available only along the centerline of the tunnel and along two further rows of points close to the left and right hand side of the tunnel. By using this information in the integration of the Euler equations, the result is that the initial flow angle measured at the above three stations persists over the straight portion of the duct. The streamline pattern obtained in the two cases described above is quite different, see Fig. 11. A boundary layer calculation just at the entrance to the S-bend, i.e. at the station X3320/Z0660, based on the two different streamline patterns, also gives quite different results with respect to the measured data, see Fig. 12a and 12b. A much more satisfactory result is thus obtained, as can be seen from this comparison, by assuming the flow to be parallel to the side walls at least at the entrance to the tunnel.

3.2   Influence of inflow boundary conditions
The importance of taking into account inflow boundary conditions in a 3D B.L. calculation of the flow at the last stations of the "Left" path in Fig. 1 has already been mentioned. The specification document of the T1 test case did not contain measurement information allowing such conditions to be determined, since the aim of the T1 test case was predictability of B.L. and N.S. calculation methods including turbulence models incorporated in these methods. Most B.L. calculations submitted covered therefore only those measurement points within the domain of dependance defined by the initial conditions upstream. However, a few B.L. calculations were submitted for the stations of present concern where the additional boundary conditions required were ignored. In retrospect, it is interesting to estimate the effect of incorporating inflow boundary conditions which can be done as follows.

To obtain a correct solution the nodes used for calculating the discretized partial derivatives at each station must lie outside of the zone of dependance for the node to be calculated. This requirement cannot be fulfilled along the left hand outermost streamline along which the crosswise velocity is directed towards the inside of the domain to be calculated. However, if experimental values are known along any curve situated to the left of this streamline, the zone of dependence of any point along it is always situated inside of known node values. Hence, the sensitivity with respect to the side boundary conditions can be estimated. This has been done by evaluating from the experimental data approximate inflow boundary conditions to determine the flow at the point X5005/Z0850. The turbulence model used in the calculations is the transport model in Lindberg (1).

As can bee seen from Figs. 13a and 13b, a significant improvement in the predictions is obtained in the mean-flow as well as in the Reynolds stress distributions in comparison with the case where the inflow velocity derivatives are set equal to zero. For this turbulence model inflow conditions for the dissipation ϵ is also needed. Since no experimented data is available for ϵ, the small crossflow approximation was used to supply the missing ϵ information.

It appears from this calculation that the inflow boundary conditions obtained from the experimental data is not of sufficient quality to reproduce all details, in particular, the irregular behaviour of $\overline{uv}/U_o^2$ and of the normal stresses. It appears that a much more detailed map of measurement would be called for to determine accurate inflow conditions.

The domain which is amenable to calculation can be extended substantially when the experimental data are used as discussed above, see Fig. 14, where the broken line indicates the left hand limit for valid calculations as opposed to the dotted line which represents the left hand characteristic emanating from the same point along the initial data line.

4. TURBULENCE MODELLING EFFECTS

As has been demonstrated by the various contributions, the differences in the shape of the external streamlines explain some of the discrepancies in the results at the stations for the cases where the flow has been calculated by a B.L. method and by means of equivalent turbulence models. To clearly distinguish the spurious effects of interpretational or numerical origin (i.e. to eliminate all factors not related to the turbulence model that can have an influence on the results), a set of additional calculations have been performed using the same numerical method with the same grid and the same shape of the external streamlines. These calculations have also been performed with the same convergence level/criteria. In the calculations four turbulence models have been employed, i.e. i) the standard mixing length model of Cebeci-Smith, ii) the low Re k-ϵ model, iii) the anisotropic low Re k-ϵ model, see Lindberg (1), and iv) the J-K anisotropic model, see Lindberg (2).

At the first station chosen for these comparisons, i.e. X4700/Z0730, see Figs. 15a and 15b, it is evident that neither a local model nor the J-K transport model is adequate. A drastic improvement in the distribution of $\overline{uv}/U_o^2$ and $\overline{uw}/U_o^2$ and all the normal stresses are obtained by using either of the two k-ϵ models. Still the $\overline{vw}/U_o^2$ is not well captured, and there are differences between measured and predicted data in the U/U_o-profile. However, the flow at this point is to some extent influenced by inflow conditions which have been ignored in these calculations. As shown in Fig. 13a, inflow data help to improve deficiencies in the U/U_o-profile and $\overline{uv}/U_o^2$-profiles. The difference between the isotropic and the anisotropic transport model does not play a significant role in these calculations. Certainly, the predicted shape of the $(\gamma_g - \gamma_t)$-curve in the anisotropic model is similar to the measured shape. However, whereas the measured data show a maximum difference of 20°, the calculations show a difference of 5°. Consequently, the calculated T−factor is small, and anisotropy does not show up in the calculated results except to a very small degree in the $\overline{vw}/U_o^2$ and $\overline{uw}/U_o^2$ distributions. (Compare also all these results with the N.S. calculations in Figs. 9a and 9b). At the next station chosen for the comparisons, i.e. X4700/Z1050, close to the right hand sidewall, see Figs.16a and 16b, the local mixing length model produces reasonable results and no major improvement is obtained by using either of the two k-ε models, which confirms our earlier observations.

5. CONCLUSIONS

In retrospect, the test case T1 has required much more effort on the experimental side than originally anticipated. Many, if not all, of the goals defined in advance have been achieved and others can still be reached by closer examination of the data already available or by new measurements in the existing facility. In spite of the relatively modest turning of the flow (which has encouraged some to try the small crossflow approximation), the three-dimensional effects have proved to be severe with respect to certain flow variables in certain areas of the duct. In the flow region upstream of the inflection point in the duct walls, the inviscid streamline turning is monotonic with a well developed crossflow. The strong anisotropy in the deduced eddy viscosity confirms that observed in previous experiments (East, van den Berg et al.), but it is surprising and disappointing that the newer prediction methods, taking full account of the flow history, could not predict this effect. Ad hoc corrections to the local eddy viscosity give better agreement than the more sophisticated models. Downstream of the inviscid streamline deflection points, pronounced crossover profiles developed as expected and the in advance measured shear stress distribution was found to be different from that calculated from any theory in use at present.

It appears worthwhile to use the data now available to develop new models in better agreement with the experimental information. It was hoped originally that a new initial data line could be defined (from measurements) in the crossover region and that one, on this basis, could study the flow relaxing from an initially complex three-dimensional state to a near two-dimensional flow at the downstream end of the duct. But there is little point in doing this at present, considering the deficiencies of the available turbulence models in the crossover regime.

A disappointing feature of the present setup is that the region available for study (when the domain of influence is properly considered) is rather small part of the total flow region. In a future test case, the appropriate side inflow conditions should be measured and supplied with the initial data. It would also be desirable to specify the full outer boundary conditions (rather than just the pressure distribution) to avoid discrepancies between predictions not related to boundary layer method or turbulence model. An advantage of the present setup is the finite domain which allows full Navier-Stokes solutions to be obtained for comparison with the boundary layer results. This comparison will not be that easy for the GARTEUR-wing or any other experiment involving a 3-D TBL flow in a medium of infinite extent and with strong interaction effects.

It would be of interest to make use of the present test case to develop new and better experiments for the purpose of developing new and improved prediction methods and models. But it is not obvious how this can be done. The secondary flows and the corner flows which have been the primary obstacles in the design of the present experiment, are an intrinsic part of any duct flow and can only be eliminated at high cost in time and effort.

6. REFERENCES

T.V.Truong: *Specifications for the test case: Boundary layer in a "S"-shaped channel [Problem T1]*, EPFL Ercoftac Pilot Centre, December 1989, Switzerland.

T.V.Truong, P.Dengel, V.Moreau, A.Nakkasyan, A.Drotz & I.L.Ryhming: *The Euroexpt series of duct flows tunnel at the EPFL, Interim Report No:2, Specialist Group 2,* EPFL/IMHEF, Switzerland

In this Workshop Proceedings:

M.Bettelini & T.K.Fanneløp: *Case T1: a boundary layer computation*

D.S.Clarke: Problem T1: *Predictions of the boundary layer flow in an S-shaped duct*

D.Dutoya & Ph.Poncelin de Raucourt: *A 3D compressible Navier-Stokes code for a boundary layer in a S-shaped channel calculation*

M.Gabillard, G.Marty, P.L.Viollet: *Problem T1: Boundary layer in a "S"-shaped channel*

C.Gleyzes, R.Houdeville, C.Mazin, J.Cousteix: *Calculation of test case T1 at Onera/Cert/ Derat*

P-Å. Krogstad & P.E. Skåre: *"S"channel Calculations*

P-Å.Lindberg: *An anisotropic eddy viscosity model, using the k-ε model or an extended Johnson & King model for solving the boundary layer equations*

J.P.F.Lindhout & B.van den Berg: *Calculation results of the NLR BOLA Method for the Ercoftac test case T1: 3D turbulent boundary layer in an S-shaped duct*

A.Nakkasyan: *A new statistical Approach to predict turbulence*

A.Nakkasyan, T.S.Prahlad, P.Gilquin: *An algebraic model based on the velocity-profile skewing*

W.Rodi & J.Zhu: *Calculation of three-dimensional turbulent flow in an S-shaped channel*

A.M.Savill, T.B.Gatski & P-Å.Lindberg: *A pseudo-3D extension to the Jonhson & King model and its application to the Euroexpt S-duct geometry*

T.V.Truong & M.Brunet: *Test case T1: Boundary layer in a "S"-shaped channel*

J.Wu, U.R.Müller, E.Krause: *Computation of the 3D turbulent boundary layer in an S-shaped channel*

CONTRIBUTOR(S)	CODE BASIS: B.L. or N.S.	TURBULENCE MODEL	NUMERICAL DETAILS	REMARKS
Bettelini & Fannelop	B.L., p measured	Mixing length (Cebeci-Smith-Michel)	Finite diff., Levy-Lees transformation external streamline coordinates	
Gleyzes, Houdeville, Mazin & Cousteix (1)	B.L., p measured	Mixing length (Cebeci-Smith-Michel)	Integral method	No useful results
Gleyzes, Houdeville, Mazin & Cousteix (2)	B.L., p measured	Mixing length (Cebeci-Smith-Michel)	Fully implicit, finite diff., local streamline directions	
Krogstad & Skåre (1)	B.L., p calculated Euler	Mixing length (Cebeci-Smith-Michel)	Small cross flow, finite diff., Levy-Lees transf. external streamline coordinates	Important differences with (2)
Krogstad & Skåre (2)	B.L., p measured	Mixing length (Cebeci-Smith-Michel)	Small cross flow, finite diff., Levy-Lees transf. external streamline coordinates	
Krogstad & Skåre (3)	B.L., p measured	Mixing length (Cebeci-Smith-Michel)	ADI scheme, Levy-Lees trans., external streamline coordinates	
Lindberg (1)	B.L., p measured	k-ε + ASM + transport eq. for T: no wall function	Finite diff., modified version of Krogstad ADI	
Lindberg (2)	B.L., p measured	J-K + transport eq. for T	Finite diff., modified version of Krogstad ADI	
Lindberg (3)	B.L., p measured	Low Re k-ε, no wall function	Finite diff., modified version of Krogstad ADI	
Lindhout & van den Berg	B.L., p measured	Mixing length (Cebeci-Smith-Michel)		
Nakkasyan, Prahlad, Gilquin	B.L., p measured	T ≠ 0 algebraic	Finite diff., modified version of Krogstad ADI	
Savill, Gatski & Lindberg	B.L., p measured	3D J-K, T ≠ 0	Same as Lindberg (2)	
Wu, Müller & Krause	B.L., p measured	Low Re k-ε, wall function	Finite diff., implicit, curvilinear coordinates	
Clarke & Wilkes (1)	N.S.	Low Re k-ε, (FHKE)	Harwell FLOW 3D, SIMPLE p-v coupling	
Clarke & Wilkes (2)	N.S.	DSM (FUDS)	Harwell FLOW 3D, SIMPLE p-v coupling	
Dutoya & Poncelin de Raucourt	N.S.	Mixing length	MATHILDA-3D unsteady, compressible N.S. solver	
Gabillard, Marty & Viollet	N.S.	Low Re k-ε	ESTET-3D, fractional step method	
Hecht, Maire, Pares & Pironneau	N.S.	Low Re k-ε		Pressure distribution obtained
Rodi, Zhu	N.S.	Low Re k-ε, wall function approach	SIMPLE p-v coupling, standard central/upwind hybrid differences	

Table I : Brief overview of the contributions; T is the turbulence anisotropy parameter (T = 0 isotropic)

Table II : Selected stations and groups of contributions for comparison

Station	Group
X4125 Z0800	Group I, Group II, Group III
X4700 Z0730	Group IV, Group V
X4700 Z1050	Group I, Group II, Group III, Group IV, Group V
X5255 Z0850	Group V
X5500 Z1198	Group I, Group II, Group III
X5500 Z1328	Group IV

Group I   : [Bettelini/Fanneløp],[Gleyzes/Houdeville/Mazin/Cousteix]
            [Krogstad/Skåre (2)], [Krogstad/Skåre (3)]
Group II : [Nakkasyan/Prahlad/Gilquin], [Lindberg (2)],
            [Savill/Gatski/Lindberg]
Group III : [Lindberg (3)], [Wu/Müller/Krause], [Lindhout/vandenBerg]
Group IV : [Gabillard/Marty/Viollet], [Clarke/Wilkes (1)], [Rodi/Zhu]
Group V   : [Lindberg (1)], [Clarke/Wilkes (2)], [Dutoya/Poncelin]

See also legend of the Table III

Table III : Comparison of external flow and boundary layer profile / wall parameters, i.e.:

$$Q_{ext}, \alpha_{ext}, \delta_{995}, \tau_{wall}, u_{\tau}, c_f, \gamma_{wall} \text{ , and H}$$

between the experimental data and the computational results from:

- B/F		: Bettelini/Fanneløp
- G/H/M/C		: Gleyzes/Houdeville/Mazin/Cousteix
- K/S	(1)	: Krogstad/Skåre [p calculated Euler, small cross flow]
- K/S	(2)	: Krogstad/Skåre [p measured, small cross flow]
- K/S	(3)	: Krogstad/Skåre [p measured, ADI scheme]
- L	(1)	: Lindberg [k-ε + ASM + transport equation for T]
- L	(2)	: Lindberg [J-K + transport equation for T]
- L	(3)	: Lindberg [low Re k-ε, no wall function]
- Lh/vdB		: Lindhout/vandenBerg
- N/P/G		: Nakkasyan/Prahlad/Gilquin
- S/G/L		: Savill/Gatski/Lindberg
- W/M/K		: Wu/Müller/Krause
- C/W	(1)	: Clarke/Wilkes [low Re k-ε, (FHKE)]
- C/W	(2)	: Clarke/Wilkes [DSM, (FUDS)]
- D/P		: Dutoya/Poncelin
- G/M/V		: Gabillard/Marty/Viollet
- R/Z		: Rodi/Zhu

The (*) mark in the tables denotes that the computational results were given by the contributor at the station close to the experimental station . The α angle values are shown with the conventional sign while the γ angle values with the inverse sign as defined in the specifications of the test case.

Station X2945Z0660

Parameter	Q_{ext} [m/s]	α_{ext} [o]	δ_{995} [mm]	τ_{wall} [N/m2]	u_τ [m/s]	$100C_f$ [-]	γ_{wall} [o]	H [-]
Experiment	20.25	3.10	75.3	0.687	0.872	0.371	-4.10	1.30
B/F	20.72	2.44	75.1	0.735	0.793	0.293	-5.03	1.29
G/H/M/C	20.66	4.40	83.0	0.805	1.159	0.314	-6.92	1.28
K/S (1)	18.19	1.80	61.1	0.511	0.676	0.276	-4.86	1.33
K/S (2)	20.69	2.70	55.9	0.545	0.793	0.294	-5.49	1.31
K/S (3)	20.67	2.70	55.0	0.514	0.770	0.278	-5.45	1.30
L (1)	21.04	2.71	75.3	0.68	0.75	0.35	-2.46	1.25
L (2)	21.04	2.71	64.6	0.65	0.74	0.33	-4.60	1.46
L (3)	21.04	2.71	75.3	0.68	0.75	0.35	-2.49	1.25
Lh/vdB		0.3				0.388	-3.2	1.28
N/P/G	20.7	1.19	72.	0.690	0.777	0.283	-5.67	1.28
S/G/L	21.04	2.71	71.1	0.93	0.88	0.48	-4.35	1.50
W/M/K	20.64	2.43	63.6	0.653	0.754	0.267	-5.19	1.29
C/W (1)	21.11		61.3	0.800	0.963	0.416	-0.84	
C/W (2)	21.35		65.1	0.818	0.984	0.425	-1.00	
D/P	19.28	0.88	91.5		0.87	0.409		
G/M/V					0.74		-2.13	
R/Z	20.35				0.834	0.336	-0.03	

Station X3320Z0340

Parameter	Q_{ext} [m/s]	α_{ext} [o]	δ_{995} [mm]	τ_{wall} [N/m2]	u_τ [m/s]	$100C_f$ [-]	γ_{wall} [o]	H [-]
Experiment	20.36	3.50	89.5	0.656	0.857	0.356	- 5.3	1.28
B/F	20.41	5.16	86.3	0.670	0.757	0.275	-10.6	1.29
G/H/M/C	20.41	6.76	73.7	0.734	1.106	0.294	-11.4	1.29
K/S (1)	17.42	4.50	76.5	0.465	0.617	0.251	-12.9	1.35
K/S (2)	20.41	4.30	69.1	0.501	0.751	0.271	- 9.9	1.31
K/S (3)	20.40	4.30	74.0	0.520	0.765	0.281	-10.3	1.31
L (1)	20.77	4.43	76.9	0.63	0.73	0.33	- 5.0	1.27
L (2)	20.77	4.43	63.6	0.60	0.71	0.31	-10.0	1.50
L (3)	20.77	4.43	76.8	0.63	0.71	0.33	- 5.1	1.27
Lh/vdB		3.0				0.350	- 8.4	1.30
N/P/G	20.4	2.95	75.	0.630	0.743	0.265	-10.4	1.30
S/G/L	20.77	4.43	71.2	0.89	0.86	0.46	- 9.2	1.54
W/M/K	20.40	4.76	74.8	0.603	0.725	0.252	- 9.5	1.31
C/W (1)	20.18		72.6	0.683	0.850	0.355	- 5.1	
C/W (2)	20.54		78.2	0.722	0.890	0.375	- 5.4	
D/P	18.83	3.77	104.6		0.82	0.375		
G/M/V					0.66		- 8.3	
R/Z	20.77				0.871	0.352	- 8.9	

Station X2945Z0340

Parameter	Q_{ext} [m/s]	α_{ext} [o]	δ_{995} [mm]	τ_{wall} [N/m2]	u_τ [m/s]	$100C_f$ [-]	γ_{wall} [o]	H [-]
Experiment	19.76	3.30	82.9	0.673	0.843	0.364	-6.80	1.30
B/F	20.25	3.93	82.5	0.694	0.770	0.289	-7.68	1.28
G/H/M/C	20.25	5.56	74.5	0.756	1.122	0.307	-9.22	1.29
K/S (1)	17.68	1.80	72.3	0.487	0.641	0.263	-5.18	1.34
K/S (2)	20.27	3.30	65.5	0.564	0.791	0.304	-7.16	1.30
K/S (3)	20.27	3.30	64.2	0.543	0.776	0.293	-7.06	1.30
L (1)	20.61	3.28	72.6	0.66	0.74	0.34	-3.30	1.26
L (2)	20.61	3.28	62.2	0.62	0.72	0.32	-7.09	1.48
L (3)	20.61	3.28	72.6	0.66	0.74	0.34	-3.35	1.26
Lh/vdB		1.9				0.375	-5.4	1.28
N/P/G	20.2	1.68	72.	0.659	0.759	0.282	-7.19	1.29
S/G/L	20.61	3.28	68.7	0.87	0.85	0.45	-7.11	1.52
W/M/K	20.26	3.74	63.6	0.631	0.741	0.267	-6.24	1.29
C/W (1)	20.29		67.6	0.709	0.871	0.369	-1.02	
C/W (2)	20.51		73.4	0.738	0.898	0.384	-1.58	
D/P	18.85	1.37	91.5		0.83	0.384		
G/M/V					0.68		-2.70	
R/Z	20.75				0.866	0.348	+0.05	

Station X2945Z0500

Parameter	Q_{ext} [m/s]	α_{ext} [o]	δ_{995} [mm]	τ_{wall} [N/m2]	u_τ [m/s]	$100C_f$ [-]	γ_{wall} [o]	H [-]
Experiment	20.31	5.20	59.6	0.674	0.866	0.364	-3.70	1.30
B/F	20.46	3.18	78.8	0.709	0.779	0.290	-6.38	1.29
G/H/M/C	20.43	5.27	82.3	0.775	1.137	0.309	-8.85	1.28
K/S (1)	17.94	2.00	71.0	0.553	0.693	0.299	-5.22	1.33
K/S (2)	20.46	3.00	66.2	0.622	0.839	0.336	-6.44	1.30
K/S (3)	20.44	3.00	67.4	0.588	0.814	0.317	-6.47	1.30
L (1)	20.81	3.16	74.2	0.66	0.75	0.34	-3.19	1.26
L (2)	20.81	3.16	64.7	0.63	0.73	0.33	-6.30	1.47
L (3)	20.81	3.16	74.2	0.66	0.74	0.34	-3.24	1.26
Lh/vdB		1.3				0.380	-4.9	1.28
N/P/G	20.4	1.6	74.	0.660	0.760	0.276	-7.14	1.29
S/G/L	20.81	3.16	69.9	0.91	0.87	0.47	-6.03	1.51
W/M/K	20.42	3.37	63.6	0.642	0.748	0.268	-6.69	1.29
C/W (1)	20.57		65.1	0.749	0.907	0.389	-1.29	
C/W (2)	20.77		67.6	0.769	0.929	0.400	-1.60	
D/P	19.05	1.3	91.5		0.85	0.395		
G/M/V					0.70		-2.72	
R/Z	20.54				0.849	0.342	+0.01	

Station X4125Z0646

Parameter	Q_{ext} [m/s]	α_{ext} [°]	δ_{995} [mm]	τ_{wall} [N/m²]	u_τ [m/s]	$100C_f$ [-]	γ_{wall} [°]	H [-]
Experiment	21.93	13.5	87.2	0.773	1.004	0.419	-25.2	1.30
B/F	22.29	15.4	92.3	0.865	0.860	0.297	-30.1	1.27
G/H/M/C	22.25	17.5	94.8	0.945	1.255	0.318	-30.0	1.25
K/S (1)	18.86	17.4	87.7	0.589	0.752	0.318	-32.6	1.34
K/S (2)	22.28	15.0	77.3	0.683	0.957	0.369	-26.7	1.31
K/S (3)	22.23	15.0	72.1	0.627	0.915	0.338	-26.4	1.31
L (1)	22.65	14.9	85.5	0.77	0.80	0.40	-11.6	1.24
L (2)	22.66	14.9	68.8	0.85	0.84	0.44	-27.2	1.51
L (3)	22.65	14.9	84.6	0.77	0.80	0.39	-11.8	1.24
Lh/vdB *		13.8				0.468	-28.2	1.27
N/P/G	22.3	13.3	89.	0.824	0.849	0.291	-30.3	1.26
S/G/L	22.65	14.9	82.0	1.08	0.95	0.56	-28.5	1.59
W/M/K	22.21	15.7	80.4	0.791	0.829	0.279	-29.4	1.28
C/W (1)	21.81		75.8	0.849	1.024	0.441	-21.3	
C/W (2)	22.29		83.0	0.891	1.073	0.463	-21.6	
D/P	20.56	14.8	118.5		0.99	0.461		
G/M/V					0.83		-26.1	
R/Z	21.78				1.014	0.434	-28.9	

Station X4125Z0800

Parameter	Q_{ext} [m/s]	α_{ext} [°]	δ_{995} [mm]	τ_{wall} [N/m²]	u_τ [m/s]	$100C_f$ [-]	γ_{wall} [°]	H [-]
Experiment	22.56	13.9	81.8	0.835	1.071	0.451	-24.9	1.29
B/F	23.09	15.2	87.5	0.935	0.894	0.300	-29.3	1.27
G/H/M/C	23.07	17.3	101.4	0.992	1.286	0.311	-30.8	1.24
K/S (1)	19.58	17.4	76.0	0.517	0.732	0.279	-33.2	1.34
K/S (2)	23.08	15.2	64.8	0.608	0.935	0.328	-27.3	1.31
K/S (3)	23.05	15.2	66.4	0.645	0.962	0.348	-27.5	1.32
L (1)	23.47	14.9	82.7	0.79	0.81	0.41	-12.3	1.24
L (2)	23.48	14.9	64.9	0.83	0.83	0.43	-26.0	1.50
L (3)	23.47	14.9	82.6	0.79	0.81	0.41	-12.5	1.24
Lh/vdB *		13.4				0.494	-28.5	1.26
N/P/G	23.1	13.3	75.	0.911	0.893	0.299	-29.2	1.26
S/G/L	23.47	14.9	73.0	1.14	0.98	0.59	-25.2	1.53
W/M/K	22.98	15.4	83.3	0.836	0.853	0.276	-30.1	1.27
C/W (1)	22.80		75.0	0.929	1.120	0.483	-20.7	
C/W (2)	23.29		79.8	0.957	1.161	0.497	-21.3	
D/P	21.18	14.8	118.5		1.05	0.489		
G/M/V					0.90		-25.6	
R/Z	21.90				1.024	0.437	-28.4	

Station X3320Z0500

Parameter	Q_{ext} [m/s]	α_{ext} [°]	δ_{995} [mm]	τ_{wall} [N/m²]	u_τ [m/s]	$100C_f$ [-]	γ_{wall} [°]	H [-]
Experiment	20.64	3.60	74.2	0.698	0.896	0.377	-7.6	1.27
B/F	20.84	5.08	83.0	0.727	0.788	0.286	-10.7	1.28
G/H/M/C	20.84	7.15	85.9	0.771	1.133	0.296	-12.3	1.28
K/S (1)	17.93	4.70	75.5	0.551	0.692	0.298	-12.2	1.33
K/S (2)	20.85	5.20	68.2	0.615	0.849	0.332	-10.7	1.30
K/S (3)	20.85	5.20	69.9	0.595	0.835	0.321	-10.8	1.30
L (1)	21.22	5.19	78.1	0.67	0.75	0.34	-5.2	1.26
L (2)	21.22	5.17	64.4	0.64	0.73	0.33	-9.9	1.48
L (3)	21.22	5.19	78.1	0.67	0.75	0.34	-5.2	1.26
Lh/vdB *		3.3				0.391	-9.2	1.28
N/P/G	20.9	3.71	78.	0.674	0.768	0.271	-12.2	1.29
S/G/L	21.22	5.19	72.4	0.92	0.88	0.47	-9.5	1.52
W/M/K	20.76	5.16	71.3	0.672	0.765	0.272	-10.4	1.29
C/W (1)	20.73		68.8	0.755	0.918	0.392	-5.4	
C/W (2)	21.05		73.4	0.780	0.947	0.405	-5.7	
D/P	19.32	3.84	104.6		0.87	0.402		
G/M/V					0.72		-8.1	
R/Z	20.67				0.868	0.353	-9.4	

Station X3320Z0660

Parameter	Q_{ext} [m/s]	α_{ext} [°]	δ_{995} [mm]	τ_{wall} [N/m²]	u_τ [m/s]	$100C_f$ [-]	γ_{wall} [°]	H [-]
Experiment	20.98	3.40	78.2	0.750	0.944	0.405	-7.4	1.26
B/F	21.36	4.66	79.4	0.785	0.819	0.294	-10.1	1.28
G/H/M/C	21.34	6.69	85.5	0.829	1.175	0.303	-11.8	1.27
K/S (1)	18.44	4.30	66.1	0.513	0.686	0.277	-11.5	1.32
K/S (2)	21.35	5.00	58.7	0.553	0.825	0.299	-10.5	1.30
K/S (3)	21.37	5.00	58.2	0.528	0.807	0.285	-10.7	1.30
L (1)	21.73	4.98	76.5	0.71	0.77	0.37	-5.0	1.25
L (2)	21.73	4.98	63.0	0.67	0.75	0.35	-9.6	1.46
L (3)	21.73	4.98	76.5	0.71	0.77	0.37	-5.1	1.25
Lh/vdB *		2.7				0.417	-8.7	1.27
N/P/G	21.3	3.48	73.	0.742	0.806	0.286	-11.2	1.28
S/G/L	21.73	4.98	69.4	0.97	0.90	0.50	-9.1	1.50
W/M/K	21.25	4.58	71.4	0.709	0.786	0.274	-9.7	1.29
C/W (1)	21.49		65.1	0.831	0.998	0.432	-5.1	
C/W (2)	21.85		71.3	0.849	1.026	0.441	-5.5	
D/P	19.84	3.59	104.6		0.92	0.431		
G/M/V					0.76		-7.6	
R/Z	20.55				0.864	0.354	-9.1	

Station X412SZ0960

Parameter	Q_{ext} [m/s]	α_{ext} [o]	δ_{995} [mm]	τ_{wall} [N/m2]	u_τ [m/s]	$100C_f$ [-]	γ_{wall} [o]	H [-]
Experiment	23.40	14.2	68.5	0.892	1.150	0.483	-25.9	1.29
B/F	24.04	15.4	73.7	1.016	0.932	0.301	-30.3	1.27
G/H/M/C	24.03	17.8	103.8	1.082	1.343	0.312	-32.6	1.24
K/S (1)	20.44	18.2	63.3	0.554	0.790	0.299	-33.5	1.34
K/S (2)	24.04	15.4	80.1	0.545	0.922	0.294	-29.6	1.29
K/S (3)	24.03	15.4	87.2	0.539	0.917	0.291	-30.0	1.29
L (1)	23.20	14.8	79.2	0.88	0.86	0.45	-12.6	1.23
L (2)	23.35	14.9	63.8	0.97	0.90	0.50	-26.9	1.51
L (3)	23.20	14.8	79.1	0.88	0.86	0.45	-12.8	1.23
Lh/vdB *		13.4				0.541	-29.5	1.25
N/P/G	24.0	12.7	132.	0.935	0.904	0.283	-33.2	1.22
S/G/L	23.29	14.9	69.3	1.28	1.03	0.66	-26.4	1.56
W/M/K	23.97	15.9	80.2	0.908	0.889	0.275	-31.1	1.27
C/W (1)	23.92		76.6	1.017	1.229	0.528	-20.7	
C/W (2)	24.26		77.4	1.041	1.262	0.541	-21.7	
D/P	21.99	15.5	118.5		1.13	0.525		
G/M/V					0.96		-25.7	
R/Z	22.11				1.049	0.450	-29.0	

Station X47002Z0890

Parameter	Q_{ext} [m/s]	α_{ext} [o]	δ_{995} [mm]	τ_{wall} [N/m2]	u_τ [m/s]	$100C_f$ [-]	γ_{wall} [o]	H [-]
Experiment	22.36	15.8	85.4	0.791	1.029	0.424	-24.7	1.31
B/F								
G/H/M/C *	23.07	19.4	110.9	0.979	1.278	0.307	-25.5	1.23
K/S (1)	19.22	18.8	90.7	0.469	0.684	0.253	-27.7	1.33
K/S (2)	23.09	16.3	76.4	0.557	0.895	0.301	-22.7	1.31
K/S (3)	23.08	16.3	79.0	0.632	0.953	0.341	-22.9	1.31
L (1)	23.50	16.4	80.6	0.82	0.83	0.42	-5.8	1.24
L (2)	23.50	16.4	59.1	0.84	0.84	0.43	-16.0	1.45
L (3)	23.50	16.4	80.0	0.82	0.83	0.42	-5.97	1.24
Lh/vdB *		15.8				0.457	-24.8	1.28
N/P/G	23.1	14.8	78.	0.818	0.846	0.268	-24.2	1.28
S/G/L	23.50	16.4	65.3	1.27	1.03	0.65	-14.6	1.49
W/M/K								
C/W (1)	22.67		83.0	0.887	1.088	0.461	-23.0	
C/W (2)	23.32		92.2	0.945	1.155	0.491	-22.6	
D/P	21.07	16.5	133.2		1.01	0.461		
G/M/V					0.89		-25.3	
R/Z	22.20				1.044	0.442	-26.0	

Station X47002Z0730

Parameter	Q_{ext} [m/s]	α_{ext} [o]	δ_{995} [mm]	τ_{wall} [N/m2]	u_τ [m/s]	$100C_f$ [-]	γ_{wall} [o]	H [-]
Experiment	22.79	15.4	95.0	0.817	1.076	0.446	-25.1	1.30
B/F								
G/H/M/C *	23.51	18.6	80.4	1.156	1.388	0.348	-21.8	1.21
K/S (1)	19.70	19.7	88.5	0.463	0.697	0.250	-27.6	1.31
K/S (2)	23.54	16.0	52.7	0.628	0.969	0.339	-20.6	1.29
K/S (3)	23.53	16.0	52.0	0.801	1.095	0.433	-21.6	1.26
L (1)	23.94	15.6	81.4	0.78	0.81	0.40	-5.8	1.23
L (2)	23.94	15.6	50.6	0.70	0.77	0.36	-13.9	1.44
L (3)	23.94	15.6	81.1	0.78	0.81	0.40	-6.0	1.23
Lh/vdB *								
N/P/G	23.5	14.8	64.	0.989	0.930	0.314	-24.7	1.25
S/G/L	23.94	15.6	56.1	1.05	0.93	0.54	-13.6	1.47
W/M/K								
C/W (1)	23.1		86.2	1.007	1.182	0.523	-24.1	
C/W (2)	23.72		97.2	1.068	1.249	0.555	-23.7	
D/P	21.57	16.9	133.2		1.09	0.514		
G/M/V					0.96		-25.4	
R/Z	23.35				1.187	0.517	-26.3	

Station X47002Z1050

Parameter	Q_{ext} [m/s]	α_{ext} [o]	δ_{995} [mm]	τ_{wall} [N/m2]	u_τ [m/s]	$100C_f$ [-]	γ_{wall} [o]	H [-]
Experiment	21.77	16.2	90.0	0.726	0.966	0.394	-25.1	1.33
B/F								
G/H/M/C *	22.56	20.0	122.2	0.836	1.180	0.273	-27.9	1.25
K/S (1)	18.68	20.7	82.6	0.418	0.627	0.226	-29.2	1.38
K/S (2)	22.57	17.2	71.1	0.476	0.809	0.257	-24.8	1.34
K/S (3)	22.64	17.2	67.3	0.579	0.895	0.313	-23.0	1.32
L (1)	23.0	17.3	87.2	0.76	0.80	0.39	-7.1	1.26
L (2)	23.0	17.3	67.7	0.82	0.83	0.42	-18.0	1.53
L (3)	23.0	17.3	86.7	0.76	0.80	0.39	-7.4	1.26
Lh/vdB *		16.0				0.413	-25.8	1.29
N/P/G	22.6	15.7	91.	0.717	0.792	0.245	-27.0	1.29
S/G/L	23.0	17.3	72.3	1.27	1.03	0.65	-17.6	1.53
W/M/K	22.60	18.4	90.3	0.778	0.823	0.266	-27.2	1.29
C/W (1)	22.54		79.8	0.823	1.042	0.427	-23.0	
C/W (2)	23.06		91.0	0.873	1.098	0.454	-23.0	
D/P	20.50	17.0	133.2		0.93	0.414		
G/M/V					0.81		-26.0	
R/Z	21.10				0.919	0.379	-26.8	

Station X4700Z1250

Parameter	Q_{ext} [m/s]	α_{ext} [o]	δ_{995} [mm]	τ_{wall} [N/m2]	u_τ [m/s]	$100C_f$ [-]	γ_{wall} [o]	H [-]
Experiment	20.71	17.8	78.9	0.628	0.854	0.340	-30.2	1.37
B/F	21.71	19.1	79.9	0.596	0.714	0.216	-28.3	1.37
G/H/M/C *	21.75	21.5	139.9	0.662	1.050	0.233	-31.3	1.29
K/S (1)	17.73	23.0	73.6	0.370	0.560	0.200	-29.7	1.46
K/S (2)								
K/S (3)								
L (1)	22.82	17.6	115.8	0.64	0.73	0.33	-9.1	1.25
L (2)	22.82	17.7	93.0	0.79	0.81	0.41	-18.4	1.54
L (3)	22.82	17.6	113.9	0.64	0.73	0.33	-9.5	1.26
Lh/vdB								
N/P/G	22.2	15.3	164.	0.613	0.733	0.217	-30.5	1.26
S/G/L	22.81	17.7	104.9	1.18	0.99	0.61	-18.4	1.55
W/M/K								
C/W (1)	21.54		85.4	0.752	0.952	0.391	-24.7	
C/W (2)	21.99		100.2	0.756	0.975	0.393	-25.2	
D/P	19.43	18.6	148.8		0.80	0.341		
G/M/V					0.62		-28.0	
R/Z	19.34				0.753	0.303	-29.5	

Station X500SZ0850

Parameter	Q_{ext} [m/s]	α_{ext} [o]	δ_{995} [mm]	τ_{wall} [N/m2]	u_τ [m/s]	$100C_f$ [-]	γ_{wall} [o]	H [-]
Experiment	23.06	12.4	99.1	0.805	1.075	0.435	-17.6	1.30
B/F								
G/H/M/C	23.40	15.2	84.3	1.081	1.342	0.329	-14.4	1.22
K/S (1)	19.33	14.0	94.5	0.423	0.653	0.229	-14.9	1.30
K/S (2)	23.44	13.0	59.2	0.637	0.972	0.344	-13.5	1.29
K/S (3)	23.39	13.0	68.3	0.773	1.069	0.418	-13.5	1.30
L (1)	23.82	13.0	108.6	0.57	0.69	0.30	-1.0	1.25
L (2)	23.82	13.1	65.8	0.51	0.65	0.26	-3.3	1.48
L (3)	23.82	13.0	108.4	0.57	0.69	0.29	-1.0	1.25
Lh/vdB								
N/P/G	23.3	11.8	60.	0.870	0.873	0.280	-13.5	1.29
S/G/L	23.82	13.0	79.8	0.75	0.79	0.39	-4.3	1.50
W/M/K								
C/W (1)	22.85		87.8	0.891	1.099	0.463	-17.2	
C/W (2)	23.55		100.2	0.955	1.173	0.496	-16.4	
D/P	21.20	12.9	133.2		1.02	0.462		
G/M/V					0.90		-17.0	
R/Z	22.98				1.114	0.470	-16.7	

Station X500SZ1000

Parameter	Q_{ext} [m/s]	α_{ext} [o]	δ_{995} [mm]	τ_{wall} [N/m2]	u_τ [m/s]	$100C_f$ [-]	γ_{wall} [o]	H [-]
Experiment	22.25	13.2	89.5	0.706	0.972	0.382	-16.4	1.31
B/F								
G/H/M/C	22.61	16.2	111.7	0.899	1.224	0.293	-17.0	1.23
K/S (1)	18.51	13.9	94.2	0.397	0.606	0.214	-14.3	1.33
K/S (2)	22.56	12.5	69.8	0.500	0.829	0.270	-12.8	1.31
K/S (3)	22.59	12.5	72.5	0.567	0.884	0.306	-13.5	1.31
L (1)	23.01	13.9	102.6	0.60	0.71	0.31	-0.3	1.25
L (2)	23.01	13.9	73.0	0.60	0.71	0.31	-4.9	1.48
L (3)	23.01	13.9	102.6	0.60	0.71	0.31	-0.3	1.25
Lh/vdB *		12.4				0.408	-14.8	1.29
N/P/G	22.6	12.5	76.	0.740	0.805	0.253	-13.6	1.30
S/G/L	23.01	13.9	84.0	0.86	0.85	0.44	-3.7	1.53
W/M/K								
C/W (1)	22.19		89.4	0.792	1.006	0.411	-16.7	
C/W (2)	22.88		98.2	0.851	1.076	0.442	-16.2	
D/P	20.38	13.1	133.2		0.92	0.405		
G/M/V					0.81		-16.9	
R/Z	21.55				0.955	0.393	-16.6	

Station X500SZ1150

Parameter	Q_{ext} [m/s]	α_{ext} [o]	δ_{995} [mm]	τ_{wall} [N/m2]	u_τ [m/s]	$100C_f$ [-]	γ_{wall} [o]	H [-]
Experiment	21.36	13.5	89.7	0.631	0.885	0.343	-15.1	1.33
B/F								
G/H/M/C	21.80	16.4	124.0	0.743	1.113	0.260	-17.1	1.26
K/S (1)	17.68	14.8	86.3	0.365	0.555	0.197	-12.6	1.37
K/S (2)	21.78	14.8	72.7	0.455	0.763	0.246	-15.0	1.34
K/S (3)	21.80	14.8	83.9	0.591	0.871	0.319	-18.1	1.32
L (1)	22.20	14.7	113.8	0.56	0.69	0.29	-0.2	1.27
L (2)	22.20	14.7	86.7	0.62	0.72	0.32	-4.0	1.51
L (3)	22.20	14.7	112.8	0.56	0.68	0.29	-0.3	1.28
Lh/vdB *		12.7				0.355	-14.4	1.31
N/P/G	21.8	13.2	100.	0.606	0.728	0.223	-14.6	1.31
S/G/L	22.20	14.7	92.4	0.90	0.87	0.46	-3.7	1.54
W/M/K	21.77	14.6	97.6	0.666	0.761	0.245	-15.3	1.32
C/W (1)	21.60		83.0	0.705	0.924	0.366	-16.1	
C/W (2)	22.26		97.2	0.767	0.994	0.399	-16.2	
D/P	19.78	13.4	148.8		0.85	0.368		
G/M/V					0.70		-16.6	
R/Z	20.21				0.823	0.332	-16.3	

Station X525SZ1000

Parameter	Q_{ext} [m/s]	α_{ext} [o]	δ_{995} [mm]	τ_{wall} [N/m2]	u_τ [m/s]	$100 C_f$ [-]	γ_{wall} [o]	H [-]
Experiment	21.76	10.0	91.5	0.656	0.917	0.356	- 9.9	1.31
B/F								
G/H/M/C	22.53	12.8	104.9	0.914	1.234	0.300	- 9.5	1.23
K/S (1)	18.42	10.2	99.1	0.392	0.599	0.211	- 5.8	1.31
K/S (2)	22.57	10.5	84.5	0.538	0.860	0.290	- 7.8	1.30
K/S (3)	22.52	10.5	97.0	0.488	0.817	0.264	- 6.2	1.33
L (1)	22.94	10.5	103.5	0.60	0.71	0.31	+3.6	1.26
L (2)	22.94	10.5	70.9	0.62	0.72	0.32	+2.5	1.49
L (3)	22.94	10.5	102.5	0.60	0.71	0.31	+3.7	1.26
Lh/vdB *		9.1				0.415	- 6.6	1.29
N/P/G	22.5	9.1	94.	0.732	0.800	0.252	- 5.2	1.28
S/G/L	22.94	10.5	41.8	1.84	1.24	0.95	+3.4	1.54
W/M/K								
C/W (1)	22.01		92.6	0.764	0.981	0.397	-10.7	
C/W (2)	22.76		101.2	0.829	1.056	0.431	-10.2	
D/P	20.39	10.2	148.8		0.92	0.404		
G/M/V					0.78		- 9.0	
R/Z	21.69				0.963	0.394	- 8.9	

Station X525SZ1150

Parameter	Q_{ext} [m/s]	α_{ext} [o]	δ_{995} [mm]	τ_{wall} [N/m2]	u_τ [m/s]	$100 C_f$ [-]	γ_{wall} [o]	H [-]
Experiment	20.80	10.5	97.4	0.604	0.841	0.327	- 8.8	1.32
B/F								
G/H/M/C	21.73	13.4	123.3	0.754	1.121	0.266	-10.1	1.26
K/S (1)	17.63	10.6	98.2	0.381	0.565	0.206	- 2.8	1.35
K/S (2)	21.74	11.6	75.4	0.504	0.802	0.272	- 7.7	1.32
K/S (3)	21.75	11.6	90.6	0.556	0.843	0.300	- 8.5	1.37
L (1)	22.12	11.4	106.4	0.55	0.68	0.28	+3.7	1.27
L (2)	22.12	11.4	76.2	0.54	0.67	0.28	+3.4	1.52
L (3)	22.12	11.4	106.3	0.55	0.68	0.28	+3.8	1.27
Lh/vdB *		9.2				0.360	- 6.2	1.31
N/P/G	21.0	9.6	159.	0.528	0.680	0.210	- 3.4	1.30
S/G/L	22.12	11.4	41.1	1.64	1.17	0.84	+4.0	1.55
W/M/K	21.69	11.5	92.4	0.702	0.782	0.260	- 7.5	1.31
C/W (1)	21.56		92.6	0.699	0.919	0.363	- 9.9	
C/W (2)	22.19		100.2	0.763	0.988	0.397	- 9.9	
D/P	19.74	10.3	148.8		0.85	0.368		
G/M/V					0.70		- 8.5	
R/Z	20.35				0.837	0.338	- 8.2	

Station X5005Z1300

Parameter	Q_{ext} [m/s]	α_{ext} [o]	δ_{995} [mm]	τ_{wall} [N/m2]	u_τ [m/s]	$100 C_f$ [-]	γ_{wall} [o]	H [-]
Experiment	20.45	14.2	76.7	0.582	0.813	0.316	-15.8	1.36
B/F	20.93	14.8	87.9	0.542	0.681	0.211	-13.6	1.37
G/H/M/C	20.98	17.4	128.5	0.636	1.029	0.241	-17.3	1.29
K/S (1)	16.80	15.5	79.3	0.370	0.531	0.200	- 8.8	1.42
K/S (2)								
K/S (3)								
L (1)	21.99	14.8	116.1	0.58	0.70	0.30	- 0.8	1.28
L (2)	21.99	14.8	92.1	0.70	0.77	0.36	- 3.1	1.52
L (3)	21.99	14.8	115.9	0.58	0.70	0.30	- 1.0	1.28
Lh/vdB *								
N/P/G	21.3	12.7	165.	0.527	0.679	0.202	-15.5	1.29
S/G/L	21.99	14.8	96.8	1.10	0.96	0.57	- 3.9	1.52
W/M/K	20.98	15.9	99.3	0.627	0.739	0.248	-15.6	1.35
C/W (1)	20.64		85.4	0.673	0.863	0.350	-15.4	
C/W (2)	21.27		100.2	0.684	0.897	0.356	-16.3	
D/P	18.87	13.7	148.8		0.75	0.32		
G/M/V					0.58		-15.6	
R/Z	18.75				0.702	0.280	-16.0	

Station X525SZ0850

Parameter	Q_{ext} [m/s]	α_{ext} [o]	δ_{995} [mm]	τ_{wall} [N/m2]	u_τ [m/s]	$100 C_f$ [-]	γ_{wall} [o]	H [-]
Experiment	22.66	8.9	99.7	0.728	1.003	0.394	-11.8	1.30
B/F								
G/H/M/C	23.38	11.7	69.4	1.059	1.329	0.323	- 7.7	1.25
K/S (1)	19.21	10.0	91.5	0.323	0.567	0.175	- 3.8	1.30
K/S (2)	23.41	8.4	44.3	0.495	0.856	0.267	- 3.9	1.30
K/S (3)	23.35	8.4	47.7	0.571	0.917	0.308	- 6.5	1.28
L (1)	23.80	8.9	96.8	0.55	0.68	0.28	+3.1	1.26
L (2)	23.80	8.9	57.3	0.46	0.62	0.24	+4.5	1.52
L (3)	23.80	8.9	98.0	0.55	0.68	0.28	+3.3	1.26
Lh/vdB *								
N/P/G	23.4	8.5	68.	0.838	0.856	0.269	- 5.0	1.29
S/G/L	23.64	9.4	28.2	1.34	1.06	0.69	+3.2	1.54
W/M/K								
C/W (1)	22.74		92.6	0.860	1.075	0.447	-11.2	
C/W (2)	23.47		104.2	0.913	1.143	0.475	-10.4	
D/P	21.12	10.0	133.2		1.0	0.451		
G/M/V					0.88		- 9.1	
R/Z	23.14				1.120	0.469	- 9.0	

Station X5255Z1300

Parameter	Q_{ext} [m/s]	α_{ext} [o]	δ_{995} [mm]	τ_{wall} [N/m2]	u_τ [m/s]	$100C_f$ [-]	γ_{wall} [o]	H [-]
Experiment	20.08	11.0	80.3	0.545	0.777	0.300	- 8.0	1.35
B/F	20.93	11.1	95.7	0.562	0.693	0.219	- 4.9	1.36
G/H/M/C	20.93	13.8	127.0	0.647	1.038	0.246	- 9.1	1.29
K/S (1)	16.88	10.7	91.8	0.400	0.554	0.216	+ 1.8	1.39
K/S (2)	20.96	12.0	94.0	0.412	0.699	0.222	- 7.6	1.35
K/S (3)	20.96	12.0	113.0	0.362	0.655	0.195	- 2.4	1.38
L (1)	22.00	11.4	86.0	0.70	0.77	0.36	+ 3.9	1.27
L (2)	21.82	11.5	74.9	0.69	0.76	0.36	+ 6.4	1.54
L (3)	22.00	11.4	86.3	0.70	0.77	0.36	+ 4.0	1.28
Lh/vdB								
N/P/G	21.0	9.6	159.	0.528	0.680	0.210	- 3.4	1.30
S/G/L	21.68	11.5	89.1	1.01	0.92	0.52	+ 4.7	1.53
W/M/K	20.93	12.0	91.4	0.609	0.729	0.242	- 5.8	1.35
C/W (1)	20.52		78.1	0.677	0.860	0.352	- 8.0	
C/W (2)	21.46		100.2	0.715	0.925	0.372	- 9.2	
D/P	18.94	10.4	148.8		0.77	0.328		
G/M/V					0.61		- 6.3	
R/Z	19.06				0.733	0.296	- 6.9	

Station X5500Z1068

Parameter	Q_{ext} [m/s]	α_{ext} [o]	δ_{995} [mm]	τ_{wall} [N/m2]	u_τ [m/s]	$100C_f$ [-]	γ_{wall} [o]	H [-]
Experiment	21.03	7.5	90.5	0.654	0.887	0.355	- 1.7	1.32
B/F	22.10							
G/H/M/C	22.10	9.4	108.8	0.851	1.191	0.291	+ 4.1	1.24
K/S (1)	18.04	6.2	109.1	0.384	0.581	0.208	+ 4.4	1.33
K/S (2)	22.10	7.5	82.8	0.529	0.835	0.286	- 1.5	1.31
K/S (3)	22.12	7.5	93.3	0.521	0.830	0.281	- 1.4	1.32
L (1)	22.50	7.6	84.5	0.72	0.78	0.37	+ 5.5	1.27
L (2)	22.50	7.6	64.8	0.65	0.74	0.33	+ 8.5	1.53
L (3)	22.50	7.6	83.4	0.72	0.78	0.37	+ 5.7	1.27
Lh/vdB *		6.0				0.375	- 0.6	1.31
N/P/G	22.1	6.3	90.	0.656	0.758	0.235	+ 2.3	1.32
S/G/L	22.5	7.6	83.6	0.84	0.84	0.43	+10.7	1.59
W/M/K								
C/W (1)	21.68		96.2	0.726	0.942	0.377	- 5.2	
C/W (2)	22.42		104.2	0.793	1.017	0.412	- 5.2	
D/P	20.04	7.6	148.8		0.88	0.384		
G/M/V					0.74		- 3.0	
R/Z	21.20				0.912	0.370	- 2.9	

Station X5500Z11198

Parameter	Q_{ext} [m/s]	α_{ext} [o]	δ_{995} [mm]	τ_{wall} [N/m2]	u_τ [m/s]	$100C_f$ [-]	γ_{wall} [o]	H [-]
Experiment	20.37	7.8	92.8	0.620	0.837	0.338	- 1.1	1.33
B/F								
G/H/M/C	21.53	9.7	124.3	0.739	1.110	0.266	- 3.3	1.26
K/S (1)	17.52	6.7	101.9	0.386	0.565	0.208	+ 5.9	1.34
K/S (2)	21.53	7.5	71.8	0.511	0.799	0.276	- 0.6	1.32
K/S (3)	21.56	7.5	95.7	0.614	0.878	0.331	- 0.7	1.36
L (1)	21.92	8.3	87.4	0.68	0.75	0.35	+ 6.4	1.28
L (2)	21.92	8.3	69.4	0.66	0.74	0.34	+ 8.4	1.52
L (3)	21.92	8.3	87.4	0.67	0.75	0.35	+ 6.7	1.28
Lh/vdB *		6.0				0.344	+ 0.9	1.32
N/P/G	21.5	6.8	95.	0.626	0.740	0.236	+ 2.0	1.32
S/G/L	21.92	8.3	84.5	0.81	0.82	0.42	+12.6	1.63
W/M/K	21.51	8.4	91.2	0.709	0.786	0.267	- 2.0	1.31
C/W (1)	21.47		95.7	0.699	0.915	0.363	- 4.5	
C/W (2)	22.15		104.2	0.774	0.993	0.402	- 4.8	
D/P	19.54	7.6	148.8		0.83	0.360		
G/M/V					0.69		- 1.9	
R/Z	20.17				0.825	0.334	- 2.0	

Station X5500Z1328

Parameter	Q_{ext} [m/s]	α_{ext} [o]	δ_{995} [mm]	τ_{wall} [N/m2]	u_τ [m/s]	$100C_f$ [-]	γ_{wall} [o]	H [-]
Experiment	19.86	8.0	84.5	0.586	0.793	0.319	+ 0.8	1.34
B/F	20.98	7.5	103.9	0.583	0.706	0.226	+ 2.7	1.35
G/H/M/C	20.98	9.8	127.2	0.672	1.058	0.255	- 1.8	1.28
K/S (1)	17.03	6.3	97.5	0.437	0.585	0.236	+ 8.6	1.36
K/S (2)	20.97	8.0	89.5	0.443	0.725	0.239	- 0.5	1.33
K/S (3)	21.02	8.0	107.8	0.508	0.778	0.274	+ 3.5	1.36
L (1)	21.52	8.4	111.5	0.61	0.72	0.32	+ 6.7	1.28
L (2)	21.52	8.4	88.5	0.68	0.75	0.35	+13.1	1.51
L (3)	21.52	8.4	111.4	0.61	0.71	0.31	+ 6.9	1.28
Lh/vdB								
N/P/G	21.0	6.5	137.	0.577	0.711	0.230	+ 2.3	1.30
S/G/L	21.52	8.4	94.0	0.99	0.91	0.51	+12.7	1.55
W/M/K	20.94	8.6	92.1	0.621	0.735	0.246	- 0.9	1.34
C/W (1)	20.60		84.1	0.733	0.899	0.381	- 2.8	
C/W (2)	21.60		104.2	0.756	0.957	0.393	- 4.0	
D/P	19.03	7.5	148.8		0.79	0.341		
G/M/V					0.62		+ 0.1	
R/Z	19.21				0.754	0.309	- 0.8	

Stations	Measured Data α_{ext} [o]	Greatest difference between measured and computed data value [o]	Greatest difference between measured and computed data contribution from	Smallest difference between measured and computed data value [o]	Smallest difference between measured and computed data contribution from
X4125Z0800	13.9	+ 3.5	K/S (1) - G/H/M/C	- 0.6	N/P/G
X4700Z0730	15.4	+ 4.3	K/S (1)	- 0.6	N/P/G
X4700Z1050	16.2	+ 4.5	K/S (1)	- 0.5	N/P/G
X5255Z0850	8.90	+ 2.8	G/H/M/C	- 0.5	K/S (2),(3)
X5500Z1198	7.80	+ 1.9	G/H/M/C	- 1.1	K/S (1)
X5500Z1328	8.0	+ 1.9	G/H/M/C	- 1.7	N/P/G

Stations	Q_{ext} [m/s]	value [%]	contribution from	value [%]	contribution from
X4125Z0800	22.56	+ 3.9	L (2)	- 13.2	K/S (1)
X4700Z0730	22.79	+ 5.1	L (2) - L (3)	- 13.6	K/S (1)
X4700Z1050	21.77	+ 5.9	C/W (2)	- 14.2	K/S (1)
X5255Z0850	22.66	+ 5.0	L (2) - L (3)	- 15.2	K/S (1)
X5500Z1198	20.37	+ 8.8	C/W (2)	- 14.0	K/S (1)
X5500Z1328	19.86	+ 8.8	C/W (2)	- 14.2	K/S (1)

Stations	γ_w [o]	value [o]	contribution from	value [o]	contribution from
X4125Z0800	- 24.9	12.6	L (1)	- 5.9	G/H/M/C
X4700Z0730	- 25.1	19.3	L (1)	- 2.5	G/H/M/C
X4700Z1050	- 25.1	18.0	L (1)	- 4.1	K/S (1)
X5255Z0850	- 11.8	16.3	L (2)	+ 0.6	C/W (1)
X5500Z1198	- 1.1	13.7	S/G/L	- 3.7	C/W (2)
X5500Z1328	+ 0.8	12.3	L (2)	- 4.8	C/W (2)

Stations	u_τ [m/s]	value [%]	contribution from	value [%]	contribution from
X4125Z0800	1.071	+ 20	G/H/M/C	- 32	K/S (1)
X4700Z0730	1.076	+ 29	G/H/M/C	- 35	K/S (1)
X4700Z1050	0.966	+ 22	G/H/M/C	- 35	K/S (1)
X5255Z0850	1.003	+ 32.5	G/H/M/C	- 43.5	K/S (1)
X5500Z1198	0.837	+ 33	G/H/M/C	- 32.5	K/S (1)
X5500Z1328	0.793	+ 33	G/H/M/C	- 26	K/S (1)

Table IV : Differences between the measured and the computed data at selected stations chosen for comparison

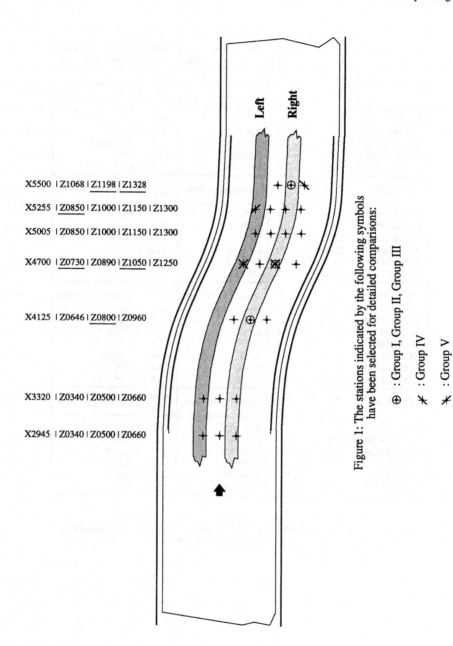

X5500 | Z1068 | Z1198 | Z1328

X5255 | Z0850 | Z1000 | Z1150 | Z1300

X5005 | Z0850 | Z1000 | Z1150 | Z1300

X4700 | Z0730 | Z0890 | Z1050 | Z1250

X4125 | Z0646 | Z0800 | Z0960

X3320 | Z0340 | Z0500 | Z0660

X2945 | Z0340 | Z0500 | Z0660

Figure 1: The stations indicated by the following symbols
have been selected for detailed comparisons:

⊕ : Group I, Group II, Group III

⫲ : Group IV

✳ : Group V

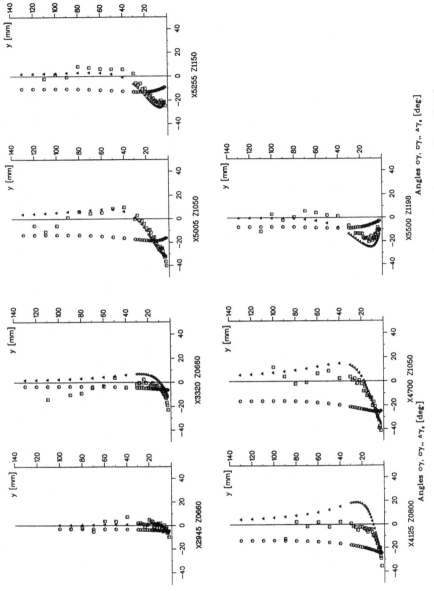

Figure 2: Development of the angles γ, γ_τ and γ_g as functions of the wall distance along the streamline path "R"

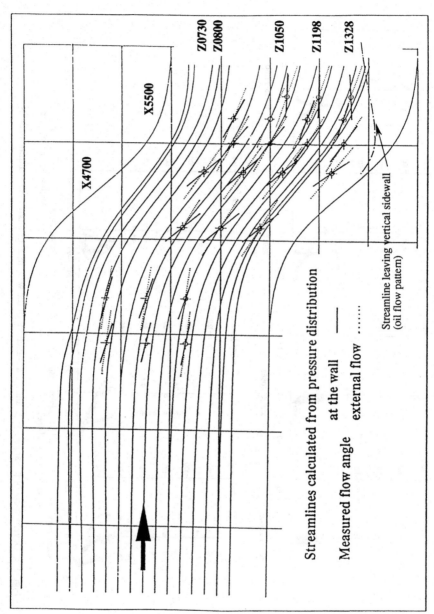

Figure 3: Calculated external streamlines and streamline angles at different locations on the test plate

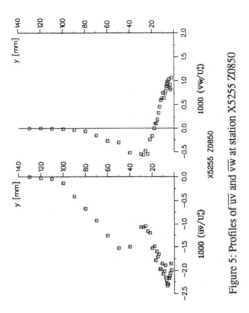

Figure 5: Profiles of $\overline{uv}$ and $\overline{vw}$ at station X5255 Z0850

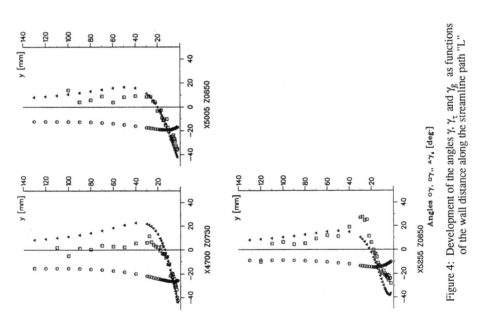

Figure 4: Development of the angles γ, γ_τ and γ_g as functions of the wall distance along the streamline path "L"

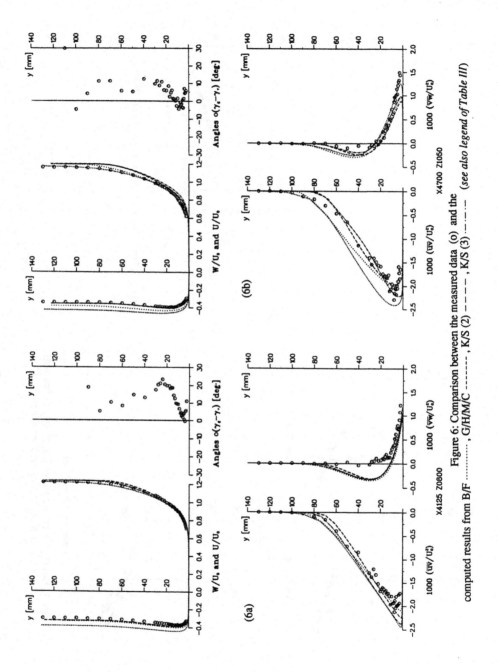

Figure 6: Comparison between the measured data (o) and the computed results from B/F ·········· , G/H/M/C — - — , K/S (2) – – – – , K/S (3) – · – · –   (see also legend of Table III)

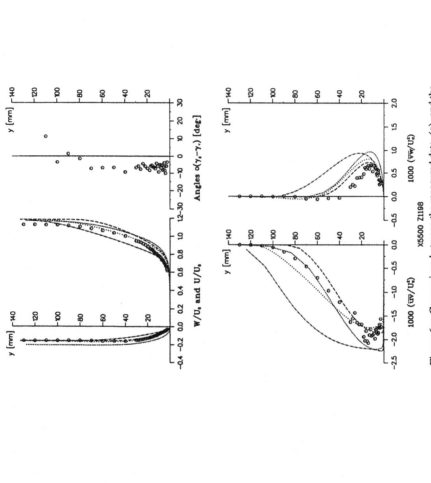

Figure 6c : Comparison between the measured data (o) and the
computed results from B/F .... , G/H/M/C ----- , K/S (2) ---- , K/S (3) --·--·-- (*see also legend of Table III*)

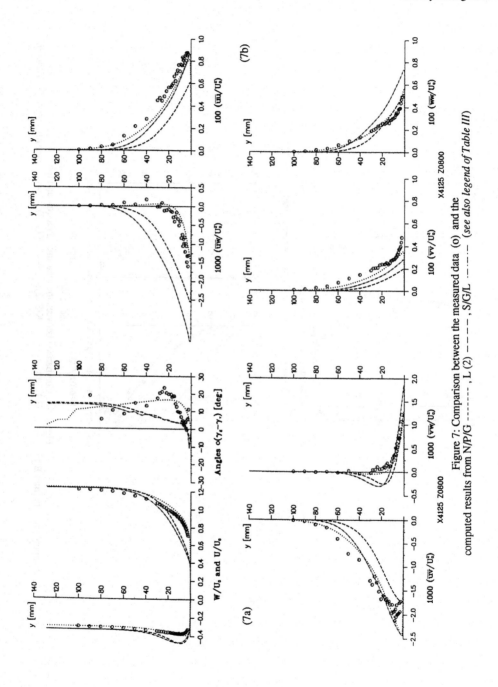

Figure 7: Comparison between the measured data (o) and the
computed results from N/P/G ——— , L (2) – – – – , S/G/L ·–·–·– (see also legend of Table III)

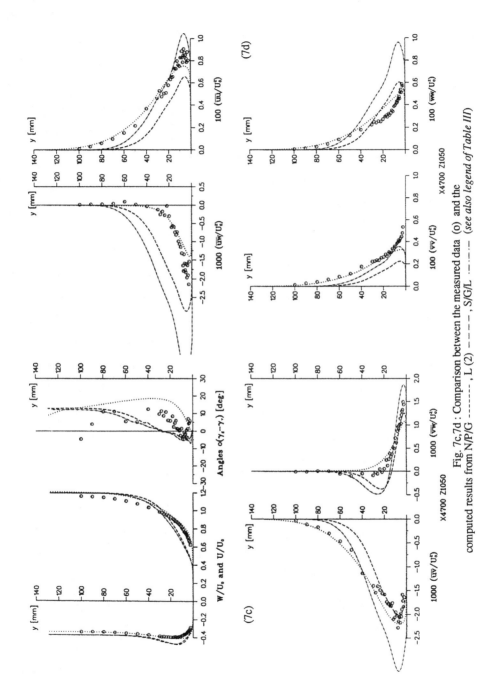

Fig. 7c,7d : Comparison between the measured data (o) and the
computed results from N/P/G ---·--- , L (2) – – – – , S/G/L ·––·–· *(see also legend of Table III)*

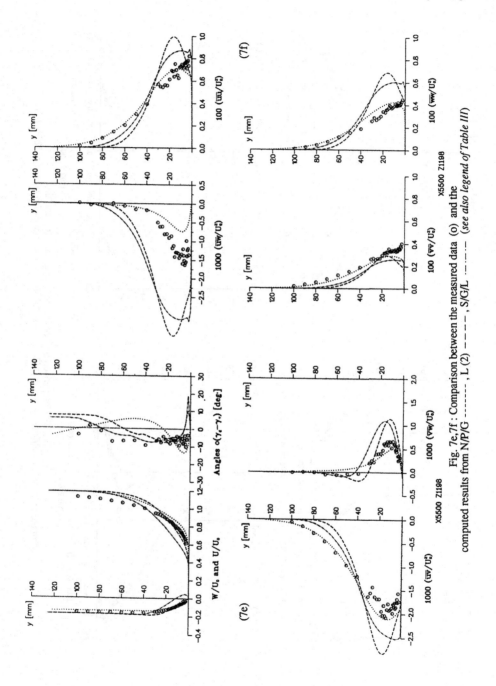

Fig. 7e,7f : Comparison between the measured data (o) and the
computed results from N/P/G ·······, L (2) − − − − , S/G/L −·−·−·− (see also legend of Table III)

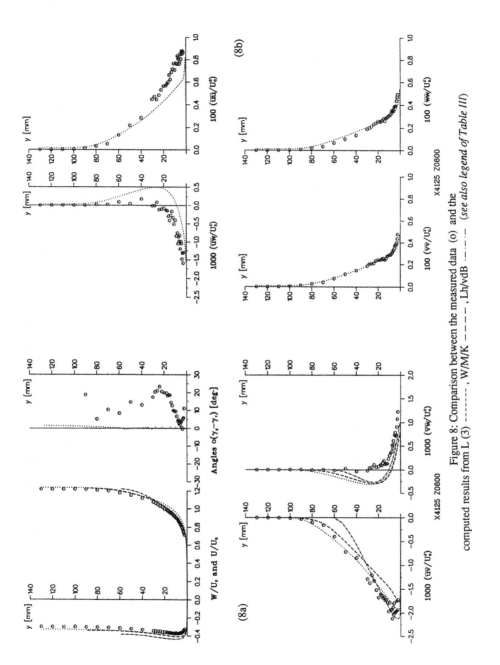

Figure 8: Comparison between the measured data (o) and the .... , W/M/K – – – – , Lh/vdB ·–·–·– (*see also legend of Table III*) computed results from L (3) ·········· and the

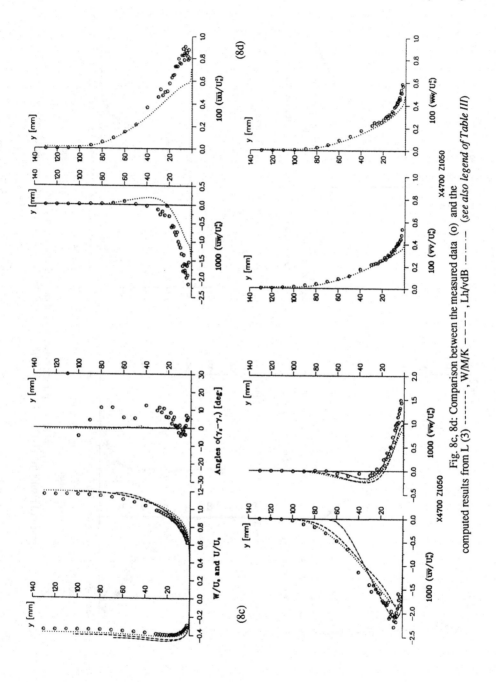

Fig. 8c, 8d: Comparison between the measured data (o) and the
computed results from L (3) ――――― , W/M/K ― ― ― ― , Lh/vdB ‧‧‧‧‧‧ (*see also legend of Table III*)

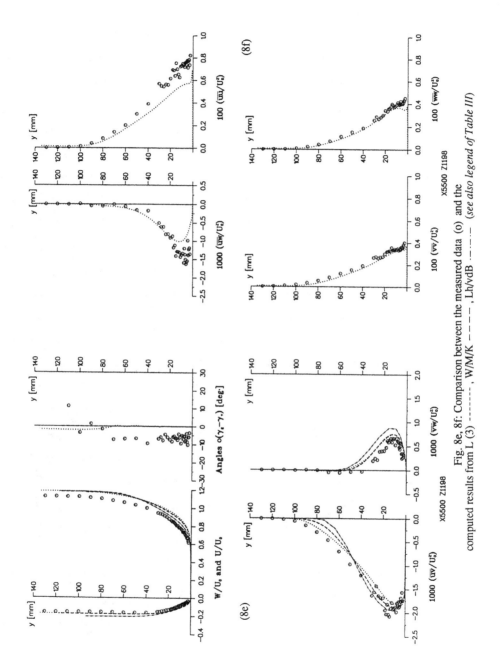

Fig. 8e, 8f: Comparison between the measured data (o) and the computed results from L (3) ——————, W/M/K – – – – , Lh/vdB ·–·–·– (see also legend of Table III)

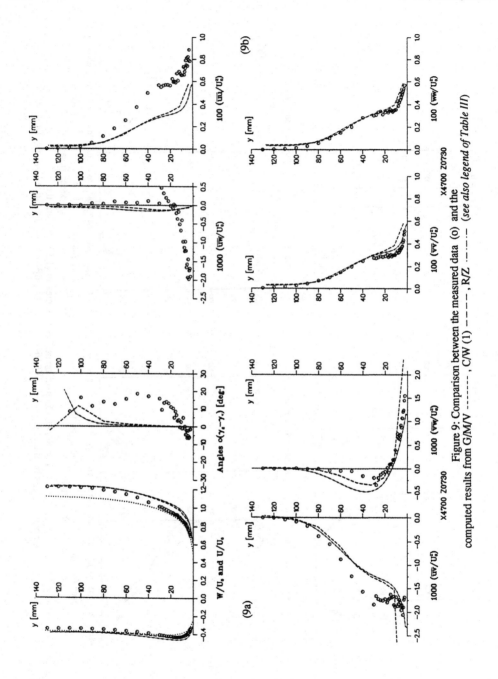

Figure 9: Comparison between the measured data (o) and the
computed results from G/M/V ------ , C/W (1) ----- , R/Z –··–··– (see also legend of Table III)

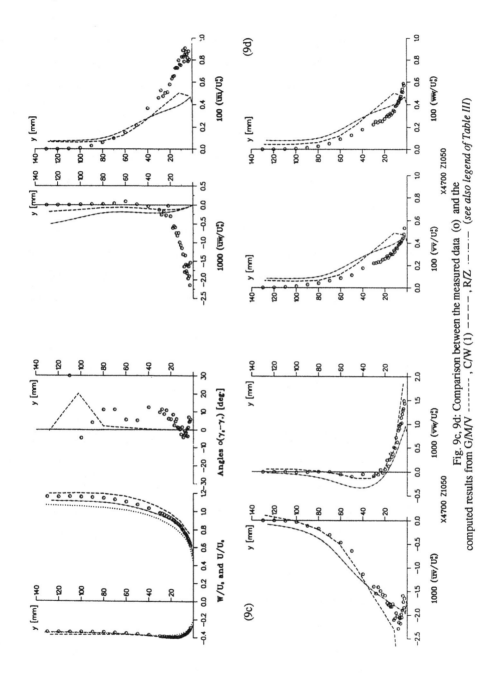

Fig. 9c, 9d: Comparison between the measured data (o) and the computed results from G/M/V ------- , C/W (1) ---- , R/Z —·—·— *(see also legend of Table III)*

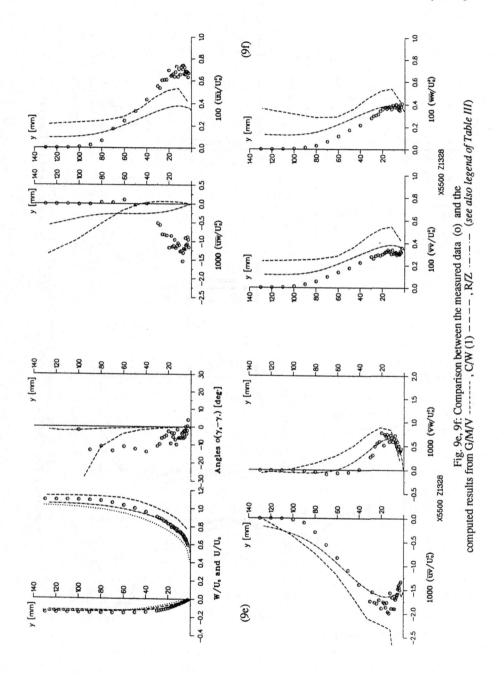

Fig. 9e, 9f: Comparison between the measured data (o) and the computed results from G/M/V ------, C/W (1) -----, R/Z ----- (see also legend of Table III)

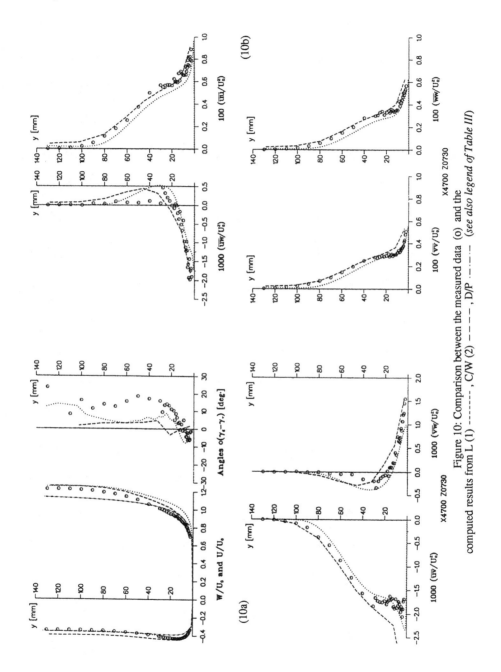

(10a)

(10b)

X4700 Z0730

X4700 Z0730

Figure 10: Comparison between the measured data (o) and the
computed results from L (1) ———, C/W (2) – – – , D/P ⋯–⋯– (see also legend of Table III)

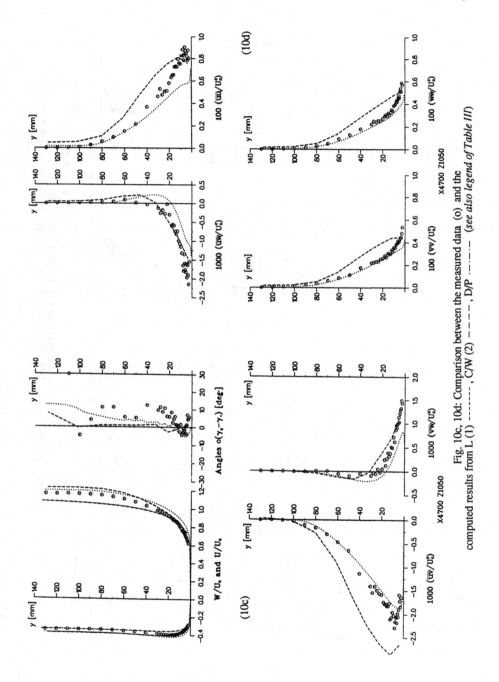

Fig. 10c, 10d: Comparison between the measured data (o) and the computed results from L (1) ⸺⸺, C/W (2) – – – –, D/P ·–·–·– (see also legend of Table III)

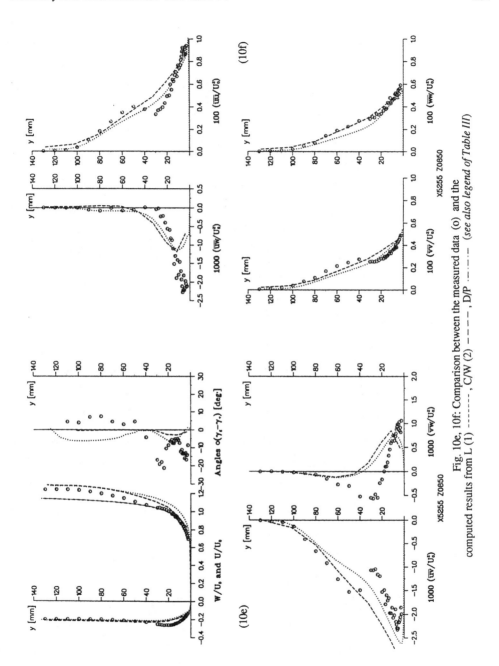

Fig. 10e, 10f: Comparison between the measured data (o) and the computed results from L (1) ———— , C/W (2) – – – – , D/P ·–·–·– (see also legend of Table III)

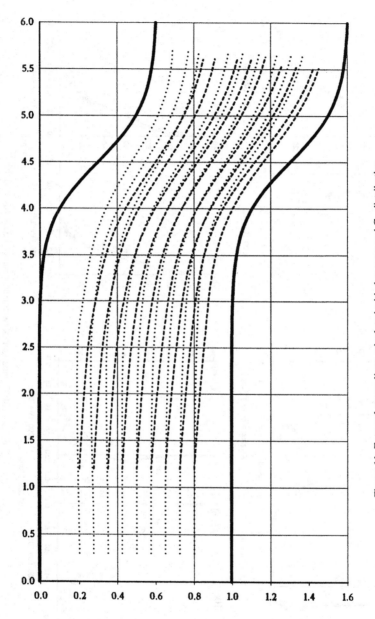

Figure 11: External streamlines calculated with the measured C_p distribution ········· , and with the measured C_p and the mean flow direction -------- .

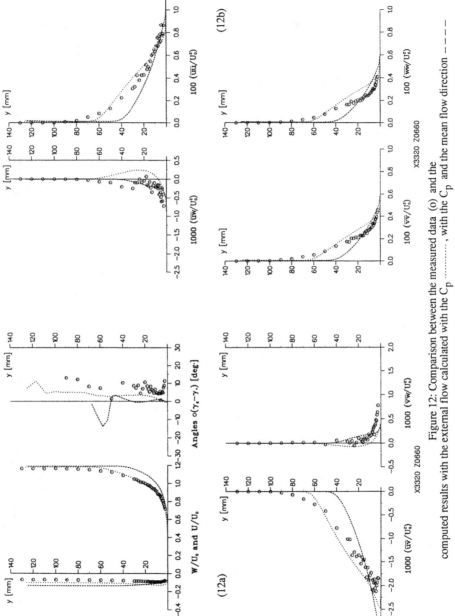

Figure 12: Comparison between the measured data (o) and the
computed results with the external flow calculated with the C_p ...., with the C_p ——— and the mean flow direction — — —

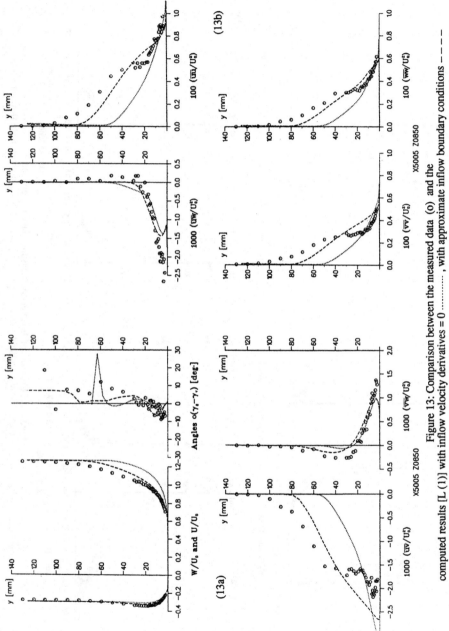

Figure 13: Comparison between the measured data (o) and the
computed results [L (1)] with inflow velocity derivatives = 0 ⋯⋯⋯⋯ , with approximate inflow boundary conditions – – – –

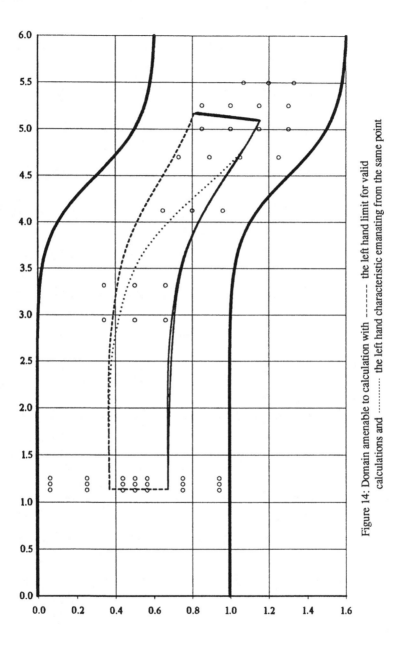

Figure 14: Domain amenable to calculation with ------ the left hand limit for valid calculations and ·········· the left hand characteristic emanating from the same point

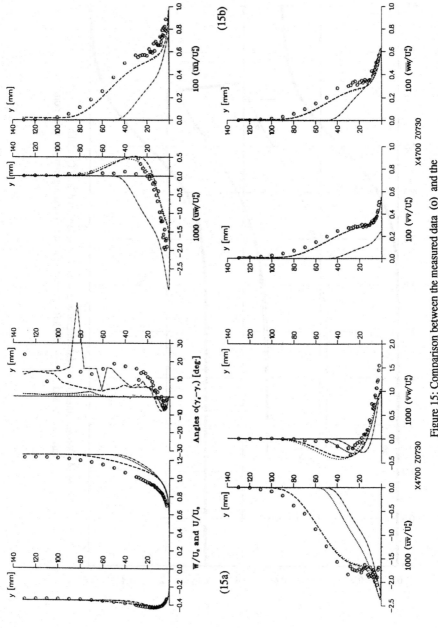

Figure 15: Comparison between the measured data (o) and the computed results with the model: algebraic ——— , low Re k-ε ·········· , anisotropic low Re k-ε – – – – , J-K anisotropic ——·——

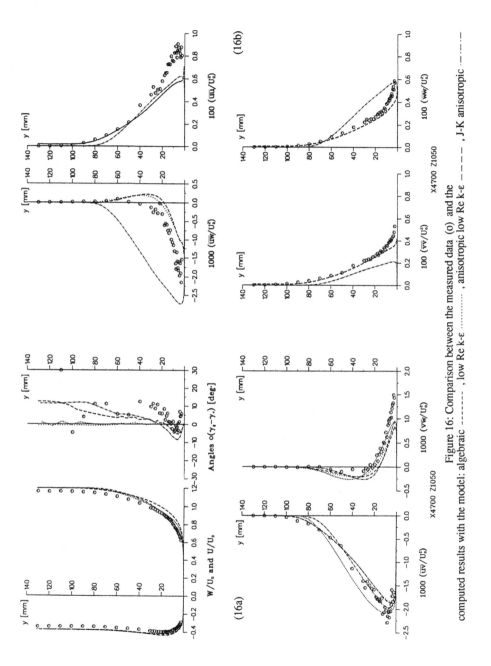

Figure 16: Comparison between the measured data (o) and the computed results with the model: algebraic ------- , low Re k-ε ·········· , anisotropic low Re k-ε ----- , J-K anisotropic ---·---

Test Case T2:
Array of Cylinders

Parallel Spectral Element Solutions of Eddy-Promoter Channel Flow

Paul F. Fischer
Center for Research on Parallel Computation
California Institute of Technology
Pasadena, CA 91125 USA

Anthony T. Patera
Department of Mechanical Engineering
Massachussetts Institute of Technology
Cambridge, MA 02139 USA

1 Introduction

We consider solution of the two-dimensional Navier-Stokes equations in the periodic domain of Fig. 1:

$$\frac{\partial \mathbf{u}}{\partial t} + \mathbf{u} \cdot \nabla \mathbf{u} = -\nabla p + \tfrac{1}{Re}\nabla^2 \mathbf{u} + f(t)\hat{e}_1 \quad in\ \Omega \tag{1}$$

$$\nabla . \mathbf{u} = 0 \quad in\ \Omega,$$

subject to,

$$\mathbf{u} = 0 \quad on\ \partial\Omega_1,\ \partial\Omega_2,\ and\ \partial\Omega_3 \quad ;$$

$$\mathbf{u}(0,y,t) = \mathbf{u}(L,y,t) \quad ; \tag{2}$$

$$p(0,y,t) = p(L,y,t) \quad .$$

Equation (1) is nondimensionalized with respect to channel half-height, h, and the centerline velocity which would arise in laminar plane Poiseuille flow in the absence of the cylinder for the same given flow rate, resulting in a Reynolds number defined as:

$$Re \equiv \frac{\tfrac{3}{2}\bar{V}h}{\nu} \quad , \tag{3}$$

where

$$\bar{V} \equiv \frac{1}{2h}\int_0^{2h} u_1(x,y,t)\,dy \quad . \tag{4}$$

246

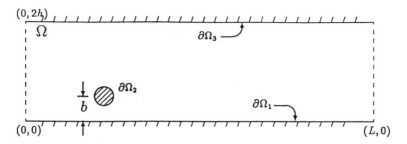

Figure 1: Eddy-promoter flow geometry: $L = 6.666h$, $b = 0.5h$, cyl. dia. $= 0.4h$.

The forcing term, $f(t)\hat{\mathbf{e}}_1$, corresponds to a mean pressure gradient driving the flow in the $\hat{\mathbf{e}}_1$ $(+x)$ direction. The functional form of $f(t)$ is chosen to enforce $\bar{V}$ to be constant with respect to time.

2 Numerical Formulation

Temporal discretization of (1) follows the fractional step method of [1,2], which leads to a series of elliptic problems to be solved at each time step by conjugate-gradient iteration. The method has been developed in such a way that it is readily implemented on distributed- or shared-memory parallel computers; we discuss these considerations in Section 4.

The fractional-step formulation is comprised of three computational steps. Beginning with explicit treatment of the nonlinear terms, compute $\hat{\mathbf{u}}$:

$$\frac{\hat{\mathbf{u}} - \mathbf{u}^n}{\Delta t} = \mathbf{C}^n + f^{n+1}\hat{\mathbf{e}}_1 \quad . \tag{5}$$

Here, $\mathbf{C}^n$ is taken to be an explicit representation of the nonlinear convective terms, in this case given by a 3rd order Adams-Bashforth discretization:

$$\mathbf{C}^n = \frac{23}{12}\mathbf{u}^n \cdot \nabla\mathbf{u}^n - \frac{16}{12}\mathbf{u}^{n-1} \cdot \nabla\mathbf{u}^{n-1} + \frac{5}{12}\mathbf{u}^{n-2} \cdot \nabla\mathbf{u}^{n-2} \quad . \tag{6}$$

Next, compute the pressure correction:

$$\begin{aligned}
\frac{\hat{\hat{\mathbf{u}}} - \hat{\mathbf{u}}}{\Delta t} &= -\nabla p & in \ \Omega \\
\nabla^2 p &= \tfrac{1}{\Delta t}\nabla \cdot \hat{\mathbf{u}} & in \ \Omega \\
\nabla p \cdot \hat{\mathbf{n}} &= \tfrac{1}{\Delta t}\hat{\mathbf{u}} \cdot \hat{\mathbf{n}} & on \ \partial\Omega.
\end{aligned} \tag{7}$$

Finally, compute the viscous correction:

$$\frac{u^{n+1} - \hat{u}}{\Delta t} = \frac{1}{Re} \nabla^2 u^{n+1} \quad . \tag{8}$$

The method is first order accurate in time [3]. The most numerically intensive steps are the (iterative) elliptic solves, with the Neumann problem for pressure (7) having significantly slower convergence than the Helmholtz equations (8) for the viscous terms.

To ensure that the constant mass flux condition is satisfied at t^{n+1}, we recognize that, as formulated in (5-8), u^{n+1} is linear in f^{n+1}. Therefore, we can decompose (5-8) into two sub-problems, one for which $f^{n+1} = 0$, and one for which $u^n, C^n = 0$. Thus, we first solve (5-8) for:

$$u^* = u^*(u^n = 0, C^n = 0, f^{n+1} = 1) \quad . \tag{9}$$

Second, we compute:

$$\bar{u}^{n+1} = \bar{u}^{n+1}(u^n, C^n, f^{n+1} = 0) \quad . \tag{10}$$

Finally, we update u^{n+1} as:

$$u^{n+1} = \bar{u}^{n+1} + \alpha u^* \quad , \tag{11}$$

where α is a scalar such that the desired mean flow, $\bar{V}$, is obtained at t^{n+1}. Note that the advantage of the splitting (9-11) is that (9) is independent of time (though not independent of Δt), and needs only be solved for u^* once in a preprocessing step, which can then be stored.

In all cases, we take as our initial condition:

$$u_1(x, y, t = 0) = y(2 - y) \tag{12}$$
$$u_2(x, y, t = 0) = 0 \quad .$$

This initial condition is not divergence free due to the presence of the cylinder, but it offers the advantage of avoiding thin shear layers near the walls during startup which would necessitate additional spatial resolution.

Spatial discretization is based upon the spectral element method [4,5] which is a high-order weighted residual technique in which the domain is broken up into relatively few, macro- (spectral) elements, and the geometry, data, and solution within each element are approximated by tensor-product polynomial basis functions. A typical decomposition of the eddy-promoter geometry is shown in Fig. 2. C^0 continuity is imposed across element interfaces and convergence is achieved through increased (polynomial) order of approximation within each element.

The above temporal discretization (5-8) leads to a series of elliptic sub-problems to be solved at each time step which are cast in the weak form:

Find $u \in \mathcal{H}_0^1(\Omega)$ such that

$$\int_\Omega \nabla\phi\nabla u + \lambda^2\phi u \, d\Omega = \int_\Omega \phi f \, d\Omega \quad \forall\phi \in \mathcal{H}_0^1(\Omega), \qquad (13)$$

where the space $\mathcal{H}_0^1$ is the space of all functions which have a square integrable first derivative and are zero on the boundary.

The spectral element method proceeds by subdividing the domain, Ω, into K elements, Ω^k, which are mapped to the square $(x,y)\big|_{\Omega^k} \rightarrow (r,s) \in [-1,1]^2$. The solution, data, and test functions are expressed as tensor-product polynomials in (r,s) of degree $\leq N$ with respect to each variable. Thus, the discrete represenation of u takes the form:

$$u(x,y)\big|_{\Omega^k} = \sum_{p=0}^{N}\sum_{q=0}^{N} u_{pq}^k h_p(r)h_q(s) \quad , \qquad (14)$$

The Legendre spectral element method in $\mathbb{R}^2$ employs Lagrangian interpolant bases $h_i(\xi)$ satisfying $h_i(\xi_j) = \delta_{ij}$, where the grid points ξ_j are the Gauss-Lobatto-Legendre points. The polynomial coefficients $u_{pq}^k = u^k(r_p, s_q)$ are therefore the grid values of u in element k. Gauss-Lobatto quadrature assures accurate approximation to the integrals in (13). Further details of the spatial discretization and formulation of the elliptic problems can be found in [5].

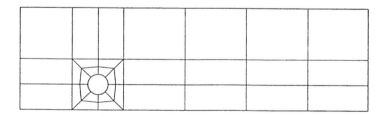

Figure 2: Spectral element mesh ($K = 33$) for the eddy-promoter geometry of Fig. 1

3 Iterative Solvers

The above discretization leads to a linear system of equations of the form:

$$\mathbf{A}\mathbf{u} = \mathbf{B}\mathbf{f} \ , \tag{15}$$

where $\mathbf{A}$ is taken to be the discrete Laplacian (for the case $\lambda = 0$), $\mathbf{B}$ the diagonal mass matrix, $\mathbf{u}$ and $\mathbf{f}$ the (global) discrete representation of the solution and data respectively. Due to memory, operation count, and parallelization considerations, iterative methods are used to solve the system (15). Such methods are dependent upon repeated evaluation of matrix-vector products of the form $\mathbf{r} = \mathbf{A}\mathbf{u}$, where $\mathbf{u}$ and $\mathbf{r}$ are intermediate vectors associated with successive iterations.

For spectral element problems in higher space dimensions $\mathbf{R}^d$ the linear operators have large bandwidth and, if formed explicitly, are non-sparse with $O(KN^{2d})$ entries. The subsequent operation count and memory requirements can be significantly reduced if the matrix-vector product $\mathbf{A}\mathbf{u}$ is evaluated element by element, using a factored form in which the discrete derivatives associated with the gradient operators in (13) are applied in a sequential fashion. A typical term in the elemental matrix-vector product $A^k u^k$ for the case of $d = 2$ is:

$$\sum_{p=0}^{N} \rho_{pj} D_{pi} \left(\sum_{q=0}^{N} D_{pq} u_{qj}^k \right) \quad \forall i,j \in \{0, ..., N\}^2 \tag{16}$$

where ρ_{ij} is the quadrature weight associated with the point (r_i, s_j), and D_{ij} is the derivative operator,

$$D_{ij} \equiv \left. \frac{dh_j(r)}{dr} \right|_{r=r_i} \tag{17}$$

The residual evaluation is completed via direct stiffness summation wherein intermediate residual values at nodes shared by multiple elements are summed and redistributed to the elemental data structures. The factored evaluation of $\mathbf{A}\mathbf{u}$ requires only $O(KN^d)$ storage and $O(KN^{d+1})$ operations for general isoparametric discretizations.

Our current implementation employs standard Jacobi- (diagonal) preconditioned conjugate gradient iteration [6]. The preconditioned $\mathbf{A}$ system has condition $O(K_1^2 N^2)$ implying an iteration count of $O(K_1 N)$, where K_1 is the number of elements in a single spatial direction. The majority of the computational effort is associated with evaluation of $\mathbf{A}\mathbf{u}$, as all other terms have an operation count of $O(KN^d)$ or less. In addition, two operations require communication of information between elements, namely, direct stiffness summation and inner-product evaluations of the form $\mathbf{r}^t\mathbf{r}$. While these steps require only $O(KN^{d-1})$ and $O(KN^d)$ operations, respectively, they represent the leading order *communication* terms in the parallel implementation discussed below.

4 Parallel Implementation

The spectral element discretizations, bases, and iterative solvers of the previous sections are constructed so as to admit a native, geometry-based parallelism, in which each spectral element (or group of spectral elements) is mapped to a separate processor/memory, with the individual processor/memory units being linked by a relatively sparse communications network. This conceptual architecture is naturally suited to the spectral element discretization in that it provides for tight, structured coupling within the dense elemental constructs, while simultaneously maintaining generality and concurrency at the level of the unstructured macro-element skeleton. The locally-structured/globally-unstructured spectral element parallel paradigm is closely related to the concept of domain-decomposition by substructured finite elements.

Our methods are implemented in an essentially machine-independent fashion. First, we construct a spectral element code in a standard high-level language in which each spectral element is treated as a "virtual parallel processor". In particular, each spectral element is treated as a separate entity, and all data structures and operations are defined and evaluated at the elemental level. The data and code are descended to M processors, each operating asynchronously. The only procedures which require communication are, by construction, the direct stiffness summation associated with residual evaluation, and vector reduction, which are relegated to special subroutines to effect data transfer. Processor synchronization is imposed at each iteration by the communication steps.

The residual calculation $\mathbf{Au}$ is the most complex operation in our parallel Navier-Stokes algorithm. Because of the variational formulation, the required action to update the residual along element interfaces is to first compute an intermediate residual vector, $A^k u^k$, within each element, and then sum correspondant edge values between elements. The parallelism in this procedure is quite evident; computation of local residuals requires $O(KN^{d+1})$ operations, while the communication between elements (processors) requires less than $O(KN^{d-1})$ words to be transmitted, resulting in a favorable computation to communication ratio of $O(N^2)$.

The conjugate-gradient iteration requires evaluation of two inner products which also require communication. These can be evaluated at a communication cost proportional to $\log_2 M$ on most topologies and any wormhole-connected communication system. The impact of this term is dependent on the work local to the processors, the ratio of computation to communication speed of the particular hardware, and the number of processors. Typically it is not significant, but it can be a source of performance degradation for small problems when M is large, and ultimately limits the speedup possible for a particular problem [7].

We have implemented our algorithms on the Intel iPSC/2-VX hypercube which is typical of the class of architectures for which the parallel spectral element method is well suited. The iPSC is a distributed memory, message passing, parallel processor consisting of $M = 2^D$ independent processor/memories, or nodes, arranged on a D-dimensional hypercube communication network. The iPSC/2-VX is an upgraded version of Intel's original iPSC hypercube which incorporates improved "worm-hole" message routing allowing data to be transferred between non-nearest-neighbor processors with minimal degradation in data transfer rate, as well as a twenty-fold reduction in message transfer times, resulting in a single word transfer rate of $\Delta(1) \simeq 300\mu sec$ and an asymptotic rate of $\Delta(\infty) \simeq 1.4\mu sec$ / word. The nodes are based upon Intel 80386/80387 processor/coprocessors with a floating point execution rate of roughly 0.1 MFLOPS, coupled with attached vector processors which achieve 3-4 MFLOPS on standard vector operations and 10-12 MFLOPS on matrix-matrix products. Typical performance for the spectral element code is 2-3 MFLOPS per node, including communication overhead.

5 Results

We have considered three classes of problems: *(i)* two-dimensional periodic geometry (e.g. Fig. 1), *(ii)* two-dimensional flow past 9 successive cylinders with parabolic velocity profile at inflow and standard Neumann boundary conditions at outflow, *(iii)* three-dimensional extensions of Fig.1 with and without endwalls. We briefly discuss the results of the class *(i)* and *(ii)* problems, and present computing times for all three classes.

Numerous simulations have been carried out for the two-dimensional, streamwise periodic flows, ranging from $Re = 0$ to $Re = 600$ [8]. In the range $Re > 0$ to $Re = Re_c$, the initially transient flow field evolves to a steady state. For subcritical Reynolds numbers close to Re_c the predominant part of the transient takes the form of an exponentially decaying sine wave, as seen for example in Fig. 3a, which is a time history of the transverse component of velocity at a point 3 cylinder diameters downstream of the center of the cylinder for $Re = 100$. As the Reynolds number approaches Re_c, the decay rate decreases; by extrapolating to zero decay rate, we estimate that $Re_c = 136$. We believe this number to be good to within 5%. However, non-splitting calculations should be undertaken before passing final judgement.

Above Re_c, the flow no longer settles to a steady state. Rather, it transitions

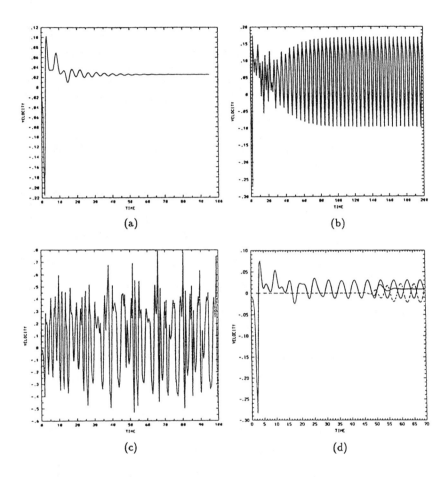

Figure 3:   Time history traces of transverse velocity component at a point three
diameters downstream of the cylinder center: (a) $Re=100$, (b) $Re=200$, (c) $Re=600$.
Fig. (d) is the comparison between the periodic and inflow-outflow calculation for
$Re = 140$ at a point 4.166 diameters downstream (of the 6th cylinder for the inflow-
outflow case). The dashed curve represents the difference between the two traces; the
convective nature of the instability can be observed by the sudden termination of the
disturbance for the inflow-outflow configuration at a time $t_c \simeq 50$.

a steady periodic state for Re near Re_c, as evidenced by the time history trace of Fig. 3b for $Re = 200$. A fit of the saturation amplitude versus Reynolds number for several Reynolds numbers slightly greater than Re_c, showed the saturation amplitude is to vary as $(Re - Re_c)^{.59}$ [9]. Fig. 4 shows a typical streamline pattern at this Reynolds number, indicating the presence of a travelling wave moving from left to right, with wavelength $\lambda = L/2$. Over the range $Re = 100$ to $Re = 200$, the Strouhal number was found to be nearly constant, with a value of $St = \frac{2fh}{3\bar{V}} = 0.188$.

At $Re = 600$, the flow ceases to exhibit the single mode behavior found at Reynolds numbers near Re_c; Fig. 3c shows the erratic behavior of the velocity signal at the point of the previous figures.

Experimental results have suggested that the observed two-dimensional instability is of a convective nature. That is, the disturbance will grow (to a saturated state) in a frame of reference moving with the flow, but will not be self-sustaining at a fixed location in a laboratory setup, if there is not sufficient "noise" in the inlet profile. To examine this possibility, a second set of simulations have been carried out in which the domain is no longer periodic, but of fixed length, $L_9 = 9.5L$, with an array of nine successive cylinders to provide a periodic geometric disturbance to the otherwise plane Poiseuille channel flow, as depicted in Fig. 5. A parabolic inlet profile was specified, with the usual Neumann outflow boundary conditions on velocity at the exit [9].

With initial conditions identical to those for the periodic case, time history traces behind each of the cylinders was found to be identical (as shown in Fig. 3d) to the periodic results up to a time $t_c \simeq l/\bar{V}$, at which point the signal would fall off

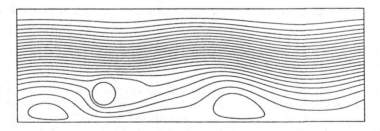

Figure 4: Streamlines for the supercritical Reynolds number $Re = 200$ reveal the travelling wave nature of the instability, with wavelength $\lambda = L/2$.

Figure 5: Multi-cylinder geometry ($K = 315$) and resultant streamlines for $Re = 160$.

rapidly to a constant value. Here, l is the distance downstream of the first cylinder. It is, in principle, possible to increase l to the point where t_c will be greater than the time required to reach a saturated periodic state. Thus, one can conclude that the periodic simulations discussed above provide relevant analysis for this physically realizable situation. Furthermore, in the presence of noise, experimental results [10] suggest that the periodic solution appears relatively quickly downstream.

6 Timings

We present in Table 1 the run time required for several of the test cases considered. All problems were computed on the Intel iPSC/2-VX, save case (a), which was solved on a DECStation 3100. Case (a) is a steady Stokes flow calculation which is discretized using consistent spaces for the velocity and pressure and which is solved using the Uzawa algorithm [5]. Case (e) is the nine cylinder calculation. For this problem, the CPU time per time-step, per processor, is roughly 15 times that required for the corresponding single cylinder case, (c). This is consistent with the increased number of degrees of freedom and increased condition number of the system matrices, as well as the counter-balancing effect of increased efficiency obtained due to the larger amount of work per processor required for the multi-cylinder case.

Case	Re	d	K	N	t_{final}	#time-steps	$\frac{seconds}{time-step}$	#-procs	CPU (hrs)
a	0	2	39	10	∞	1	228.	DEC	.063
b	100	2	39	10	95.5	26920	2.45	8	18.3
c	160	2	33	8	200.0	40000	2.50	8	28.0
d	600	2	39	10	200.0	79280	2.85	8	62.7
e	160	2	315	8	82.5	16540	18.8	16	86.5
f	150	3	99	8	200.0	40000	25.2	32	280.

Table 1: Computer performance for the eddy promoter problem.

Acknowledgements

We wish to thank Michael Schatz, Randall Tagg, and Harry Swinney of the University of Texas at Austin for their assistance and for the insight provided through their experimental work. This work was supported by the ONR and DARPA under contracts N00014-85-K-0208, N00014-88-K-0188, and N00014-89-J-1610, by NSF under Grants DMC-8704357, ASC-8806925, and Cooperative Agreement No. CCR-8809615, and by Intel Scientific Computers.

References

[1] S.A. Orszag and L.C. Kells (1980): Transition to turbulence in plane Poiseuille flow and plane Couette flow. *J. Fluid Mech.* 96, 159-205

[2] N.K. Ghaddar, G.E. Karniadakis, and A.T. Patera (1986): A conservative isoparametric spectral element method for forced convection; application to fully developed flow in periodic geometries. *Num. Heat Trans.* 9, 277-300

[3] R. Temam (1977): *Navier-Stokes Equations* (North-Holland, Amsterdam)

[4] A.T. Patera (1984): A spectral element method for fluid dynamics; Laminar flow in a channel expansion. *J. Comput. Phys.*, 54, 468-488.

[5] Y. Maday and A.T. Patera (1989): "Spectral element methods for the Navier-Stokes equations," in *State of the Art Surveys on Computational Mechanics* Eds. A.K. Noor and J.T. Oden, (ASME, New York), 71-143

[6] G.H. Golub and C.F. Van Loan (1983): *Matrix Computations*, (John Hopkins University Press, Baltimore, Maryland)

[7] P.F. Fischer and A.T. Patera (1991): Parallel Spectral Element Solution of the Stokes Problem. *J. Comput. Phys.*, 92

[8] G.E. Karniadakis, B.B. Mikic, and A.T. Patera (1988): Minimum-dissipation transport enhancement by flow destabilization: Reynolds' analogy revisited. *J. Fluid Mech.*, 192, 365-391

[9] M.F. Schatz, R.P. Tagg, H.L.Swinney, P.F. Fischer, and A.T. Patera (1990): Supercritcal transition in plane channel flow with spatially periodic perturbations. submitted to Phys. Rev. Lett.

[10] H. Kozlu (1989): *Experimental Investigation of Optimal Heat Removal From a Surface*. Ph.D. Thesis, Massachusetts Institute of Technology

2D and 3D simulations of the periodic flow around a cylinder between 2 walls

Marc BUFFAT

Laboratoire de Mécanique des Fluides et Acoustique, ECL,
Av. Guy de Collongues, 69131 Ecully, France

1   Introduction

We study the flow around a periodic array of cylinders between two parallel walls as described by Pironneau (this book). We solve the incompressible Navier-Stokes equations with a velocity pressure formulation in two and three dimensions. The reference velocity V is the maximum laminar velocity in the channel for a given flow rate Q ($V = Q/2h$) and the reference length h is half of the distance between the walls.

2   Numerical method

To solve the Navier-Stokes equations, we use the finite element code "Nadia" developed at the laboratory for the prediction of internal subsonic flows and we refer to [1] for a complete description of the numerical method. Nadia is a time marching code and it uses a semi-implicit time integration with first order accuracy in time to calculate stationary flows or a corrected implicit scheme with second order accuracy to simulate precisely unsteady flows. For the space discretization, we use the $P^1/P^1 iso P^2$ element with a piecewise linear pressure and a piecewise linear velocity on a grid twice finer. To solve the linear system of coupled equations arising from the discretization, an iterative Uzawa algorithm is used. To accelerate the speed of convergence of this algorithm a minimal residual method with preconditioning is implemented. The non-symmetric linear systems are solved using a Conjugate

Gradient Square method with Incomplete LU preconditioning. For the space discretization of the convective term at high Reynolds number. we use either a stream line upwinding technique or a modified finite volume / finite element approximation with discontinuous flux splitting [2]. The periodic boundary conditions on the velocity $\vec{u}$ are imposed explicitly with the first order scheme or implicitly with the second order scheme by imposing at the entrance the velocity profile $\vec{u}$ calculated at the exit. For the pressure, we impose a reference value in the outlet section.

3 Numerical precision

To check the numerical precision of our simulations, we have used two triangular meshes: a regular coarse mesh with 1602 nodes and a locally refined mesh with 4434 nodes. From the plots of the amplitude of the vertical ve-

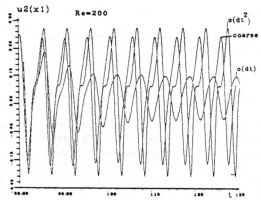

Figure 1: vertical velocity u_2 at $x_1 = 2d$ with different schemes

locity u_2 at the point $x_1 = 2d$ behind the cylinder (figure 1), we notice that the first order scheme is too much diffusive and is unable to predict correctly the amplitude. The difference in amplitude between the first order and the second order scheme reaches 70%, and is higher than the difference between the coarse mesh and the fine mesh results (14%). The discrepancies on global quantities, as the time average skin friction coefficient C_f, are lower

(20% between the first order and the second order scheme, 12% between the coarse mesh and the fine mesh). For this problem the error in time is then much higher than the error in space, and we need second order accuracy in time to predict precisely such unsteady flows. For a Reynolds number higher than 600, we have used an upwind formulation and the modified discontinuous flux splitting yields the better results (i.e the less diffusive). For the non-stationary test cases, we also found that the time-integration must be carried out during a long time (typically 100 sec. or more) leading to extensive computations. The chosen time step varies from 0.1 to 0.05 sec. (50 to 100 iterations per period) and the number of iterations ranges from 1000 to 3000, which corresponds to about 20 times the average travel time of the domain. The code runs on a Sun 4/360 workstation and on a parallel vector computer Alliant FX80 with 8 processors and 64 Mbytes of core memory. The 2D code is not optimized for a vector computer and the mean cpu time per time step (Δt=0.05, N=4434) is 146 sec. on the Alliant with one ACE (10 Mflops peak) and 80 sec. on the Sun (3 Mflops). The 3D code is parallelized and vectorized, and the mean cpu time per time step (Δt=0.1, N=33642) is 370 sec. on the Alliant with 4 ACE (40 Mflops peak).

4   2D calculations

The mean pressure drop C_p and the mean friction drag C_f obtained with the fine mesh are listed in the table below.

Re	0	100	200	600	1200
C_p	6.53	0.098	0.073	0.042	0.044
C_f	4.92	0.969	0.726	0.307	0.126

Re=0: The Stokes flow is obtained with a low Reynolds number (Re= 10^{-2}). The flow is stationary with no attached eddies behind the cylinder. The flow around the cylinder is symmetrical in comparison with the y axis, and the stagnation point is located at 7 degree below the horizontal line, leading to a lift force on the cylinder.

Re=100: Starting from the Stokes flow, we observe a damped oscillatory movement with very low vertical velocity (figure 2). The frequency of these damped periodic oscillations is equal to the fundamental frequency f_1 of the problem. This fundamental frequency is about 0.2 hz ($f_1 = 0.189$) , which leads to a wave length L_1 equal to one half of the distance between the cylinder: $L_1 = V_m/f_1$ where $V_m = \frac{2}{3}V$ is the mean velocity. At the stationary state, the flow has one attached eddy behind the cylinder.

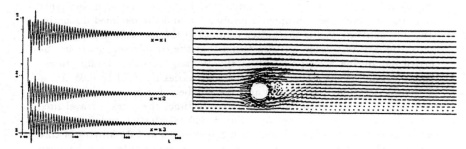

Figure 2: velocity u_2 versus time at x_1 and velocity field for Re=100

Re=200: The flow is perfectly periodic with a frequency equal to f_1 (figure 3). We observed the periodic vortex shedding behind the cylinder, which maintains the global oscillatory movement. From the study of Karniadakis et al. [3] on the stability of this eddy-promoter flow, this steady periodic flow corresponds to nonlinear saturation of unstable Tollmien-Schlichting-like travelling waves, with two wavelengths per periodicity length L and a frequency $f = 0.185$ which is equal to the frequency $f_1 = 0.189$ to within 2%. Let us defined a cylinder Reynolds number R_c, based on the diameter of the cylinder and on the mean velocity near the cylinder $V_c = \frac{3}{4}V$: $R_c = \frac{V_c * d}{\nu} = 0.3Re$, then the Strouhal number St: $St = \frac{f_1 * d}{V_c} = 0.53 f_1$, is equal to 0.1. The first transition (Hopf bifurcation) occurs between Re=100 (damped periodic) and Re=200 (periodic). Using the Landau theory [4] for the growth of the amplitude of the perturbation near the threshold, we found a critical Reynolds number of 114.

Re=600: the flow is unsteady with large vortices which are created periodically on the lower wall behind the cylinder and are convected by the mean flow (figure 4). Vortex shedding still occurs at the frequency f_1, but now the flow is not mono-periodic but quasi-periodic. Looking at the power spectrum of the vertical velocity u_2, we found that two basic frequencies f_1 and f_2 are present together with their harmonics and integer combinations. The second frequency f_2 is equal to 0.34hz and the associated Strouhal num-

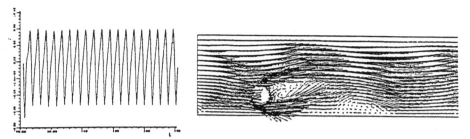

Figure 3: velocity u_2 versus time at x_1 and velocity field for Re=200

ber is $St = 0.18$, closed to the typical Strouhal number of the Karman vortex
street $St = 0.2$. This second frequency is then related to the frequency of
the vortex shedding behind an isolated cylinder.

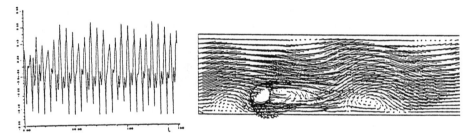

Figure 4: velocity u_2 versus time at x_1 and velocity field for Re=600

Re=1200: the flow is unsteady, with large vortices on the lower wall and
also on the upper wall. The power spectrum of the vertical velocity is more
complicated, but the fundamental frequency f_1 is still present.

5   3D calculations

For the 3D simulations, we use a mesh with 33642 velocity nodes. The
chosen width is 2, and we apply Neuman boundary conditions on the lateral

planes. The mean pressure drop C_p, the mean friction drag C_f and the mean frequency f_1 are listed in the table below.

Re	100	200	600
C_p	0.0998	0.070	0.0455
C_f	0.990	0.602	0.235
f_1 (hz)	—	0.20	0.18

For Re=100 and 200, the values are close to the values found in the 2D simulations with a similar grid size (coarse mesh) indicating that the flow remains globally two dimensional. For Re=100 the flow is stationary and 2D and for Re=200 the flow is periodic with a fundamental frequency f_1 of 0.2 hz. However, for Re=600 the flow becomes very unsteady with 3D instabilities appearing near the walls (figure 5).

6 Conclusion

The numerical simulation of unsteady flows is a severe test case to check the precision of a numerical code. The accuracy of the time integration is very important to predict the instabilities, which develop in such flows. From a physical point of view, the study of unsteady flows with numerical simulations is a promise of a better understanding of transition to the turbulence. While the 2D simulations allow the analysis of the first transition, further 3D simulations are needed to characterize the 3D instabilities in this eddy promoter flow.

References

[1] M. BUFFAT. to appear in the IJNMF, 1991.

[2] M. BUFFAT. thèse, ECL, 1991.

[3] G. KARNIADAKIS, B. MIKIC, A. PATERA. JFM,1988, vol 2, 365-391.

[4] L.D. LANDAU, E.M. LIFSHITZ.Pergamon press, 1987.

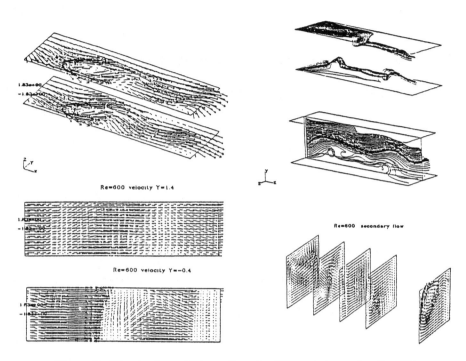

Figure 5: 3D velocity and streamlines in different planes for Re=600

N3S (Release 2.1) 2D-Computations, Test T2.

L. JANVIER, J.P. CHABARD, T. BIDOT

EDF - Laboratoire National d'Hydraulique, 6 quai Watier, 78400 Chatou.

1 NUMERICAL METHOD

1.1 Governing Equations:

The N3S finite element code (release 2.1) was used for both laminar and turbulent 2-D computations. N3S solves the Navier-Stokes equations for an incompressible fluid.

$$\frac{\partial \vec{U}}{\partial t} + \vec{U} . \nabla \vec{U} = -\frac{1}{\rho} \nabla P + \nabla . \left\{ (v + v_t) \nabla \vec{U} \right\}$$

$$\nabla . \vec{U} = 0 \tag{1}$$

For the turbulent part, a standard k-ε model was used. It consist in two advection-diffusion equations for the kinetic energy k and its dissipation rate ε, the eddy viscosity is related to k and ε.

$$\frac{\partial k}{\partial t} + \vec{U} . \nabla k = \nabla . \left((v + \frac{v_t}{\sigma_k}) \nabla k \right) + P - \varepsilon \tag{2}$$

$$\frac{\partial \varepsilon}{\partial t} + \vec{U} . \nabla \varepsilon = \nabla . \left((v + \frac{v_t}{\sigma_\varepsilon}) \nabla \varepsilon \right) + \frac{\varepsilon}{k} (C_{\varepsilon 1} P - C_{\varepsilon 2} \varepsilon) \tag{3}$$

with the production term due to shear flow and the the eddy viscosity expressed by :

$$P = v_t \left(\frac{\partial u_i}{\partial x_k} + \frac{\partial u_k}{\partial x_i} \right) \frac{\partial u_i}{\partial x_k} \qquad v_t = C_\mu \frac{k^2}{\varepsilon} \tag{4, 5}$$

The set of constants used is :

σ_ε	σ_κ	$C_{\varepsilon 1}$	$C_{\varepsilon 2}$	C_μ
1.3	1.0	1.44	1.92	0.09

264

1.2 Wall Boundary Conditions

No-slip condition for the velocity for laminar flows laminar ; if turbulent, the finite element boundary mesh is assumed to be at a distance $y \neq 0$ from the wall (Figure 1). We assume that a logarithmic profile is valid at the distance y of the wall. A Reichardt law has been chosen instead of the logarithmic law, because it gives a continuous transition from the viscous sublayer to the logarithmic region [Chabard(88)]. Thus it can be used for y^+ within the range [5,200]. It degenerates to linear law for $y^+ < 10$ and to the logarithmic law for $y^+ > 30$. Let the velocity be known from the previous time step, then using y, (6) gives the friction velocity u_*. Then we can check if the value of y^+ is admissible. If not, y is modified using the corresponding upper or lower limit, and a new value of u_* is computed.

(Figure 1)

$$\left\{ \begin{array}{l} y^+ = \dfrac{\rho \, y \, u_*}{\mu} \\[2mm] \dfrac{|v|}{u_*} = 2.5 \ln(1+0.4y^+) + 7.8 \left(1 - \exp(-\tfrac{y^+}{11}) - \tfrac{y^+}{11} \exp(-0.33 y^+)\right) \end{array} \right.$$

(6)

The boundary conditions for the velocity are given by the impermeability condition and the tangential friction stress (7). If n denotes the unit outer normal vector to the wall, τ the tangential direction the friction stress is assumed to act along the τ direction. The boundary conditions for k and ϵ (8) express the equilibrium between production and dissipation at the wall (K is the Karman constant).

$$v \cdot n = 0 \qquad \text{and} \qquad \sigma \, \tau = -\rho \, u_*^2 \, \tau \tag{7}$$

$$k = \frac{u_*^2}{\sqrt{C_\mu}} \qquad \text{and} \qquad \epsilon = \frac{u_*^3}{K \, y} \tag{8}$$

1.3 Method of solution:

The resolution of the momentum equations is split in three steps :

/1/ Advection of all variables (9), C is k, ϵ or any component of the velocity, $\overline{C}$ is then the result of the advection step. If Σ_c is the right hand side term associated with the variable C, $\Sigma_c = S_{cc} + S_{cd}$, a part of it ,S_{cc}, is taken into account while we solve the

advection stage (for k and ε) and S_{cd} in the second step (for temperature T, if necessary)
This step is performed using a characteristics method.
/2/ Diffusion (except for velocity components) (10), each linear system is solved by a
conjugate gradient method preconditioned by an incomplete Choleski decomposition.
/3/ Stokes problem (velocity field and pressure) (11) : $\overline{C} = \overline{U}_i$ after the advection step.

/1/
$$\frac{\overline{C} - C^n}{\Delta t} + \vec{U} \cdot \nabla C = S_{cc} \tag{9}$$

/2/
$$\frac{C^{n+1} - \overline{C}}{\Delta t} - \nabla \cdot \left\{ \kappa_c \, \nabla C \right\} = S_{cd} \tag{10}$$

/3/
$$\begin{cases} \dfrac{\vec{U}^{\,n+1} - \vec{U}}{\Delta t} + \dfrac{1}{\rho} \nabla \, P^{n+1} + \nabla \cdot \left\{ (\nu + \nu_t^n) \, \nabla \overrightarrow{U^{n+1}} \right\} = 0 \\ \nabla \cdot \vec{U}^{n+1} = 0 \end{cases} \tag{11}$$

The wall boundary conditions may couple the velocity components. This coupling will
appear in the velocity matrix through the terms involving the tangential shape functions on
the wall. Thus the matricial form of the Stokes problem can be written as follows (let W be
the tangential component of the velocity at the wall boundary) :

$$\begin{bmatrix} A_{uu} & 0 & A_{uw}^t & B_u^t \\ 0 & A_{vv} & A_{vw}^t & B_v^t \\ A_{uw} & A_{vw} & A_{ww} & B_w^t \\ B_u & B_v & B_w & 0 \end{bmatrix} \begin{bmatrix} U \\ V \\ W \\ P \end{bmatrix} = \begin{bmatrix} S_u \\ S_v \\ S_w \\ 0 \end{bmatrix} \tag{12}$$

To solve this problem an Uzawa method is employed. We use a preconditionning matrix C
defined as $C = (\alpha \, (C_{lap})^{-1} - \nu \, M_p{}^{-1})^{-1}$). Several choices are available for the
operator C. For the first part of C : $C_{lap} = C_{incomp} = \Delta$, or,
$C_{lap} = C_{comp} = B \, (DiagM_v)^{-1} \, B^t$, where M_v is the velocity mass matrix, B the
divergence matrix, for the second one the pressure mass matrix may not be taken into
account.

At each iteration of the algorithm a linear system on the velocity has to be solved. It is an
implicit system on the velocity components because of coupling terms. It is solved by the
ICCG method and a symmetric compact storage is used for the matrix.The Uzawa
algorithm iterates directly on the divergence of the velocity field. Thus the residual
divergence can be rendered as small as desired. (Of course the smaller the criterion, the
greater the number of iterations.)

2 MESH

The element which has been used is the P1-isoP2 triangle.Two meshes have to be considered: the standard one for the pressure and a refined one, for the velocity's components (figure 2), in each triangle, 4 sub-elements are obtained:

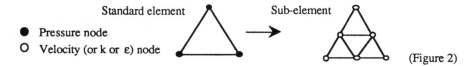

● Pressure node
O Velocity (or k or ε) node

(Figure 2)

An unstructured mesh was generated by SUPERTAB™. Identical distributions of nodes at lines x=0m and x=20/3m on the one hand, and at lines x=1.5m and x= 20/3+1.5 m on the other hand. The right boundary is at the distance 1.5m from the boundary considered in the proposed case and so, it includes half a part of the next cylinder. Doing so, we can impose the periodic conditions (figure 3). The mesh contains 9647 velocity nodes (9189 inner nodes, 388 boundary nodes, 70 nodes on the inlet and outlet), 2471 pressure nodes (40 on the cylinder), 4705 triangles. The size of the element near the wall of the cylinder is 0.0314 m, the maximum size of any element is nearly 0.30 m

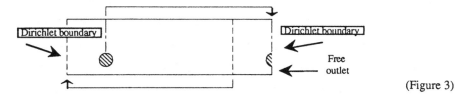

(Figure 3)

3 BOUNDARY CONDITIONS

3.1 Periodic Boundary Conditions

At the first time step given values (Dirichlet) are used at the inlet , and at the outlet (above the cylinder, below the flow is let free), then we impose the peridiodic conditions as follows: the inside flow computed at x= 20/3 is carried back at the inlet (x= 0), the inside flow computed at x= 1.5   is carried forward at the outlet x=20 / 3 + 1.5 .The same procedure is used for k and ε for the last value of the Re number.

3.2 Wall boundary conditions (channel and cylinders):

Laminar case : both the components of the velocity are set to 0 ; turbulent case : logarithmic profile is assumed for the tangential velocity (see 1.2).

4 FLOW CONDITIONS

The flow conditions (increasing Re) are imposed by diminishing the values of the viscosity. Let V be the mean velocity at the inlet : $V= 1.ms^{-1}$, a parabolic profil is first imposed with Vmax=1.5 ms^{-1} ($V(y)= 3y - 3/2\ y^2$), then Re = 3/2 ((hV)/ν) or Re = 3/4 (Q / ν) and, if Q is the flow rate, $Q = 2.m^2s^{-1}$. While setting the periodic conditions at the inlet, we check the flowrate, in fact a corrective coefficient 0.999 to 1.002 was used. For the last case we set the value of k = 2,5 10^{-3} and ε = 3,4 10^{-3} in the whole channel.

5 PERFORMANCE OF THE METHOD

The computer used was a CRAY YMP 464 (System : UNICOS 5.0 ; Fortran compiler : cft77 version 3.0.2.10). The "CPU " is given for the same required precision for the divergence (less than 10^{-8}).

	Reynolds number	Time step (s)	CPU per time step	Number of time steps
A	0 (0.01)	0.1s	33.0	33
B	100	0.1s	32.5	180
C	200	0.1s	24.4	180
D	600	0.1s	16.4	180
E	1200	0.1s	12.4	180
F	10 000 (k-ε)	0.025s	15.2	653

Memory (Mwords), add about 0.5 Mw for the load module :
Central Memory   laminar case   0.46   turbulent case   0.77
Disk Space   laminar case   0.33   turbulent case   0.46

6 RESULTS

	Reynolds number	period (if any, the last three are used)	Strouhal fd/V	head loss $h\ \Delta P / (\rho\ V^2\ L)$.	viscous friction coefficient $\overline{Cf}$
A	0 (0.01)	none	---------	648.9	- 4735.0
B	100	none	---------	0.0945	- 0.859
C	200	2.31s	0.17	0.0555	- 0.469
D	600	2.36s	0.17	0.0321	- 0.214
E	1200	2.40s	0.17	0.0252	- 0.131
F	10 000	2.04s	0.196	0.0195	-0.038

7 COMMENTS.

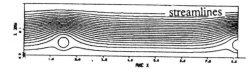

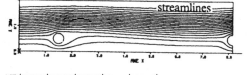

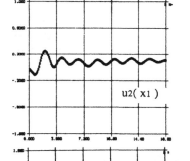

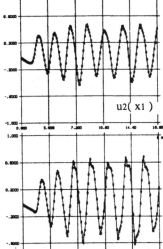

Re = 0 :   Figure 4
The computed flow shows no variation
at all after a few time steps.
Although we can dropp the advection
term, this was approximated by setting
Re = 0.01 in order to have the same
simulation conditions than the ULYSSE
computations [Laurence-Goutorbe-
Sebag, this book].

Re = 100 :   Figure 5
No oscillations appear.

Re = 200 :   Figures 6 and 7
Small oscillations appear, decreasing in
time.

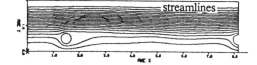

Re = 600 :   Figures 8 and 9
Vortex shedding is visible on the
streamlines. Magnitude of oscillations
is about $u_2 = 40\%$.

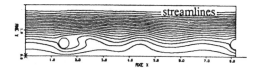

Re = 1200 :   Figure 10 and 11
Same as Re = 600, but velocity
oscillations become dissymetrical .

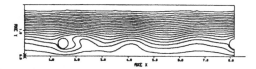

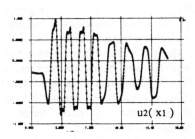

u2(x1)

Re = 10 000   Figure 12

Figure 13

Figure 14

Turbulence modelling k-ε. We don't reach any stabilized periodic flow within the space of time computed. See [Laurence-Goutorbe-Sebag, this book].to compare with the simular computations made with the ULYSSE code and a R_{ij} - ε model . The kinetic energy produced at the wall is higher, which explains the higher level reached by the viscous friction coefficient for that case .

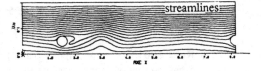

streamlines

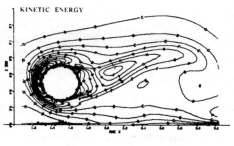

KINETIC ENERGY

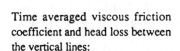

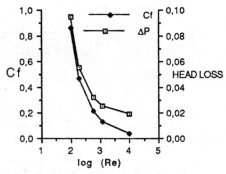

Time averaged viscous friction coefficient and head loss between the vertical lines:

          1/ X=0

and   2/ X= 20/3

versus Re

(Figure 15)

8 REFERENCES

[Chabard(88)]
J.P. Chabard: Projet N3S Manuel théorique - version 2.0, Rapport EDF HE-41/88.09

ULYSSE (Release 4.0) 2D-Computations, Test T2.

D. LAURENCE, T. GOUTORBE, S. SEBAG

EDF - Laboratoire National d'Hydraulique,
6 quai Watier, BP 49 78401, Chatou.Cedex, FRANCE

1 NUMERICAL METHOD

1.1 Governing Equations:
A finite volume curvilinear 2-D code (ULYSSE-4.0) was used for both laminar and turbulent computations. For the turbulent part, two models are available: standard k-ε model using the Eddy viscosity concept, and the Reynolds Stress Model (RSM) [1]:
- k-ε model :

$$\frac{\partial k}{\partial t} + \overline{U}_i \frac{\partial k}{\partial x_i} = \frac{\partial}{\partial x_i}\left(\nu_t \frac{\partial k}{\partial x_i}\right) + P - \varepsilon$$

$$\frac{\partial \varepsilon}{\partial t} + \overline{U}_i \frac{\partial \varepsilon}{\partial x_i} = \frac{\partial}{\partial x_i}\left(\frac{\nu_t}{\sigma_\varepsilon} \frac{\partial \varepsilon}{\partial x_i}\right) + C_{\varepsilon 1} \frac{\varepsilon}{k} P - C_{\varepsilon 2} \frac{\varepsilon^2}{k}$$

with: $P = \nu_t \left(\frac{\partial \overline{U}_i}{\partial x_k} + \frac{\partial \overline{U}_k}{\partial x_i}\right) \frac{\partial \overline{U}_i}{\partial x_k}$

The set of constants used is :

σ_ε	σ_K	$C_{\varepsilon 1}$	$C_{\varepsilon 2}$	C_μ
1.3	1.0	1.44	1.92	0.09

- Reynolds Stress Model :

$$\frac{\partial \overline{u_i u_j}}{\partial t} + \overline{U}_k \frac{\partial \overline{u_i u_j}}{\partial x_k} = P_{ij} + \Phi_{ij,1} + \Phi_{ij,2} + d_{ij} - \frac{2}{3}\varepsilon \delta_{ij}$$

$$\frac{\partial \varepsilon}{\partial t} + \overline{U}_k \frac{\partial \varepsilon}{\partial x_k} = C_{\varepsilon 1} \frac{\varepsilon}{k} P - C_{\varepsilon 2} \frac{\varepsilon^2}{k} + C_\varepsilon \frac{\partial}{\partial x_k}\left(\frac{k}{\varepsilon} \overline{u_k u_l} \frac{\partial \varepsilon}{\partial x_l}\right)$$

In this expression, except the production term P :

$$P_{ij} = -\left(\overline{u_i u_k}\frac{\partial \overline{U}_j}{\partial x_k} + \overline{u_j u_k}\frac{\partial \overline{U}_i}{\partial x_k}\right)$$

all R.H.S. terms are modelized according to Launder, Reece, Rodi [1] :

$$d_{ij} = C_s \frac{\partial}{\partial x_k}\left(\frac{k}{\varepsilon} \overline{u_k u_l} \frac{\partial \overline{u_i u_j}}{\partial x_l}\right)$$

$$\Phi_{ij,1} = -C_1 \frac{\varepsilon}{k}\left(\overline{u_i u_j} - \frac{2}{3}k \delta_{ij}\right)$$

$$\Phi_{ij,2} = -C_2 \left(P_{ij} - \frac{2}{3}P \delta_{ij}\right)$$

$C_{\varepsilon 1}$, $C_{\varepsilon 2}$, C_ε, C_1, C_2, Cs are model constants.

σ_ε	$C_{\varepsilon 1}$	$C_{\varepsilon 2}$	c_ε	c_1	c_2	c_s
1.3	1.44	1.92	0.15	1.8	0.6	0.22

1.1 Space discretization:

A 2D structured non-orthogonal body-fitted grid is generated and serves for storage of velocity vectors (both components at same location). Since cells are liable to be very distorted and irregular, finite volume approach is prefered to finite differences. A simple way to define derivatives (divergence: D(u) , gradient: G(p)) is to write :

$$G(u) = \int_M \text{div } \vec{u} \, dv = \int_{\partial M} \vec{u} \cdot \vec{n} \, .ds \quad , \text{ et } G(p) = \int_M \overrightarrow{\text{grad}} \, p \, . \, dv = \int_{\partial M} p \, . \, \vec{n} \, .ds$$

Where M is a cell defined by 4 velocity nodes and of border ∂M. Straightforward integration assuming linear variations between nodes yields, e. g. for divergence of velocity integrated over 2D controle volume V:

$$D(u) = \frac{1}{V} \int_{\partial M} \vec{u} . \vec{n} \, ds = \frac{1}{V} \int_{\partial M} \vec{u} \wedge \vec{dl} \quad \vec{K}$$

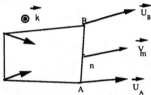

Note that derivatives are defined as averages over the 4 node cells and thus attributed to the corresponding node of a staggered grid (middles of primary-grid cells). Examining the equations we notice that there is one order of derivation between: velocity and pressure or Reynolds stresses, Reynolds stresses and velocity, and between pressure and velocity (when one writes the Poisson equation).

1.3 Solution procedure:

ULYSSE-4 code [2] solves transport and diffusion equations for an arbitrary number of unknowns in addition to the Navier Stokes equations (so RSTM was conveniently implemented). It is a time marching code and the resolution of the unsteady equations is split in two (three for velocity) steps:

a) Advection of any variable f by a Lagrangian method using third order interpolation.

$$\frac{\tilde{f} - f^n}{\Delta t} = - U \cdot \vec{\nabla} f \qquad \Rightarrow \qquad \tilde{f}(X) = f^n(X - \int_0^{\Delta t} U^n \, dt)$$

b) Diffusion and source terms for all the variables :

For any quantity f (i.e., k, $\overline{u_i u_j}$, ε or U_i), we introduce the increment δf :

$f^{n+1} = \tilde{f} + \delta f$.   Next the balance equations are expressed at an intermediate time t^n < $t^{n+\alpha}$ < t^{n+1} and linearized with respect to δf .

With the RSM [3] this leads to a linear system <u>coupling all variables</u> : velocity components, Reynolds Stress components and dissipation, i. e.7 unknowns.

For the components of **U**, the pressure gradient (using the last known pressure as predictor) is included in the r.h.s. . Thus one should read, in $f^{n+1} = \tilde{f} + \delta f$, $\tilde{\tilde{f}}$ instead of f^{n+1} since a pressure correction is yet to be applied before obtaining value at time t^{n+1}.

c) **Pressure -correction** of the velocity field by the pressure increment is finaly applied to ensure mass conservation.

$$\frac{\overrightarrow{u^{n+1}} - \overrightarrow{\tilde{\tilde{u}}}}{\Delta t} = -\frac{1}{\rho} . \overrightarrow{\nabla}(P^{n+1} - P^n)$$

P^{n+1} is solved through a Poisson equation obtained by taking the divergence of the discretized momentum equation. At this level, numerical action is taken to prevent numerical oscillations.

All linear systems are solved by a three-level conjugate residual algorithm. For step 2 a diagonal preconditionning is applied to ensure that residuals corresponding to velocities, stresses and dissipation (summed up while defining steepest descent parameters) are of similar order of magnitude .

1.4 Boundary Conditions:
Laminar : No-slip condition for the velocity at the wall.

Turbulent :
The law of the wall is assumed, i.e. the velocity profile at the previous time step is introduced in the logarithmic law to determine the friction velocity u_* , see JANVIER et al.

in this vol.. Also the same procedure is used to impose pseudo-periodic boundary conditions in the flow direction (inner mesh line copied onto boundary mesh line using an extended overlaping domain).

For the turbulent kinetic energy k, and its dissipation rate ε the local equilibrium assumption yields :

$$k = \frac{u_*^2}{\sqrt{C_\mu}} \qquad \varepsilon = \frac{u_*^3}{K\,y}$$

($K = 0.42$, $C_\mu = 0.09$, y : distance to wall)

For RSM, as previously the law of the wall is assumed for the velocity. Wall boundary conditions for the stresses are standard plane channel flow values, i.e. :

$$\overline{u^2} = 5.u_*^2 \; ; \; \overline{v^2} = u_*^2 \; ; \; \overline{w^2} = 2.u_*^2 \; ; \quad \overline{uv} = -u_*^2$$

These values are valid in a local referential and a rotation is applied when necessary.
For the dissipation, same expression as above (k- ε model) is used.

2.COMPUTER:
CRAY XMP 216, System: UNICOS 5.0, Fortan compiler : CFT77 version 3.0.2.10.

3.RESULTS:
Number of nodes : 8060 (= 124 x 65), Memory : 2 Mwords

Reynolds num.	CPU / time step	Num. time steps	h ΔP / (ρ V² L)	average Cdf
0. (0.01)	3.36s	846	771.	7.
100	3.37s	1484	.096	.9
200	3.33s	1756	.057	.5
600	3.30s	1163	.032	.25
1200	3.27s	1173	.027	.15
10 000 (k-ε)	4.03s	1200	.0245	.06
10 000 (RSM)	4.00s	2000	.0234	.05

4.COMMENTS

Re = 0 : usualy means dropping the advection term in the Navier Stokes equation. Here, this was approximated by setting Re = 0.01 , to test the standard version of the code.

Re = 200 : Small oscillations appear, decreasing in time down to u_2 = 5%. Strouhal number is St = 0.20.

Re = 600 : Vortex shedding is now clearly visible on the streamlines. Magnitude of oscillations is about u_2 = 40%. St= 0,19 . Blobs of vorticity also appear along the lower wall.

Re = 1200 : Same as Re = 600, only velocity oscillations become very dissymetrical (positive values of u_2 are much larger than negative values).

Re = 10 000 requires turbulence modelling. With the standard k-ε model and wall functions, the size of the mesh cell near the wall should be greater than y^+ = 10, but capturing features of small vortices require on the other hand a fine mesh near the cylinder. In the present computations y^+ ranges from 1. to 10. (use of the low Re version of the turbulence models would require refining the mesh by a decade, i.e. y^+ = 0.1)

In this last computation with the R_{ij} -ε model , we experienced as on other test cases, that the computation cannot converge if non-physical initial and inlet values are given , i.e. Reynolds Stress profiles must be in equilibrium with the given inlet velocity profiles. Hence the first 10 seconds of the computation consists in replacing the periodicity condition by standard "turbulent channel flow" profiles for all turbulent quantities at entrance. Periodicity is switched on only when this "solitary cylinder" computation is converged. This change is visible on u_2 profiles.

Compared to the k-ε computation, the R_{ij} -ε model yields a higher amplitude of oscillations for u_2 and the streamlines are perturbed on a longer scale downstream. Levels of turbulence are smaller allowing the mean flow to oscillate more energetically. Origin of the different behaviour is the lower production of energy near the stagnation point with the $R_{ij} - \varepsilon$ model.

References :

1.   Launder B.E., Reece G.J. and Rodi W. : "Progress in the development of a Reynolds stress closure.", J.F.M. vol.68-3 (1975)

2.   Laurence D. "Code ULYSSE : Note de Principe", EDF report HE41/89-32B

3   Sebag S., Laurence D. in Proc. "Int. Symp. Eng. Turbulence Modelling & Measurements", W. Rodi & E. N. Ganic , ELSEVIER 1990.

4   Sebag S., "Modèle de Turbulence au second ordre en domaines courbes" Thèse de Docteur-Ingénieur Ecole Centrale Lyon (1991)

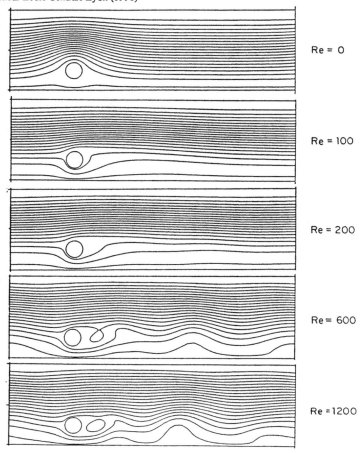

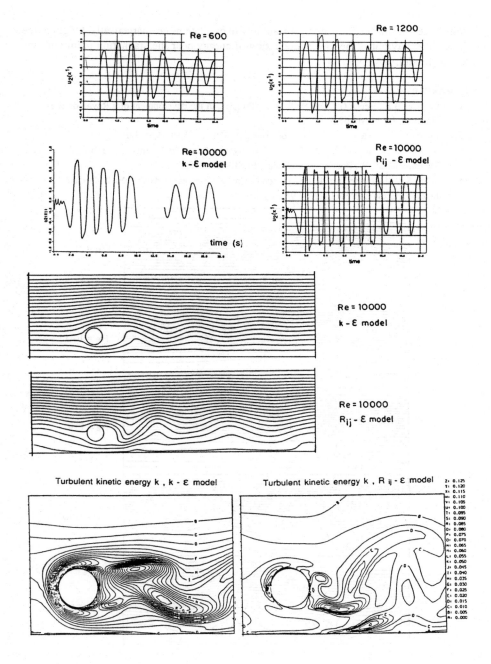

Turbulent kinetic energy k , k - ε model   Turbulent kinetic energy k , R ij - ε model

NUMERICAL SIMULATION OF A PERIODIC ARRAY OF CYLINDERS BETWEEN TWO PARALLEL WALLS[+]

J. Liou and T.E. Tezduyar

Department of Aerospace Engineering and Mechanics and Minnesota Supercomputer Institute
University of Minnesota, Minneapolis, Minnesota 55415

SUMMARY

A finite element method for two-dimensional, spatially periodic, incompressible flows governed by the vorticity-stream function formulation is used to solve this problem. We tested three cases: Stokes flow, flow with Reynolds number 100 and flow with Reynolds number 200. For the first two cases we obtained the steady-state solutions, while for the last case we found a temporally periodic solution.

THE GOVERNING EQUATIONS AND BOUNDARY CONDITIONS

Let Ω denote the two-dimensional computational domain and $(0,T)$ denote the time interval with x and t representing the coordinates associated with Ω and $(0,T)$. The vorticity-stream function formulation of the incompressible Navier-Stokes equations consists of a convection-diffusion equation for the vorticity and a Poisson equation for the stream function:

$$\partial\omega/\partial t + \mathbf{u} \cdot \nabla \omega - \nu \nabla^2 \omega = 0 \qquad \text{on } \Omega \times (0,T), \qquad (1)$$

$$\nabla^2 \psi + \omega = 0 \qquad \text{on } \Omega \times (0,T), \qquad (2)$$

At the boundaries Γ_3, Γ_4 and Γ_5 (see Figure 1), we specify no-slip conditions:

$$u_n = \mathbf{n} \cdot \mathbf{u} = -\partial\psi/\partial\tau = 0, \qquad (3)$$

$$u_\tau = \tau \cdot \mathbf{u} = \partial\psi/\partial n = 0, \qquad (4)$$

where $\mathbf{n}$ and τ are the unit normal and tangential vectors.

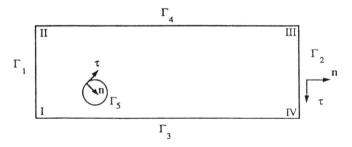

Figure 1. The problem configuration

[+] This research was sponsored by NASA-Johnson Space Center under contract NAS-9-17892 and by NSF under grant MSM-8796352.

277

From equation (3), it is known that the value of the stream function is invariant along Γ_3, Γ_4 and Γ_5. The value of the stream function at Γ_3 and Γ_4 can be specified (say ψ_3 and ψ_4) based on the flow rate between the two walls. However, the value of the stream function at the internal boundary, Γ_5, is unknown and has to be determined as part of the solution. Therefore, an additional equation is needed to determine this unknown value. For viscous flows, assuming that u_τ is invariant along the internal boundary, it can be shown that

$$\partial u_\tau/\partial t + (1/\rho)\, \partial p/\partial \tau + \nu\, \partial \omega/\partial n = 0 \quad \text{on } \Gamma_5. \tag{5}$$

Because $u_\tau = 0$, by integrating (5) from a reference point τ_0 to the point of interest τ_1, we obtain

$$(p_1 - p_0)/\rho = -\int_0^1 (\nu\, \partial \omega /\partial n)\ d\tau \tag{6}$$

Based on the single-valuedness of pressure, an additional equation, to determine the unknown stream function at Γ_5, is derived as

$$\int_{\Gamma_5} (\nu\, \partial \omega /\partial n)\ d\tau\ = 0 . \tag{7}$$

To impose the periodicity in the x_1-direction, along Γ_1 and Γ_2 the following constraints are implemented

$$\omega\,((x_1)_I, x_2, t) = \omega\,((x_1)_{IV}, x_2, t) \qquad \forall\, x_2 \in (\,(x_2)_I , (x_2)_{II}\,),$$
$$\forall\, t \in (0,T)\ , \tag{8}$$
$$\psi\,((x_1)_I, x_2, t) = \psi\,((x_1)_{IV}, x_2, t) \qquad \forall\, x_2 \in (\,(x_2)_I , (x_2)_{II}\,),$$
$$\forall\, t \in (0,T)\ . \tag{9}$$

THE FINITE ELEMENT FORMULATION

Let $\mathcal{E}$ denote the set of elements resulting from the finite element discretization of the computational domain Ω into subdomains Ω^e, $e = 1,2,...,n_{el}$ where n_{el} is the number of elements. Let Γ^e denote the boundary of Ω^e, and Γ_G denote the external boundary where both components of velocity are specified (i.e. $\Gamma_3 \cup \Gamma_4$). We associated to $\mathcal{E}$ the finite dimensional space H^{1h}, which is the piecewise bilinear finite element function space.

The trial function spaces for the vorticity and stream function are defined, respectively, as

$$\widetilde{S}^h = \{\ \omega^h \mid \omega^h \in H^{1h},\ \omega^h((x_1)_I, x_2, t) = \omega^h((x_1)_{IV}, x_2, t)$$
$$\forall\, x_2 \in (\,(x_2)_I , (x_2)_{II}\,)\ \forall\, t \in (0,T)\ \} \tag{10}$$

and

$$S^h = \{ \ \psi^h \mid \psi^h \in H^{1h}, \ \psi^h((x_1)_I, x_2, t) = \psi^h((x_1)_{IV}, x_2, t)$$

$$\forall \ x_2 \in (\ (x_2)_I, (x_2)_{II}\ ) \ \forall \ t \in (0,T),$$

$$\psi^h = \psi_3 \text{ on } \Gamma_3 \times (0,T), \ \psi^h = \psi_4 \text{ on } \Gamma_4 \times (0,T),$$

$$\partial \psi^h / \partial \tau \doteq 0 \text{ on } \Gamma_5 \times (0,T) \ \} . \tag{11}$$

The weighting function spaces are listed below:

$$V^h_* = \{ \ w^h \mid w^h \in H^{1h}, \ w^h \doteq 0 \text{ on } \Gamma_3 \cup \Gamma_4 \cup \Gamma_5,$$

$$w^h((x_1)_I, x_2) = w^h((x_1)_{IV}, x_2) \ \forall \ x_2 \in (\ (x_2)_I, (x_2)_{II}\ ) \}, \tag{12}$$

$$\hat{V}^h_* = \{ \ w^h \mid w^h \in H^{1h}, \ w^h \doteq 0 \text{ on } \Gamma_3 \cup \Gamma_4 \cup \Gamma_5,$$

$$w^h((x_1)_I, x_2) = w^h((x_1)_{IV}, x_2) \ \forall \ x_2 \in (\ (x_2)_I, (x_2)_{II}\ ) \}, \tag{13}$$

$$V^h_G = \{ \ w^h \mid w^h \in H^{1h}, \ w^h \mid_{\Omega^e} = 0 \ \forall \ \Omega^e \notin \mathcal{E}_G \ \}, \tag{14}$$

$$V^h_5 = \{ \ w^h \mid w^h \in H^{1h}, \ w^h \mid_{\Omega^e} = 0 \ \forall \ \Omega^e \notin \mathcal{E}_5 \ \}, \tag{15}$$

$$V^h_{5R} = \{ \ w^h \mid w^h \in V^h_5, \ \frac{\partial w^h}{\partial \tau} \doteq 0 \text{ on } \Gamma_5 \ \}, \tag{16}$$

where $\mathcal{E}_G$ and $\mathcal{E}_5$ are defined as

$$\mathcal{E}_G = \{\Omega^e \mid \Omega^e \in \mathcal{E}, \Gamma^e \cap \Gamma_G \neq \varnothing \} \ , \tag{17}$$

$$\mathcal{E}_5 = \{\Omega^e \mid \Omega^e \in \mathcal{E}, \Gamma^e \cap \Gamma_5 \neq \varnothing \}. \tag{18}$$

For more on the discrete variational formulations associated with (1) and (2), the interested reader can see [1] and [2].

RESULTS

We use the fully-coupled formulation [1] of the vorticity-stream function formulation to solve this problem. An implicit-explicit method [3] is used to save CPU time and memory. Eight iterations (with coefficient matrix reformation and refactorization every four iterations) are taken for each time step . The mesh contains 5,472 elements and 5,652 nodal points. We chose the first four layers of elements around the cylinder and the first layer of elements adjacent to the upper and lower solid walls as implicit elements. We tested three cases: Stokes flow, flow with Reynolds number 100, and flow with Reynolds number 200. It can be proved from Stokes's theorem that for this problem the integral of the vorticity over the entire domain has to remain constant, and that constant is zero.

All computations were performed on a CRAY-2 with 4 CPUs, 4.1 ns clock, 512 Megawords of memory and UNICOS 4.0 operating system. It takes 36 CPU seconds for each time step, and 50% of this time is spent in the subroutine for the direct solution of the linear equation system involved.

Stokes flow

For this case, we drop the convection term in equation (1), and solve for the steady-state solution. Figure 2(a) shows the vorticity contours and streamlines. Two computational domains were patched together for better visualization of the spatial periodicity.

flow with Reynolds number 100

Here a time step of 0.006 was chosen, and the problem was run for 10,000 time steps. Figure 3 shows the time history of the lift, drag, and skin friction coefficients, and the vertical component of the velocity at 2, 3 and 4 diameters downstream of the center of the cylinder, respectively. The flow reaches steady-state after t = 42. The steady-state values of C_L, C_D, and C_f are 0.234, 2.687, and 0.942, respectively, while the steady-state vertical component of the velocity at 2, 3, and 4 diameters downstream are 0.0572, 0.0187 and 0.00142. Figure 2(b) shows the vorticity contours and streamlines at various times. Again, the results are presented over two computational domains.

flow with Reynolds number 200

The same set-up was chosen for this case except for the flow rate between the walls which changes the Reynolds number. Figure 4 shows the time history of the lift, drag, and skin friction coefficients, and the time history of the vertical component of the velocity at 2, 3, and 4 diameters downstream of the center of the cylinder. Vortex shedding can be observed from these figures. The flow first reaches a temporally periodic pattern and then becomes rather unperiodic. Based on the measurements in the temporally periodic part of the solution, the Strouhal number is 0.194 and the time-averaged values of C_L, C_D, and C_f are −0.121, 1.628, and 0.507, respectively. The time-averaged values of the vertical component of the velocity at 2,3, and 4 cylinders downstream are 0.109, 0.0552 and 0.0183, respectively. Figure 2(c) shows the vorticity contours and streamlines at various times.

REFERENCES

1. T.E. Tezduyar, J. Liou, D.K. Ganjoo, and M. Behr, "Solution Techniques for the Vorticity-Stream Function Formulation of Two-Dimensional Incompressible Flows", to appear in *International Journal for Numerical Methods in Fluids*.

2. T.E. Tezduyar and J. Liou,"Computation of Spatially Periodic Flows Based on the Vorticity-Stream Function Formulation", to appear in *Computer Methods in Applied Mechanics and Engineering*.

3. T.E. Tezduyar and J. Liou," Adaptive Implicit-Explicit Finite Element Algorithms for Fluid Mechanics Problems", *Computer Methods in Applied Mechanics and Engineering*, 78 (1990), pp.15-179.

(a)

(b)

(c)

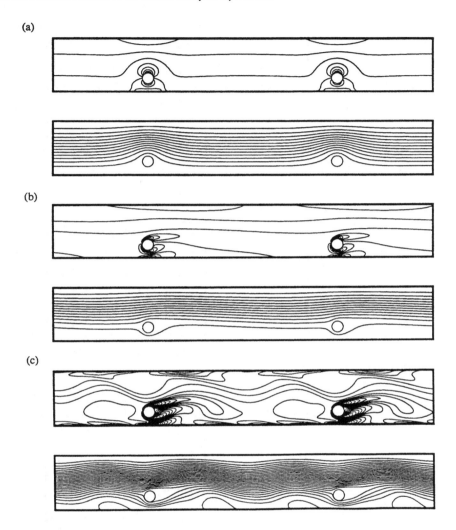

Figure 2.   Vorticity and stream function: (a) Stokes flow, (b) flow with Reynolds number
100 at t = 45.0, (c) flow with Reynolds number 200 at t = 120.0.

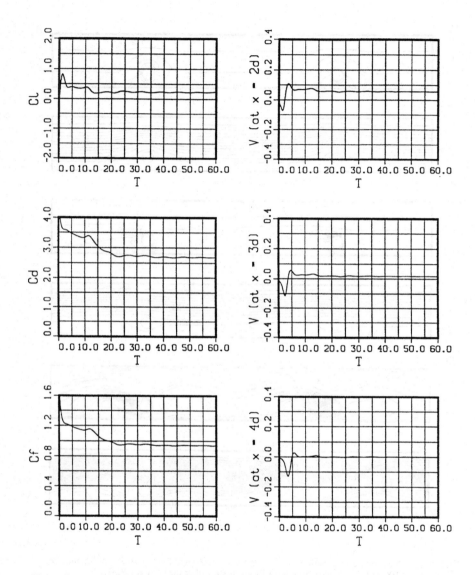

Figure 3.   Flow with Reynolds number 100: The time history of the lift, drag, and skin
friction coefficients, and the vertical component of the velocity at 2, 3, and 4
diameters downstream of the center of the cylinder.

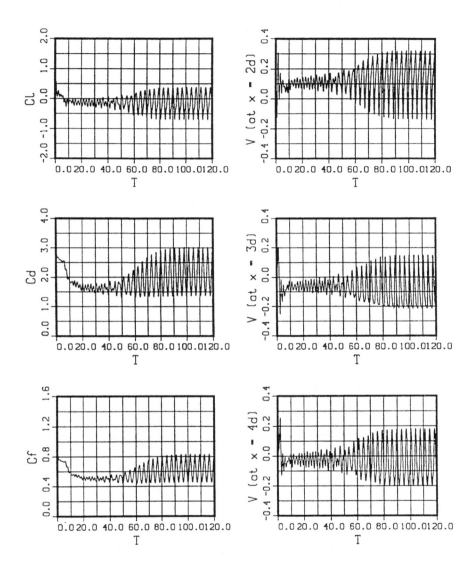

Figure 4.   Flow with Reynolds number 200: The time history of the lift, drag, and skin
            friction coefficients, and the vertical component of the velocity at 2, 3, and 4
            diameters downstream of the center of the cylinder.

ERCOFTAC WORKSHOP
Lausanne March 1990
Numerical Simulation of unsteady flows
Test Case : T2

M. Ravachol
DGT/DEA
AMD – BA
78, quai Marcel Dassault
92214 Saint Cloud, France

1 Numerical Method:

The time integration of the Navier Stokes equations is acheived through the use of an operator splitting method. In the case concerned we use a θ scheme which can be described as follows:

$u^0 = u_0$ given,

then for $n \geq 0$, starting from u^n we solve

$$\begin{cases} \frac{u^{n+\theta}-u^n}{\theta \Delta t} - \alpha\nu\Delta u^{n+\theta} + \nabla p^{n+\theta} = \beta\nu\Delta u^n - (u^n \cdot \nabla)u^n \text{ in } \Omega, \\ \\ \nabla \cdot u^{n+\theta} = 0 \text{ in } \Omega, \\ + \text{ boundary conditions}, \end{cases}$$

then

$$\begin{cases} \frac{u^{n+1-\theta}-u^{n+\theta}}{(1-2\theta)\Delta t} - \beta\nu\Delta u^{n+1-\theta} + (u^{n+1-\theta} \cdot \nabla)u^{n+1-\theta} = \alpha\nu\Delta u^{n+\theta} - \nabla p^{n+\theta} \text{ in } \Omega, \\ \\ + \text{ boundary conditions}, \end{cases}$$

and finally

$$\begin{cases} \frac{u^{n+1}-u^{n+1-\theta}}{\theta \Delta t} - \alpha\nu\Delta u^{n+1} + \nabla p^{n+1} = \beta\nu\Delta u^{n+1-\theta} - (u^{n+1-\theta} \cdot \nabla)u^{n+1-\theta} \text{ in } \Omega, \\ \nabla \cdot u^{n+1} = 0 \text{ in } \Omega, \\ + \text{ boundary conditions.} \end{cases}$$

The coefficients α and β are given by:

$$\alpha = \frac{1-2\theta}{1-\theta} \quad \text{and} \quad \beta = \frac{\theta}{1-\theta}.$$

The parameter θ is chosen in the following manner

$$\theta = \frac{1-\sqrt{2}}{2}.$$

In the above scheme, the periodic boundary conditions are enforced only during the linear steps. For the nonlinear step we use Dirichlet boundary conditions for the inflow – the value imposed being the one found in the previous step – and Newman boundary conditions for the outflow:

To solve the linear step we use a preconditioned conjugate gradient operating on the pressure. The periodicity is then obtained through the flux.

For the nonlinear step we use a nonlinear GMRES algorithm.

For the spatial approximation we use a continuous approximation for both pressure and velocity. Let us call T_h the triangulation used for the pressure and $\tilde{T}_h$ the triangulation used for the velocity. The finite element space used to approximate the pressure is

$$P_h = \{q_h | q_h \in C^0(\bar{\Omega}), \quad q_h|_T \in P_1, \forall T \in T_h\}.$$

The triangulation $\tilde{T}_h$ of Ω is obtained from T_h by joining the midpoints of the edges of $T \in T_h$ as shown in Figure 1

The finite element space used to approximate the velocity is then

$$V_h = \{v_h | v_h \in (C^0(\bar{\Omega}))^2, \quad v_h|_T \in (P_1)^2, \forall T \in \tilde{T}_h\}.$$

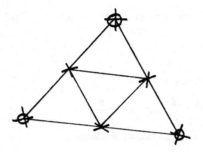

Figure 1: P1 iso P2 finite element (○ pressure nodes × velocity nodes)

2   Performance of the method

In the meshes used we have 1759 elements and 952 points for the pressure and 7063 elements and 3663 points for the velocity.

The time step used is $\Delta t = 0.1$ for Re $= 100$. and $\Delta t = 0.01$ for the other computations. The computations have been performed on an IBM 3090 200-VF used in scalar mode.

The average CPU time per time step is summarized in the following table.

Re	100	200	600	1200
CPU(s)	11.93	13.73	16.82	18.97

3   Numerical Results

In the following computation, for Re $\neq 0$, the Stokes solution has been used as the initial solution for the velocity.

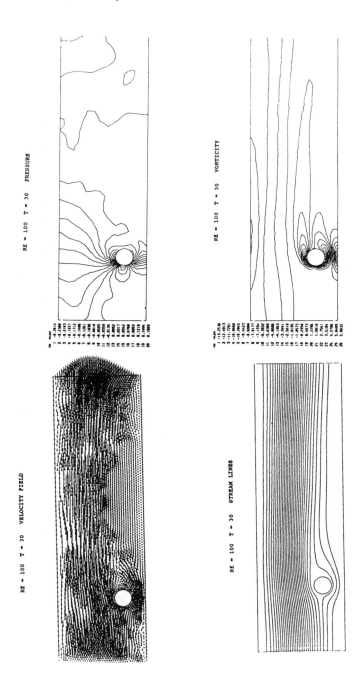

RE = 100   T = 30   PRESSURE

RE = 100   T = 30   VORTICITY

RE = 100   T = 30   VELOCITY FIELD

RE = 100   T = 30   STREAM LINES

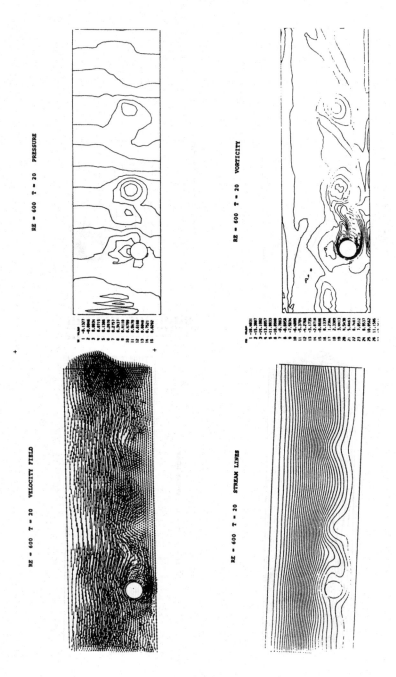

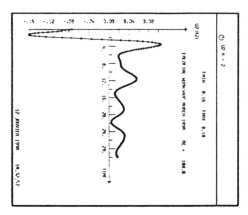
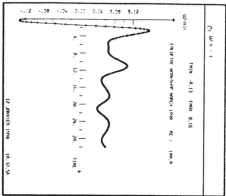

vertical velocity versus time and drag coefficient

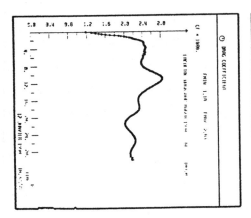
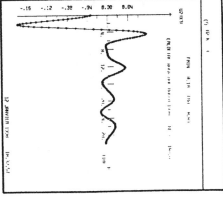

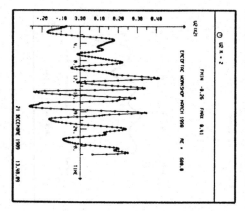
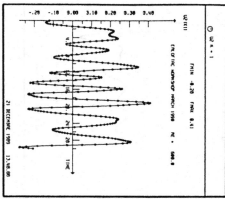

vertical velocity versus time and drag coefficient

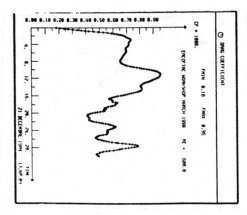
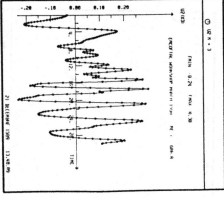

Calculation of Laminar Flow around a Circular Cylinder between Two Flat Plates

W. RODI and J. ZHU

Institute for Hydromechanics, University of Karlsruhe, D-7500 Karlsruhe 1
F.R.Germany

This paper reports on the prediction of the test problem T2 - flow around a circular cylinder between two flat plates. Only the case with steady laminar flow at Re=100 was considered because at present only a version of the code for steady calculations is available. The flow was calculated with a two-dimensional version of the general, 3D vectorized finite-volume method described briefly in the authors' paper for the test problem T1 in these proceedings. Therefore, only those details are given here which are particular to the present problem.

A grid configuration of H-type was adopted which allows one set of grid lines basically to align with the main flow direction, and avoids the troublesome branch-cut problem associated with C- or O-type configurations. The flow field was discretized by a relatively fine grid with 114×80 nodes, as shown in Figure 1(a). The grid was generated by a hybrid algebraic/differential method developed by Zhu et al. (1989). This is a two-step procedure consisting of an intermediate and a final transformation. In the intermediate transformation, the Poisson equations are solved to generate a coarse grid which can be boundary orthogonal. The final transformation involves a bicubic spline fitting on the coarse grid. An important feature of the method is that the grid spacing can be changed quickly and explicitly while retaining the overall smoothness and boundary orthogonality. Figure 1(b) shows the grid in detail near the cylinder. It can be seen that the grid is smooth and resolves quite well the high-shear layer near the body.

The governing equations for two-dimensional steady flows are given in the companion paper for test case T4. Three discretization schemes, i.e., Hybrid (Spalding, 1972), QUICK (Leonard, 1979) and SOUCUP (Zhu and Rodi, 1991) were used to approximate the convection terms of the transport equations. SOUCUP combines three simple discretization schemes, namely second-order upwind, central and (first-order) upwind differencing, with the switch from one scheme to another being automatically controlled by a convection boundedness criterion. It has been found that SOUCUP is capable of yielding low diffusive and always bounded solutions (Zhu and Rodi, 1991). For the present problem, the results obtained with QUICK and SOUCUP are basically the same, while the hybrid scheme produced some difference in the wake region near the cylinder. Only the SOUCUP calculation is presented here.

A fully developed parabolic velocity profile was used as initial field to start the calculation. The zero streamwise gradients of flow velocities were imposed at the outflow boundary. After each iteration, the flow velocities at the inlet were set to those at the outlet to satisfy the periodic boundary condition. The calculation required 597 iterations and 44 seconds of CPU-time on a Siemens VP400-EX computer.

The predicted values of the pressure difference Δp and the friction drag coefficient C_f are:

$$\Delta p = \frac{h(p_1 - p_2)}{\rho V^2 L} = 0.08689$$

$$C_f = \frac{\int_{circle} \tau_w n_2 ds}{0.5\rho V^2 d} = 0.9776$$

and the predicted streamlines are shown in Figure 2.

REFERENCES

Leonard, B. P., 1979, "A stable and accurate convective modelling procedure based on quadratic interpolation", *Comput. Methods Appl. Mech. Eng.*, 19, 59-98.

Spalding, D. B., 1972, "A novel finite-difference formulation for differential expressions involving both first and second derivatives", *Int. J. Numer. Methods Eng.*, 4, 551-559.

Zhu, J., Rodi, W., and Schönung, B., 1989, "A fast method for generating smooth grids," in *Proc. 6th Int. Cof. on Numerical Methods in Laminar and Turbulent Flow*, Swansea, 1639-1649.

Zhu, J., and Rodi, W., 1991, "A low dispersion and bounded convection scheme," to appear in *Comput. Methods. Appl. Mech. Eng.*

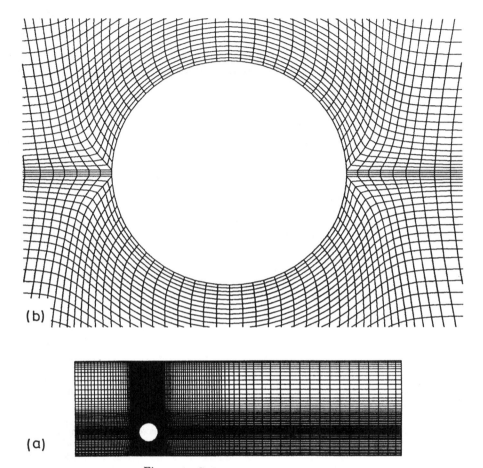

Figure 1. Computational grid

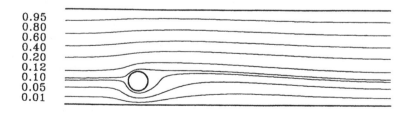

Figure 2. Predicted streamlines

SUMMARY AND CONCLUSIONS FOR TEST CASE T2

O. Pironneau

(Université Paris 6 and INRIA, 78153 Le Chesnay Cedex, France)

Abstract

We compare ten sets of computations for test case T2 (periodic flow around a cylinder between two flat plates). Results at Reynolds Re = 100 are satisfactory except some inaccuracy for the decay rate towards the steady solution. At Re = 200 a scheme second order in time was needed to compute accurately the amplitudes of the oscillations in the wake. At higher Re only 3D computations can predict the drag crisis that was observed experimentally but with todays computers it seem still too early. Turbulence models could not be validated for want of experimental data.

1. INTRODUCTION

Test case T2 was proposed for the following reasons :

i) Industrial interests : The problem arise in the computer industry for the cooling of circuit boards. As show by Karniadakis et al. [1] the cooling is more efficient at low Reynolds number.

ii) External flow : It is in a semi infinite domain and it carries some of the difficulties of external flows (like flows around a cylinder) but there are no difficulties in the modeling of boundary conditions. Most test cases on external flows are so dependant upon the type of boundary conditions applied at infinity that it is impossible to evaluate the other features of the methods.

iii) complex geometry : The geometry is sufficiently complex to refect some of the difficulties found in industrial configurations.

iv) Computable flow : The Reynolds numbers for which the flow is of industrial interest is generaly low (between 100 and 1500). Thus the flow is computable without turbulence modeling. The problem has been studied both numerically and experimentally by Karniadakis et al. [1] and they found the following critical issues :
- the flow is stationnary and stable up to $Re = Re_1^c$ around 140,
- the flow is transient but 2D up to $Re = Re_2^c$ around 200, after which 3D instabilities develop.
- there seems to be a drag crisis around $Re = Re_3^c$ around 100.

To be able to calculate numerically Rc_1^c one should have an accurate prediction of the amplitudes of the oscillations in the flow (see Fortin , Fischer-Patera) and/or of the decay rate at $Rc < Rc_1^c$. The computation of Re_2^c is more difficult and requires a full 3D computations (or 2D spanwise coupled).

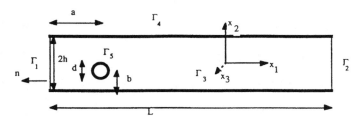

Figure 1 *Geometry of the computational domain.* $a = 1.5$ $b = 0.5$ $d = 0.4$ $h = 1$ $L = 20/3$. *The flow rate across the chanel, Q, is prescribed. No slip boundary conditions are applied on $\Gamma_3, \Gamma_4, \Gamma_5$. Periodicity is imposed on $\Gamma_1.\Gamma_2$ and on the two remaining sides if the computation is done in 3D. The Reynolds number is* $Re = \frac{3}{4}Q\nu$

2. DESCRIPTION OF THE CONTRIBUTIONS

Ten participations were received three of which were commercially available codes (FIDAP, N3S and NEKTON)

7 with the finite element method (FEM)

2 with the finite volume method (FVM)

1 with the spectral element method (SEM).

All use the physical variables u, p except one which uses the vorticity w and the stream function ψ. All contributions have an implicit solver either on the pressure alone or on all variables u and p. Only two contributions have attempted a 3D simulation at $Re = 600$ and higher; on the other hand three contributions have tried turbulence models at Reynolds numbers much higher than those asked on the proposal. A. Fortin took an interesting stand point by trying to apply the techniques of dynamical systems to study the long time behavior of the flow and to study the stability and bifurcations. Finally Rodi computed the flow at $Re = 100$ by solving directly the stationnary Navier-Stokes equations.

One of the numerical difficulty for this test case is due to the periodic boundary conditions. One either had to modify the program to treat it in an implicit fashion or treat it explicitly by feeding back at the entrance at every time steps the results of the previous time step at the exit. The explicit procedure was found to work well if there is an overlapping region (see Janvier et al) and if the computational domain is bigger than the cell.

To treat the periodicity in an implicit fashion one has to modify the Stokes or Oseen solvers by adding the pressure drop k as a variable :

Find p, u, k such that $u, p - kx_1$ are x_1- periodic and

$$\alpha u - \nu \Delta u + Bu + \nabla p = f, \quad \nabla \cdot u = 0, \quad \int_\Omega u = Q, \quad u = 0 \text{ on walls.}$$

Of course this better procedure requires to rewrite some part of the codes specifically for this workshop.

3. COMPARISON OF RESULTS AT Re = 100

Everyone found that the solution is asymptotically stationary but there are some discrepancies on the decay rates which is rather slow as seen on figure 7. One can separates the contributions in two groups :

1. Those who used a method first order accurate in time,
2. Those who used second order accuracy or more.

In group 1, the decay rate is always over estimated while in group 2 it is accurate or slightly over estimated. Streamlines and C_f (skin friction coefficients) are accurately computed in all contributions.

Name	Cf_100	Cf_200	Cf_600	Cf_1200
Buffat 2D	0.99	0.73	0.28	
Buffat 3D	0.98	0.6	0.235	0.126
Fortin				
Laurence	0.8	0.53	0.23	0.13
Janvier	0.75	0.49	0.23	0.125
Patera-Fisher	0.94	0.58		
Ravachol	0.8	0.5	0.23	0.12
Ricou				
Rodi	0.977			
Tezduyar	0.92	0.51		

Figure 2 *Values found for C_f at various Re*

4. COMPARISON OF RESULTS at Re = 200

Here again the C_f are stringingly accurate and contributions cannot be compared with this criterion.

The best criterion seems to be the vertical velocity at the point which is two diameter behing the cylinder : amplitudes and frequences of the signals.

We note the following

- First order accurate methods in time were not able to predict correctly the amplitudes (there is a lack of experimental data on this part however). Figure 8 shows several results and their discrepancies.

- Frequencies were correct but secondary frequencies were often wrong, usually because the contributor did not perform enough time steps.

Based on the amplitudes Fischer-Patera and Buffat computed Re_1^c and found 137 and 114 respectively to be compared with the experimental value 138. Streamlines were accurate by in large and displayed two eddies per cell as observed experimentally. A detailed evaluation of the eddy-shapes was not possible due to lack of experimental data. In any case this could be done only with vorticity plots.

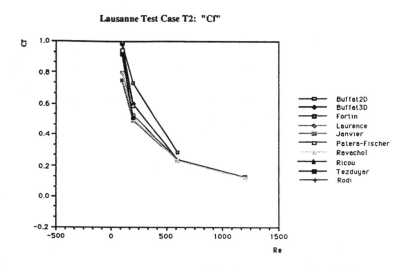

Figure 3 *Plot of C_f versus Re*

5. COMPARISON at Re = 600

In principle only Fischer-Patera and Buffat can be compared. Both found that the flow is fully 3D however I found that unless plots of the span-wise velocities at some point are provided only qualitative comparisons can be made. Figure 9 shows one of the results; vertical velocity amplitudes seem to agree as well as the Strouhal number. Computations in 2D can be compared and similar conclusions can be drawn regarding the amplitudes and frequencies (see graphs).

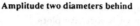

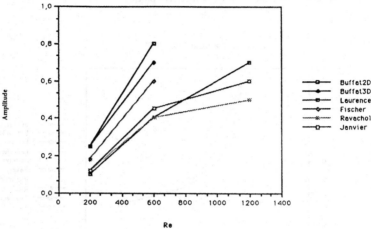

Figure 4 *Plot of $u_{max}(x = 2d)$ computed versus Re*

6. COMPARISONS at High Re

Janvier et al and Goutorbe et al ran a $k - \varepsilon$ model and a $R_{ij} - \varepsilon$ model. The $R_{ij} - \varepsilon$ model gives a more physical answer (figure 10); differences are large (~ 100 %) and again experimental data were not available to validate the models. Perhaps the computations could be coupled with the temperature equation for experimental comparisons.

Name	Um_200	Um_600	Um_1200	freq@200	Method
Buffat 2D	0.31	0.8		0.18	FEM O(h^2+δt^2)
Buffat 3D	0.25	0.7		0.2	FEM O(h^2+δt^2)
Fortin				0.1855	FEM O(h^2+δt^2)
Laurence	0.1	0.4	0.7		FEM O(h^2+δt)
Janvier	0.12	0.45	0.6	0.17	FEM O(h^2+δt)
Patera-Fisher	0.2	0.6		0.1855	SEM O(h^n+δt^n)
Ravachol	0.12	0.4	0.5	0.12	FEM O(h+δt)
Ricou	0.3	1.0			FEM O(h^2+δt)
Rodi					FVM O(h^2)

Figure 5 *Values of $u_{max}(x = 2d)$ computed*

7. COMPARISONS OF COMPUTING TIME

It is always difficult and dangerous to scale computing time to compare programs run on different machines but we made the dangerous attempt. By multiplying the computing time by the machine speed (megaflop) and the time step we computed an estimate of the time needed to reach the physical time $t = 1$ on a 1MFLOP machine. The figures are summarized on the diagram below ; these should be scaled down to an equal number of degree of freedom of everyone. Ricou used an out of core linear solver, when the memory of the machine is not sufficient (which was the case here); while it is time consuming it allows the program to run on small memory machines like IBM-AT. Liou et al. probably used an unnecessarily small time step (but it must be remembered that the emphasis was more on good results than on fast computing time). Ravachol has an unexpectedly fast algorithm based on an iterative GMRES solver.

Name	Nb U nodes	seconds per Δt	MFLOP	delta t	CPU time
Buffat 2D	4600	146	10	0.1	15000
Buffat 3D	33600	370	40	0.1	13000
Fortin	2700	10	10	0.1	10000
Laurence	8000	3	50	0.1	15000
Janvier	9600	24	100	0.1	24000
Patera-Fisher	1700	2.5	20	0.02	10000
Ravachol	3600	14	10	0.1	1400
Ricou	2700	614	5	0.02	60000
Rodi	8800	0.073	200	inst	8800
Tezduyar	5600	6	50	0.006	75000

Figure 6 *Computing time of various authors. The last column indicates the computing time needed to reach time t=1 on a 1 MFLOP machine. These numbers should also be scaled down to the same number of degree of freedoms (number of velocity nodes) for everyone*

CONCLUSION AND PROSPECTIVE

In conclusion, it is fair to say that most contributions have produced good results for the skin friction coefficients and the general features of the flow (streamlines, number of eddies...) up to Re = 200.

For more detailed informations like critical Reynolds numbers and of amplitude of oscillations only those who were second order in time (and space) could obtain them.

The 3D instabilities are still at the border of computability and we except that it will be done in the coming years. The challenge for the 90's for this problem is to be able to predict the sudden rise in drag which occurs around Re = 1000 - 1500 very much like the drag crisis of a cylinder when the boundary layer becomes turbulent.

In the mean time we expect to obtain more experimental results especially for the amplitude of the velocities at a point two diameters behing the cylinder and at various Reynolds number including $Re = 10000$.

Although high Reynolds number flow simulations were attempted with turbulence modeling we could not say much about the validity of the results for wants of experimental data.

REFERENCE

[1] G. Karniadakis, B. Mikic, A.Patera Minimum - dissipation transport enhancement by flow destabilization. J. Fluid Mech. **192** 365-391 (1988).

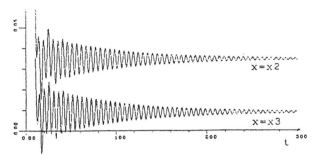

Figure 7 *Amplitude versus time of the vertical velocity component at three location behind the cylinder. Re=100 (from Buffat)*

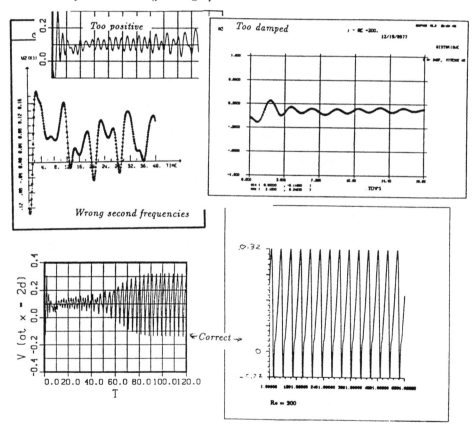

Figure 8 *Amplitude versus time found by several authors at Re=200*

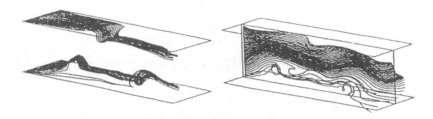

Figure 9*Stream lines of the flow at Re=600 showing the 3D character (from Buffat).*

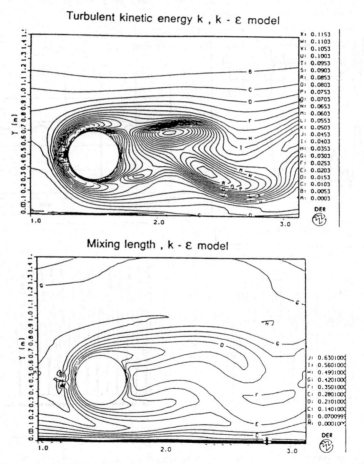

Figure 10*Level lines of k found from the k − ε model and a R_ij − ε model (from Goutorbe- Laurence).*

Invited Lecture

Transition Description and Prediction

D. ARNAL

ONERA/CERT
Aerothermodynamics Department
2 avenue E. Belin
31055 TOULOUSE CEDEX (FRANCE)

1 INTRODUCTION

Since the classical experiments performed by O. REYNOLDS (1883), the instability of laminar flows and the transition to turbulence have maintained a constant interest in fluid mechanics problems. This interest results from the fact that transition controls important hydrodynamic quantities such as drag or heat transfer. The objective of this paper is to give an overview of the transition problems, in two dimensional, incompressible flow.

An overall picture of the boundary layer development is shown on figure 1. From the leading edge to a certain distance x_T, the flow remains laminar. At x_T, turbulent structures appear and transition occurs. From x_T to x_E, there is a noticeable change in the boundary layer properties. The transition process involves a large increase in the momentum thickness θ and a large decrease in the shape factor H. The displacement thickness $\delta_1 = H\theta$ exhibits a more complex evolution. The skin friction coefficient Cf increases from a laminar value to a turbulent one, the latter being in some cases an order of magnitude larger than the former. It is obvious that the location and the extent of transition depend on a large range of parameters, such as external disturbances, vibrations, pressure gradient, roughness ...

In fact, two problems are to be studied :
a) What are the mechanisms leading to turbulence ?
b) Once the first turbulent structures are created, what will be their subsequent development up to the formation of a fully turbulent boundary layer ?

Both problems will be discussed in this paper. After a general description of the mechanisms leading to turbulence (paragraph 2), some of the practical transition prediction methods will be described (paragraph 3). Typical examples of comparisons between experiments and computations will also be given.

2 GENERAL DESCRIPTION OF THE TRANSITION MECHANISMS

2.1 Basic Concepts

When a laminar flow develops along a given body, it is strongly affected by various types of disturbances generated by the model itself (roughness, ...) or existing in the freestream (turbulence, noise, ...). These disturbances are the sources of complex mechanisms which ultimately lead to turbulence. In fact, two kinds of transition processes are usually considered :

304

a) If the amplitude of the forced disturbances is small (low freestream turbulence level for instance), one can observe at first two-dimensional oscillations developing downstream of a certain critical point. After a linear amplification of these waves, three-dimensional and non linear effects become important, inducing secondary instabilities and then transition. In these cases of "natural" transition, the transition Reynolds numbers are usually very large.

b) If the amplitude of the forced disturbances is large (high freestream turbulence level, large roughness elements), non linear phenomena are immediately observed and transition occurs a short distance downstream of the leading edge of the body. This mechanism is called a "bypass", in this sense that the linear stages of the transition process are ignored (bypassed), (see MORKOVIN, 1984).

2.2 "Natural" Transition

a) Linear stage - As it was pointed out, the instability leading to transition starts with the growth of two-dimensional disturbances, the existence of which was first demonstrated by the now classical experiments of SCHUBAUER and SKRAMSTAD, 1948. In fact, the existence of small, regular oscillations travelling in the laminar boundary layer was postulated many decades ago by Lord RAYLEIGH (1887) and PRANDTL (1921). Some years later, TOLLMIEN and SCHLICHTING worked out a linear theory of boundary layer instability, so that the waves are usually referred to as the "TOLLMIEN-SCHLICHTING waves" (TS waves). Nevertheless, the linear stability theory received little acceptance, essentially because of a lack of experimental results. The measurements of SCHUBAUER-SKRAMSTAD completely revised this opinion by demonstrating the real existence of the TS waves.

A complete account of this linear stability theory is out of the scope of this paper (see MACK, 1984, for complete information). Only some of the basic features will be briefly described.

It is assumed that the TS waves can be expressed by :

$$q = \hat{q}(y) \exp(\sigma x) \exp\left[i\,(\alpha x - \omega t)\right] \tag{1}$$

q represents a velocity or a pressure fluctuation ; σ, α and ω are the spatial amplification rate, the wavenumber and the circular frequency, respectively. Introducing (1) into the linearized NAVIER-STOKES equations leads to a system of ordinary differential equations, the combination of which gives the well known ORR-SOMMERFELD equation. Due to the homogeneous boundary conditions (the disturbances must vanish at the wall and in the freestream), the problem is an eigenvalue one. When the mean velocity profile $U(y)$ is specified, a non zero solution of the stability equations exists for particular combinations of the four real parameters R, σ, α and ω, where R is the REYNOLDS number.

The ORR-SOMMERFELD equation was solved by many authors. The results of such computations are represented on figure 2 for the BLASIUS flow. All the parameters were made dimensionless with the freestream velocity U_e and the displacement thickness δ_1, so that $R\delta_1 = U_e\delta_1/\nu$. The figure shows some curves of constant amplification rate σ in the

(α, $R\delta_1$) plane ; curves of constant frequency ω are not represented for clarity. The curve $\sigma = 0$ is called the neutral curve ; it separates the region of stable ($\sigma < 0$) from that of unstable ($\sigma > 0$) disturbances. There is a particular value of the REYNOLDS number below which all disturbances decay ; it is the critical REYNOLDS number, $R\delta_{1cr}$, which is equal to 520 for the BLASIUS flow.

b) Secondary instability, breakdown, turbulent spots - In order to illustrate the downstream evolution of the TS waves, figure 3 presents an example of smoke visualization obtained by KNAPP et al., 1966. A laminar boundary layer develops in natural conditions on an ogive nose cylinder aligned with the freestream. It can be seen that the two-dimensional TS waves take the form of concentrated bands of smoke around the cylinder (left part of the sketch). These "rings" become more distinct as they move down the body, indicating the existence of a strong amplification. When the initially weak disturbances reach a certain amplitude, their evolution begins to deviate from that predicted by the linearized theory : the waves are distorted into a series of "peaks" and "valleys". As the flow proceeds downstream, this pattern becomes more and more pronounced. CRAIK, 1971, used a weakly non linear theory in order to explain the appearance of this peak-valley system ; his model was consistent with some experimental observations, but was inoperative in other cases. More satisfactory results were obtained by HERBERT, 1985, who developed a linear secondary instability theory based on the FLOQUET theory.

Further downstream, non linear mechanisms become dominant. The peak-valley system gives rise to a vortex filament (horseshoe vortex), which breaks down into smaller vortices, which again break down into smaller vortices. The fluctuations finally take a random character and form a so-called "turbulent spot" : it is the transition onset. In the transition region (between x_T and x_E, figure 1), the turbulent spots are swept along with the mean flow ; they grow laterally and axially, overlap and finally cover the entire surface.

One has to keep in mind that the non linear phase and the breakdown process occur over a relatively short distance. For typical flat plate conditions, the streamwise extent of linear amplification covers about 75 to 85 per cent of the distance to the beginning of transition. This explains that calculation methods based on linear theory only (e^n method, paragraph 3.1) give good results for predicting the transition location.

c) Receptivity - It has been shown that the development of the TS waves can be correctly predicted by the linear stability theory. The main problem is now to explain the birth of these waves, i.e. to establish the link between their initial amplitude A_0 and the forced disturbances. The concept of receptivity, introduced by MORKOVIN (1969), describes the means by which these forced disturbances (sound, freestream turbulence) enter the laminar boundary layer and impose their signature in the disturbed flow. If they are small, they will tend to excite the TS waves, which constitute the normal modes of the boundary layer. Recent works by GOLDSTEIN (1983) and KERSCHEN (1989) have shown that the receptivity process occurs in regions of the boundary layer where the mean flow exhibits rapid changes in the streamwise direction. This happens near the body leading edge and in any region farther downstream where some local feature forces the boundary layer to adjust on a short streamwise length scale.

Up to now, receptivity studies were restricted to two-dimensional disturbances (sound, two-dimensional convected gusts ...). In many practical situations, however, the forcing disturbances are three-dimensional. In such cases, the receptivity mechanisms are unknown. The only available information come from experiments and illustrate the effect of the freestream disturbances amplitude on the transition REYNOLDS number R_{xT}. Figure 4 shows the evolution of R_{xT} as a function of the freestream turbulence level

$Tu = \tilde{u}_e/U_e$, where $\tilde{u}_e$ is the rms value of the freestream disturbances. At first sight, the experimental data seem to collapse into a single curve and it is clear that transition moves rapidly upstream when Tu increases. However, as Tu becomes small, R_{xT} reaches a constant value which depends on the experimental set-up : this value is about $2.8 \ 10^6$ for SCHUBAUER-SKRAMSTAD, 1948, and $5 \ 10^6$ for WELLS, 1967. In fact, sound component controls transition when Tu is very small and the effect of "true" freestream turbulence (vorticity fluctuations) can be only observed at values of Tu greater than $0.1 \ 10^{-2}$. On the other side, TS waves are never observed as soon as Tu exceeds 2 or $3 \ 10^{-2}$: transition becomes triggered by "bypass" mechanisms.

2.3 Bypass
There is no theory for describing the bypass processes occurring in a laminar boundary layer submitted to large disturbances ; a typical example is the T3 test case where high freestream turbulence levels cause rapid transitions. As the phenomena are essentially non linear, direct numerical simulations could be useful in such complex situations; see for instance the work done by SPALART (1988) for the problem of leading edge contamination ; see also the contribution presented by YANG and VOKE (Queen Mary College) during this Workshop.

It must be pointed out that, even in the presence of very large disturbances, transition cannot appear below a certain REYNOLDS number R_{mint} (minimum REYNOLDS number for turbulence).There is ample evidence (MORKOVIN, 1984) that boundary layers below R_{mint} are unable to sustain self-energizing turbulent motions on the scale of the shear layer thickness. For instance, when "vigorous freestream turbulence is convected toward the boundary layer ..., some mild increase in mean wall stress (is observed), but the boundary layer remains intrinsically laminar". For a two-dimensional, incompressible boundary layer developing on a flat plate, the experiments showed that R_{mint} is close to the critical REYNOLDS number R_{cr} deduced from the linear theory. In negative pressure gradient, R_{mint} becomes much lower than R_{cr}, and a bypass transition occurs generally at a REYNOLDS number between R_{mint} and R_{cr} (it is the case, for instance, on a turbine blade).

3 TRANSITION PREDICTION METHODS
The problem considered here is how to predict the position and the extent of the transition region in two-dimensional, incompressible flow. It will be assumed that the main factor acting on the transition mechanisms is the freestream turbulence level Tu. Most of the methods described in this paragraph do not claim to represent the intricate physics of the transition process : they only constitute short term answers to practical problems.

3.1 Calculation Methods Based on Stability Computations
Let us recall that the general expression of a TS wave is :

$$q = \hat{q}(y) \exp(\sigma x) \exp\left[i \ (\alpha x - \omega t)\right] \tag{1'}$$

where q stands for a velocity or a pressure fluctuation. The total amplification rate of a single dimensionless frequency $F = 2\pi f v/U_e^2$ (f is the physical frequency) is defined as :

$$A/A_0 = \exp\left[\int_{x0}^{x} \sigma \, dx\right] \tag{2}$$

A is the wave amplitude and the index 0 refers to the streamwise position where the wave becomes unstable. As an example, figure 5 shows total amplification curves corresponding to various frequencies, obtained for the BLASIUS profile. The dashed line represents the envelope of these cruves, which will be called n :

$$n = \underset{F}{\text{Max}} \,(\ln \,(A/A_0)) \text{ at a given x} \tag{3}$$

The so-called e^n method was developed independently by SMITH-GAMBERONI and by VAN INGEN (1956). SMITH and GAMBERONI compared the theoretical value of the n factor with transition locations measured on airfoils ; in all cases, transition was found to occur when $n \approx 9$; this means that turbulent spots appear when the most unstable frequency is amplified by a factor e^9. The same result was obtained by VAN INGEN, with a slightly lower value of n (7 to 8).

The e^n method is currently used for the case of natural transitions. The success of this method is certainly due to the fact that many experimental data are obtained in wind tunnels where the disturbance environment is similar ; in particular, the freestream turbulence level is usually rather low, let say $Tu \approx 0.1$ %. Let us observe that the e^n method was successfully used for predicting the transition onset on a cone in flight conditions, for Mach numbers between 0.5 and 2 (DOUGHERTY and FISCHER, 1980). The experimental data were correlated with $n \approx 10$.

MACK (1977) suggested an empirical relationship between Tu and the value of n at the transition location :

$$n = - 8.43 - 2.4 \ln Tu \tag{4}$$

For $Tu < 10^{-3}$, sound disturbances may become the factor controlling transition rather than turbulence and relation (4) may give poor results. On the other side, if $Tu = 2.98 \; 10^{-2}$, relation (4) implies that n = 0, i.e. transition occurs at the critical Reynolds number. It is the bypass limit.

3.2 Transition Criteria

In the following lines, the word criterion can be interpreted as a more or less empirical correlation between boundary layer parameters at the transition onset. These parameters may be some characteristic Reynolds number or a similarity variable, among others. The transition criteria are often used for practical applications, because they are easily introduced in engineering prediction methods such as integral methods. They often provide a fairly acceptable compromise between accuracy and simplicity.

Historically, the first proposed criteria (MICHEL, 1951 ; GRANVILLE, 1953 ; CRABTREE, 1958, for instance) took only into account the pressure gradient effects ; the freestream turbulence level was implicitly assumed to be low.

By reviewing the literature, HALL and GIBBINGS (1970), proposed the following relationship between Tu and the momentum thickness Reynolds number at the transition location :

$$R\theta_T = 190 + \exp(6.88 - 103 \text{ Tu}) \tag{5}$$

This criterion is valid for zero pressure gradient cases only. One can observe that for large values of Tu, $R\theta_T$ reaches an asymptotic value equal to 190, in agreement with the available experimental data (see paragraph 2.3).

More recently, the combined effects of Tu and of the streamwise pressure gradient were introduced into empirical correlations. In most of the cases, the pressure gradient parameter is the POHLHAUSEN parameter Λ_2 :

$$\Lambda_2 = \frac{\theta^2}{\nu} \frac{dU_e}{dx} \tag{6}$$

Such criteria were developed by VAN DRIEST and BLUMER, 1963, DUNHAM, 1972, SEYB, 1972, ABU-GHANNAM and SHAW, 1980. Figure 6 and 7 present the correlations proposed by VAN DRIEST-BLUMER and DUNHAM. In both cases, the value of $R\theta$ at the transition point is given as a function of Λ_2 and Tu. An interesting observation is that, for large values of Tu (of the order of 5 %), the transition Reynolds number no longer depends on the pressure gradient parameter. In particular, for accelerated flows, the value of $R\theta_T$ becomes much lower than the critical Reynolds number deduced from the linear stability theory. This behaviour is consistent with available experimental data (BLAIR and WERLE, 1981, for instance) and strongly supports the concept of "bypass".

Figure 8 shows a criterion developed by ARNAL et al., (1984b) by combining relation (4) and exact stability computations. The difference between the transition Reynolds number and the critical Reynolds number is given as a function of Tu and of a mean POHLHAUSEN parameter (arithmetic average of Λ_2 from the critical to the transition point). By contrast with the criteria presented on figures 6 and 7, the transition Reynolds numbers predicted by this correlation are always larger than the critical Reynolds number, i.e. this criterion cannot be applied for bypass transitions occuring in negative pressure gradients.

3.3 Transition Region Calculations Based on Intermittency Methods
Let us assume now that the transition onset is known. A second objective is to compute the transition region itself, the extent of which may be as long as the laminar region which

precedes it. An important parameter is the intermittency factor γ, which represents the fraction of the total time that the flow is turbulent. Many simple transition models are based on "intermittency methods" in which laminar and turbulent quantities are weighted by γ.

a) *Evolution of the intermittency factor* - SCHUBAUER and KLEBANOFF (1956) measured the streamwise evolution of γ for several flat plate experiments ; the length of the transition region varied from one case to another, but the intermittency distribution conserved the shape of the Gaussian integral curve. DHAWAN and NARASIMHA (1958) proposed the following "universal" distribution of γ for flat plate experiments :

$$\gamma = 1 - \exp\left[- 0.411 \ (x - x_T)^2/\lambda^2\right] \tag{7}$$

with $\lambda = x(\gamma = 0.75) - x(\gamma = 0.25)$.

Under the assumption that the turbulent spots appear randomly with a source rate density g, EMMONS, 1951, showed that the intermittency factor at a given point P could be written as :

$$\gamma = 1 - \exp\left[- \int_R gdV\right] \tag{8}$$

where R is the influence volume of P (locus of all points which influence the state of turbulence at point P). By assuming that g could be approximated by a DIRAC's delta function, CHEN and THYSON (1971) derived an analytical expression of γ valid for plane or axisymmetric flows.

In the previous expressions, γ is a function of the streamwise distance x. It is also possible to assume that the intermittency factor depends on θ/θ_T only (ARNAL et al., 1984a) where θ_T is the momentum thickness at the transition onset. This approach will be discussed below.

b) *Intermittency methods* - In these methods, the turbulent shear stress - $\overline{u'v'}$ appearing in the x-momentum equation is usually expressed as :

$$- \overline{u'v'} = \gamma \nu_t \frac{\partial U}{\partial y} \tag{9}$$

where ν_t is a turbulent viscosity given by a classical turbulence model.

In the calculation method developed at ONERA/CERT (COUSTOLS, 1983), ν_t is expressed by a mixing length scheme and the intermittency function is an analytical function of θ/θ_T which is represented on figure 9. In order to model the overshoot in the skin friction coefficient in the middle of the transition region, γ is constrained to reach a maximum value well above unity ; it becomes clear that γ does not represent the physical intermittency factor. It is only an empirical corrector type function which gives a smooth evolution between the laminar and the turbulent state.

An application of this method is shown on figure 10. The calculations are compared with an experimental configuration where transition occurs in a mild positive pressure gradient.

The evolutions of the shape factor H, of the integral thicknesses δ_1 and θ, and of the skin friction coefficient Cf are fairly well reproduced.

3.4 Transport Equations

During the last two or three decades, calculation methods using transport equations have been developed and applied to more and more complex turbulent flows. In addition, attempts have been made for extending the range of applications of such methods to the prediction of transition phenomena. For this, additional terms or empirical functions were introduced in the fully turbulent form of the equations ; they depend usually on a "turbulence Reynolds number" which represents the ratio of the turbulent to the viscous shear stress.

The numerical problem is to solve a set of parabolic partial differential equations with appropriate initial and boundary conditions. The calculation starts in laminar flow with specified initial profiles and proceeds step by step in the streamwise direction. If the turbulent quantities are amplified, a "numerical transition" may occur, in this sense that the shear stress - $\overline{u'v'}$ becomes large and modifies the mean velocity profile. These methods present the advantage that a single run is needed for the computation of the laminar, transitional and turbulent boundary layer. A second advantage is that the influence of some important factors acting on the transition mechanisms (streamwise pressure gradient, freestream turbulence level) appears naturally as boundary conditions.

Transport equations models were used by many contributors for the test case T3 ; the characteristics of these models and the corresponding results are described in separate papers. We will only present typical results obtained at ONERA-CERT (ARNAL, 1984) in order to show the advantages and the limitations of such methods.

The turbulence model is the well known k-ε model initially developed by HANJALIC-LAUNDER (1972) for high Reynolds numbers and extended by JONES-LAUNDER (1972) in order to take into account the low Reynolds number effects. Figure 11 shows a comparison between the numerical results and the experimental data obtained by TURNER (1971) on the pressure side of a turbine blade. The freestream velocity increases rapidly and almost linearly with the streamwise distance. The dashed curves represent the results of laminar calculations.

For Tu = 0.45 10^{-2}, the boundary layer remains laminar over the entire length of the blade. For Tu = 5.9 10^{-2}, transition occurs close to the leading edge. At the intermediate turbulence level, transition starts at x $\approx$ 1.5 cm and the numerical results indicate that the boundary layer is not fully turbulent at the trailing edge. In all cases, the experimental behaviour is well reproduced.

The ability of the k-ε model to predict transition onset was also checked for a <u>low value of Tu</u> and for the three velocity distributions plotted on figure 12. For x > 0.15 m, the curves 1, 2 and 3 correspond respectively to negative, zero and positive pressure gradients. The table reported in figure 12 gives the numerical values of H and Rθ at the transition location. They indicate that the transition Reynolds number increases in the adverse pressure gradient. This result is completely at variance with all available experimental results. Other computations have shown that the stabilizing effect of wall cooling cannot be reproduced by the k-ε model. These strong disagreements are due to the fact that the real transition mechanisms are linked with stability properties and the transport equations could hardly be expected to reproduce linear stability results !

4 CONCLUSIONS

Obviously, none of the presented methods is able to predict correctly transition for all practical purposes. All these techniques need the introduction of empirical data, which reduce considerably the range of applications. These data are, for instance, low Reynolds number functions (transport equations) or values of critical amplification rates (e^n method).

At low values of Tu, methods based on linear stability theory are certainly the most accurate ones, but models based on transport equations are unable to give satisfactory results. At high values of Tu, the linear stability theory no longer applies, and the e^n method, as well as criteria based on this theory, are not valid. In such cases, non linear phenomena become predominant and seem to be fairly well described by the transport equations. However, the influence of the initial profiles of the turbulent quantities (turbulent kinetic energy, dissipation, ...) need to be clarified.

REFERENCES

ABU-GHANNAM B.J., SHAW R., 1980, J. of Mech. Eng. Sci. Vol. 22, N° 5
ARNAL D., 1984, AGARD R-709, pp 2-1 to 2-71
ARNAL D., COUSTOLS E., JUILLEN J.C., 1984a, La Rech. Aérosp. N° 1984-4
ARNAL D., HABIBALLAH M., COUSTOLS E., 1984b, La Rech. Aérosp. N° 1984-2
BLAIR M.F., WERLE M.J., 1981, UTRC Report R 81-914388-17
CHEN K.K., THYSON N.A., 1971, AIAA J., Vol. 9 N° 5, pp.821-825
COUSTOLS E., 1983, Thesis presented at ENSAE, TOULOUSE
CRABTREE L.F., 1958, J. of the Royal Aeronautical Society
CRAIK A.D.D., 1971, J.F.M. Vol. 50, Part 2
DHAWAN S., NARASIMHA R., 1958, J.F.M. Vol. 3 Part 4
DOUGHERTY N.S. Jr., FISHER D.F., 1980, AIAA Paper 80-0154
DUNHAM J., 1972, AGARD Meeting "Boundary Layer in Turbomachines", PARIS
EMMONS H.W., 1951, J. Aero. Sci. N° 18, pp. 490-498
GOLDSTEIN M.E., 1983, J.F.M. Vol. 127, pp. 59-81
GRANVILLE P.S., 1953, David Taylor Model Basin Report 849
HALL D.J., GIBBINGS J.C., 1972, J. Mech. Eng. Sci. Vol. 14, N° 2
HANJALIC K., LAUNDER B.E., 1972, J.F.M. Vol. 52, Part 4
HERBERT T., 1985, AIAA Paper 85-0489
JONES W.P., LAUNDER B.E., 1972, Int. J. of Heat and Mass Transfer, Vol. 15, N° 2
KERSCHEN E.J., 1989, AIAA Paper 89-1109
KNAPP C.F., ROACHE P.J., MUELLER T.J., 1966, UNDAS-TR-866 CK
MACK L.M., 1977, Jet Prop. Lab. Publication 77-15, PASADENA
MACK L.M., 1984, AGARD R-709, pp. 3-1 to 3-81

MICHEL R., 1951, ONERA Technical Report 1/1578 A
MORKOVIN M.V., 1984, Transition in Turbines, NASA Conf. Publ. 2386
SCHUBAUER G.B., KLEBANOFF P.S., 1956, NACA Report 1289
SCHUBAUER G.B., SKRAMTAD H.K., 1948, NACA Report 909
SEYB N.J., 1971, Rolls Royce Report, Bristol Engine Division
SMITH A.M.O., GAMBERONI N., 1956, DOUGLAS Aircraft Co. Report ES 26 388
SPALART P., 1988, AGARD Conf. Proceedings N° 438
TURNER A.B., 1971, J. of Mechanical Engineering Science, Vol. 13, N° 1
VAN DRIEST E.R., BLUMER C.B., 1963, AIAA Journal Vol. 1, N° 6
VAN INGEN J.L., 1956, DELFT Univ. of Techn. Report UTH-74
WELLS C.S., 1967, AIAA Journal Vol. 5 N° 1, pp. 172-174

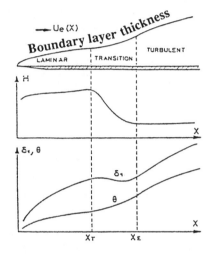

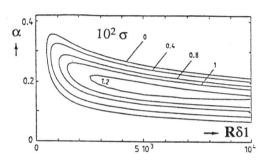

Fig. 2 - Stability diagram (BLASIUS flow)

Fig. 1 - Boundary layer development

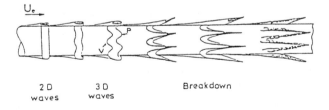

Fig. 3 - Overall picture of the transition process

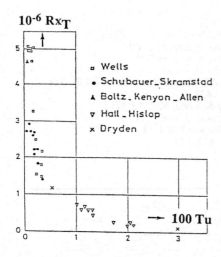

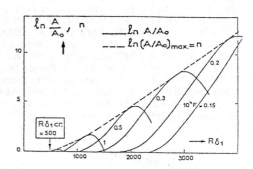

Fig. 5 - Global amplification rates for various frequencies (BLASIUS flow)

Fig. 4 - Effect of freestream turbulence on transition Reynolds number

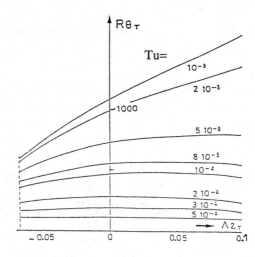

Fig. 6 - Criterion proposed by VAN DRIEST-BLUMER

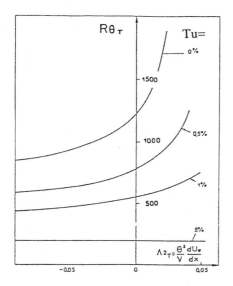

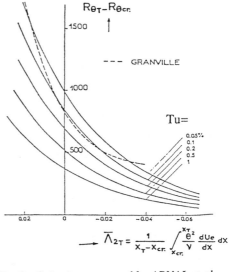

Fig. 7 - Criterion proposed by DUNHAM

Fig. 8 - Criterion proposed by ARNAL et al.

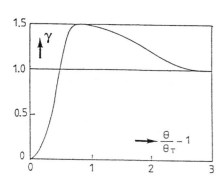

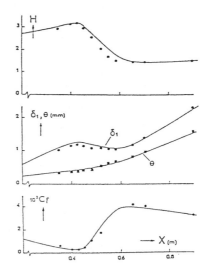

Fig. 9 - Intermittency function

*Fig. 10 - Boundary layer transition in positive pressure
            gradient - Symbols : experiments ; _____
            computations*

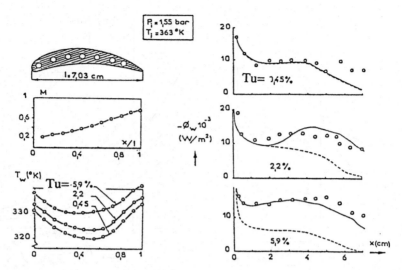

Fig. 11 - *Combined effects of negative pressure gradient and of high freestream turbulence level*
Left hand side : geometry, Mach number and temperature distributions
Right hand side : wall heat flux - Symbols : experiments ; ⎯⎯ k-ε model ; ------ laminar computations

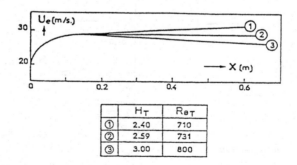

	H_T	Re_T
①	2.40	710
②	2.59	731
③	3.00	800

Fig. 12 - *Effect of pressure gradient at low Tu (k-ε model)*

Test Cases T3A, T3B:
Free-Stream Turbulence

INTRODUCTION

The ability to predict transition is becoming an ever more important requirement in many areas of fluids engineering; perhaps particularly so in aerodynamics, where the majority of internal flows (eg within gas turbines) are already largely transitional and the desire to employ laminar flow control to external surfaces (eg of aircraft) is shifting the emphasis from turbulent to transitional flows there also. Indeed transition prediction is now a pacing item for many CFD applications.

Traditionally transition (under the influence of free-steam disturbances) has been divided into two types: so-called 'natural' transition when the disturbance level is low, and 'by-pass' transition when the free-stream turbulence intensity is sufficient (>1%) to promote a much more rapid transition by 'by-passing' some of the normal transition mechanisms/stages. Recently, however, evidence has been growing that Tollmein-Schlicting waves play a significant role in both processes, which suggests there may be some chance of developing a single successful predictive strategy for all such flows. Unfortunately, although considerable advances have been made in improving the predictive capability of turbulence models over the last 20 years, rather less progress has been made in adapting and applying these to transition problems. This has been partly due to uncertainties concerning the validity of using established low-Re 'near-wall region' treatments to handle low-Re transitional flow regimes, and partly because the extensions required to account for inhomogeneity/intermittency (in the case of natural transition) and diffusion controlled flows (in the case of bypass transition) have not been a pressing requirement for other applications. In addition, whilst models have been subjected to a variety of turbulent test case validation exercises, no similar co-ordinated assessment of transition models/test case has been attempted.

In view of the need to distinguish between the ability of models to 'post-' or 'pre-'dict transition for practical engineering flows, the present Workshop has adopted the policy of supplying only the initial and boundary conditions required to make 'blind' computations. Sufficient computors were encouraged to participate to ensure that models ranging in complexity from the simplest industrial design codes to 'state-of-the-art' stress transport research models and Large Eddy Simulation codes were evaluated, and computors were advised not to employ any existing empirical correlations. In addition, at least for the T3 Test Case, the experimental data was taken specifically for industrial purposes, in a carefully set-up and controlled experimental facility, with a view to examining numerical problems and grid-refinement. Sensitivity to conditions and model constants/approximations could be examined in detail.

Besides providing the data and supporting some of the modelling, Rolls-Royce has also funded a parallel Direct Numerical Simulation of the test cases to provide a more extensive 'data'-base to improve physical understanding of the phenomena involved in such transition scenarios, to aid model refinement, and to provide additional Test Cases for later evaluation studies.

318

The Influence of a Turbulent Free-Stream on Zero Pressure
Gradient Transitional Boundary Layer Development
Part I: Test Cases T3A and T3B

P.E. ROACH & D.H. BRIERLEY

Advanced Research Laboratory, Rolls-Royce plc, PO Box 31,
Derby, DE2 8BJ, England.

ABSTRACT

This paper summarises the findings of recent zero pressure
gradient transitional boundary layer experiments. Boundary
layer measurements rely extensively on hot-wire anemometry,
and include all three components of normal stresses, two
components of shear stress, the three components of mean
velocity, wall surface shear stress, and pressure
distributions. The main conclusions of this work are:
1. A turbulent free-stream (intensity Tuo) has a first order
influence on the location of the start of the transition
process and the boundary layer state upstream, during and
downstream of the transition zone.
2. The present data indicate that with low Tuo the main
mechanism responsible for transition is the
Tollmien-Schlichting instability. This instability seems to
disappear at higher Tuo, being bypassed by some other
mechanism. It is here noted that this switch-over in
preferred mechanism could present severe problems for
prediction methods.

1 INTRODUCTION

Interest in both experimental and computational studies of
transitional boundary layers has increased in the last
decade, motivated in part by the quest for improved
performance and handling qualities of gas turbines. In
parallel with this, the blading design procedures need
validation from a large and detailed database of high
quality and reliability. In the turbine, the concept of
'aft-loading' the blade (ie producing a lift distribution
such that most of the lift occurs over the latter half of
the blade) is finding increased usage. The idea of
aft-loading is to reduce profile loss by maintaining a large
amount of laminar flow over the blade suction surface. This
must not, however, be accompanied by a large boundary layer
separation and this, in turn, depends on the position and
extent of the transition region. Similarly, in compressors
the concept of 'controlled-diffusion' blading is intended to

319

achieve increased lift whilst avoiding boundary layer
separation and its associated losses. This idea again relies
on the transition process occurring prior to the
controlled-diffusion region of the blade surface.

The development of theoretical methods for transition
prediction depends on sufficiently detailed experimental
data to describe adequately the complex physical processes
involved. Although transition has been studied experimentally by many workers, few (if any) have supplied the
degree of detail necessary to validate the Navier-Stokes
computer codes currently under development. For this reason,
the Company has initiated a series of wind tunnel
experiments to examine the laminar-to-turbulent transition
zone in developing boundary layers. The proposed experiments
cover both the zero pressure gradient flow plus pressure
distributions representative of turbine blading. The
influence of an external turbulent free-stream is of
primary interest, with the experiments performed at
turbine-blade-type Reynolds numbers.

This paper summarises the findings of the zero pressure
gradient experiments. Boundary layer measurements rely
extensively on hot-wire anemometry, and include all three
components of normal stresses, two components of shear
stress, the three components of mean velocity, wall surface
shear stress, and pressure distributions. Analysis of the
present results, together with those in the literature, is
presented in ref. 1.

2 EXPERIMENTAL METHODS

The wind tunnel used in the present experiments is of the
closed-circuit type and is described in detail in ref. 2.
Its major elements include a centrifugal fan blowing through
a large plenum fitted with turbulence-reducing honeycomb and
gauzes, a small 2-D contraction, leading to a highly
versatile working section, and a wide-angle diffuser. The
working section is of 2m length, 0.71m width and 0.26m
height with full length perspex side walls for good optical
access, see figure 1. The bottom wall can be inclined
sufficiently to produce a zero pressure gradient on the test
surface which is hung from the ceiling of the working
section. High performance flow cleaning filters are employed
in the return ducting, together with a water-cooled heat
exchanger. As a result of the careful design, the working
section free-stream turbulence intensity is 0.2+/-0.05% over
the operating velocity range of 0 to 25m/s, and the air
temperature can be controlled to +/-0.1 C degrees during the
course of a day's operation.

Measurements are made on a flat perspex test surface which is hung from the roof of the working section. The test plate has numerous static pressure tappings and surface plugs which are interchangeable and flush to within 25µm (when the tunnel is operated at a nominal 20 C). The (1700mm length, 20mm thick, 710mm wide) test surface is extremely flat and employs a small leading edge radius of 0.75mm with a 5 degree chamfer on the other side to the (lower) test surface. The test plate leading edge is mounted 40mm from the working section upper wall, and the circulation about it is controlled by a combination of a trailing edge flap and adjusting the pressure drop across the working section exit plane (by means of gauzes). Moreover, the test plate is inclined at 0.5 degree to the main flow vector which, together with the circulation control measures, ensures attached, steady leading edge flow with the stagnation streamline located on the test surface.

The turbulence generating grids used in these experiments are also detailed in ref. 2, together with the turbulence intensity characteristics. These can be located at the entrance plane of the working section, at the downstream end of the 2-D contraction. They are made of square or round bars and woven wire mesh, and are positioned 610mm upstream of the test plate leading edge. The grids have been designed according to the methods of ref. 3, and can generate test plate free-stream turbulence intensities from 0.5% to 7%, and integral length scales from 5mm to 30mm. With the grids in location, it was found that the generated turbulence was extremely homogeneous and isotropic, with a streamwise-to-normal fluctuating velocity ratio of about 1.005.

A DISA 55M series anemometer system was used to make velocity measurements in the wind tunnel. Custom designed single and X-wire probes (total wire length 1.5mm, sensing length 0.5mm, wire diameter 2.5µm) were used to measure both the free-stream turbulence properties and the boundary layer velocity profiles. Gold-plated wire probes have been used to minimise prong interference, with the small sensing length chosen to minimise spatial averaging from both the turbulent eddies and from the mean velocity shear. The primary purpose for using the single wire probes is to get sufficiently close to the test surface in order that accurate measurements of the boundary layer integral parameters could be made. In practice, the probe prongs were always placed on the test surface with the traverse proceeding away from this surface, ending at about three times the boundary layer thickness. The minimum wire height was measured to be 29µm, but because the wall proximity error was found to be significant (for y+<3), a certain number of measured points were deleted from the profiles before analysing the data.

The single wire probes also provide an initial estimate of the streamwise component of fluctuating velocities which may be compared with the X-wire results. The X-wires (both U-V and U-W components) were used to measure normal and shear stress distributions through the boundary layers. They provide measurements of all three components of normal stresses, together with two components of shear stress (uv and uw).

A number of techniques have been examined to measure surface shear stress. In the case of a turbulent boundary layer, the law of the wall has been used with the Karman and offset constants of 0.41 and 5.20 respectively. A Preston tube has also been used (making use of Patel's calibration), and comprises a square-cut circular tube of 2.49mm outside diameter, located on the surface of a test surface plug. The momentum balance approach was also used and it was found that all three techniques yielded skin friction coefficients which agreed to within 2% of each other. Attempts were made at calibrating both a heated film gauge (DISA R45) and a small (50μm height) razor blade device in both turbulent and laminar boundary layers. The aim was to produce a 'universal' calibration which could be used in both laminar and turbulent flows, and by implication in the transitional zone also. Unfortunately, these attempts failed for various reasons. As a result, it was decided to use the momentum balance technique in both the laminar and transitional zones. This was compared with the wall-slope approach in laminar boundary layers with negligible free-stream turbulence and found to agree to better than 1%.

A Hewlett-Packard HP86A desk-top computer was used to process the raw data and drive the traverse gear for the hot-wire probes. This data is then transferred via an IEEE 488 interface to a DEC PDP-11/44 mini-computer for later processing, analysis and presentation in both tabular and graphic form. The single wire measurements have been made at typically 15 streamwise locations, and up to 46 points within each boundary layer. Sample times of approximately 30 seconds per measurement have been used, with a bandwidth of 3kHz on-line, and 10kHz recorded. Recognising the value of the results and the probable requirement for further analysis, all of the raw data have been simultaneously recorded on an instrumentation FM tape recorder (SE 7000M).

3 EXPERIMENTAL RESULTS

3.1 Preliminary Considerations

When performing flat plate boundary layer experiments, one of the major problems that needs to be addressed is the flow condition near the leading edge. In the present case, it is

essential to avoid flow separation and instability since
this would corrupt later, downstream measurements. One
approach used to overcome this problem is to design a
(usually) very gradual elliptic leading edge which ensures
that flow separation is avoided and the stagnation
streamline is firmly anchored on the test surface side of
the plate. Although a 'safe' approach, it does produce a
virtual origin shift for the boundary layers developing
downstream (as in the work of Gostelow, 4), and has a first
order influence on the location of transition, ref. 5. It
can also confuse the understanding of the results since the
flow field is then a combination of the zero (or prescribed)
pressure distribution over the bulk of the test surface,
plus some (usually undefined) pressure distribution near the
leading edge. Ideally, therefore, a test surface design is
required such that the leading edge zone comprises a
negligibly small proportion of the overall length: this is
not to be overlooked, and supplies the first problem which
must be solved before proceeding with the experimental
program.

The solution adopted in the present study is to employ a
flat test plate with a small leading edge radius. This
approach can readily lead to flow separation at the leading
edge which would render any further measurements pointless.
To overcome this problem, in the wind tunnel in use it is
possible to incline both the test plate and the opposite
tunnel wall in such a way as to ensure zero pressure
gradient over the bulk of the test plate length. In
addition, use is made of the test plate trailing edge flap,
coupled with the variable pressure drop working section exit
plane gauzes, which effectively control the
circulation about the plate and thus overcome the tendency
to separate near the leading edge. This may seem a rather
complex solution, but is found to be most useful in
minimising the leading edge/test plate interaction. The
effectiveness of the approach can be judged by noting that
the bulk pressure distribution for the tests reported here
amounts to no more than 0.17% of the dynamic pressure, q,
for 200mm <X< 1500mm (the most downstream measurement
plane). Between 100mm and 200mm, the pressure falls by
0.75%q, and between 50mm (most upstream pressure tapping)
and 200mm by only 3%q. Such figures are rarely quoted in
other experiments of this nature, and highlight the quality
of the present work. In addition, single wire explorations
were made near the leading edge (+/-50mm in X) to detect any
symptoms of flow separation or unsteadiness: no such
problems were encountered. Moreover, excellent agreement of
the present laminar boundary layer data with the Blasius
solution provided further evidence that the flow about the
leading edge region is attached and steady.

Another major consideration in experiments of this nature is
the flow two-dimensionality. Measurements have been made
with the aid of a single wire probe in both laminar and
turbulent boundary layers at several streamwise planes, both
with and without the addition of a turbulent free-stream.
The first problem addressed was the growth of the
side-wall/test-plate corner flow contamination. This was
found to grow at an angular rate of 10 degrees, in excellent
agreement with the findings of Charters, ref. 6. Accounting
for the 30mm thickness of the side-wall boundary layers at
the test plate leading edge, this results in a 'potential'
flow region of 120mm width at the most downstream
measurement plane (X=1500mm). The next problem addressed is
the degree of two-dimensionality within this potential flow
region. The present measurements show scatter of the order
3% (peak-to-peak) in both skin friction and momentum
thickness (variance of 2.4%), and only 1.5% in form factor.
Considering the uncertainty in individual measurements to be
1%, these figures again highlight the good quality of the
present results, and are comparable to the results quoted in
ref. 7, for example.

Before presenting the results, it is first useful to outline
the conditions examined. Table 1 summarises the seven flow
conditions, highlighting the main parameters of interest.
The data are stored in files on 8" floppy disk and may be
accessed readily, knowing the unique file number. The file
name is constructed from the file number, preceded and
succeeded by the characters as outlined in the table notes.
The preceding characters denote the probe type: 1W being the
single wire, XV being the U-V cross-wire probe, and XW being
the U-W cross-wire probe. The code name identifies each of
the seven flow conditions examined. The ERCOFTAC workshop,
problem T3 test cases A and B are noted in this table also.

3.2 Mean Flow Data
Integral parameter plots for all seven conditions are
presented in figures 2 to 4. As seen, the start of
transition had not been reached with both the NOGRID and
SMRLO test conditions. Both the SMRHI and PRLO conditions
had just started the transition process, whilst the PRHI
(ERCOFTAC Case A), PSLO and PSHI (ERCOFTAC Case B)
conditions covered the full range of laminar, transitional
and turbulent boundary layer flow. It is also evident that
the skin friction coefficient, Cf, is increased, whilst the
form factor, H, is decreased, prior to the start of
transition. Full analysis of the present data is included in
reference 1. Both laminar and turbulent conditions are seen
to be in broad agreement with the expected curves taken from
ref. 8.

Sample mean velocity profiles are shown in figures 5 to 7, showing the PRHI flow condition (Case A), plotted in wall co-ordinates. Figure 5 shows the laminar flow conditions up to the point of the start of transition, with the free-stream U+ figures rising monotonically. Figure 6 shows the transitional data up to the end of transition, where the profiles are seen to gradually move towards the law of the wall curve. Finally, figure 7 shows the fully turbulent data, which again show the rising free-stream U+ levels. These data appear to typify the behaviour of the present transitional flow measurements, and are qualitatively similar to those in the literature, see for example ref.4.

Figure 8 shows sample laminar velocity profiles obtained at nominally equal Reynolds numbers, but with a range of free-stream turbulence levels (Tuo=0.2%, 2.2%, 5.2%). Figure 9 shows similar data obtained from fully turbulent boundary layer conditions (Tuo=1.5%, 3.8% and 4.0%). Both figures show 'fuller' profiles as the turbulence is increased, implying a reduction in form factor. Note, however, that the influence of free-stream turbulence is more pronounced in the case of the laminar boundary layers.

3.3 Fluctuating Measurements

Before examining the present X-wire measurements, it should first be noted that corrections have been applied to the data to account for the influence of velocity shear on the UW probe measurements. Even so, the actual magnitude of the results shown will be slightly in error, due to the influence of spatial averaging over the sensor length, ref. 3. It is possible that the single wire measurements of u' are more accurate than those using the cross-probes, the differences becoming greater in proportion to the velocity shear, in agreement with the findings of Hart, ref. 9. With this in mind, some of the present results for the PRHI condition (Case A) are shown in figures 10 to 12, showing the characteristics of the streamwise fluctuating velocities for the laminar, transitional and turbulent boundary layers respectively. Clearly seen is the rise in $u'(=\sqrt{(u^2)})$, which starts well upstream of the start of transition, peaking somewhere in the middle of the transition zone, then decaying before the end of transition and on into the turbulent boundary layer. The location of the peak in the u' profiles also moves from about a y+ of 20 (y/δ =0.40) in the laminar zone to y+=15 (y/δ=0.04) in the turbulent boundary layer. The spanwise and vertical stress components show somewhat different trends, in that their magnitudes are very low prior to the start of transition, rise quite rapidly through the transition zone, peaking at the end of transition, then falling only gradually in the turbulent boundary layer: more detail is given in ref. 1, including a description of the shear stress properties.

Further information is obtained from an examination of the
peak values in u' for all the present test conditions:
figure 13 shows these data, plotted against Rθ. It is
apparent that all the laminar boundary layer data form
independent curves, but that the turbulent boundary layer
data appear to form a single curve for all three conditions.
In these, the free-stream turbulence intensity, Tuo, varies
from 1% to 4%, and the data suggest no major influence of
Tuo, at least for Tuo>1%. With the SMRHI condition, however,
where Tuo=0.5%, somewhat different characteristics   are
observed, with the transitional data rising sharply above
the rest. This suggests different mechanisms are at work in
the transition process with this lower level of Tuo. This
suggestion is supported by examination of the raw signals on
an oscilloscope: just prior to transition, with low Tuo,
there is clear evidence of periodic activity in the signals,
indicative of some Tollmien-Schlichting wave   mechanism
leading to transition, ref. 8. Moreover, spectral analysis
also reveals a narrow band of excited frequencies consistent
with theory and measurements, see figures 14 and 15, and
refs. 8 and 10.

With higher Tuo, there is apparently no such activity, the
raw signals appearing almost turbulent in nature, though
the boundary layers are evidently pre-transitional. Also
included in figure 13 are some results of Hart(9), who
examined the characteristics of transitional boundary layers
on a non-lifting aerofoil. Consequently, there is   a
significant adverse pressure gradient over a large fraction
of the surface, but nevertheless, it is interesting to note
that his data are in good agreement with the present
measurements in the turbulent boundary layer zone. (His
levels of Tuo are in the range 2% to 3% and are thus similar
to the present experiments.)

These findings have far-reaching implications in modelling
such transitional boundary layers with free-stream
turbulence intensities of the order 0.5%. Although the
turbulence intensities in the turbine region of   a
gas-turbine are likely to be much higher generally, between
blades having high acceleration within the passage it is
conceivable that these high levels could be locally reduced
to levels less than 1%. If such a reduction in Tuo does
occur, then it is immediately apparent that   great
difficulties will be encountered in any modelling efforts.
Further study of this problem area is necessary in the
future.

Now, among the best measures of structural changes within a
developing boundary layer are the intensity ratios, u'/v' &
u'/w' & v'/w', and the shear correlation coefficients,
uv/u'.v' & uw/u'.w' & vw/v'.w'. Typical results from the

present work are shown in figures 16 to 19 (Case A), and are compared with Klebanoff's(11) high Reynolds number, turbulent boundary layer data (Rθ=8000). Broadly speaking, it can be seen that the present results are markedly different from those of Klebanoff in the laminar region, but rapidly approach his results through the transitional and turbulent zones. The differences observed in the turbulent zone are basically a result of Reynolds number, as confirmed in the work of Murlis et al(12). Even though poor agreement with Klebanoff's results is indicated for the UV correlation, the agreement with the results of Murlis et al is much better. It is interesting to note that the UW shear correlation coefficient is insignificant in the laminar, transitional and turbulent boundary layer zones. Both the correlation coefficients and the intensity ratios indicate the efficiency of turbulent mixing within the boundary layer. It is known that almost any turbulent eddy will contribute to the streamwise velocity fluctuation, because of the presence of a mean velocity gradient. On the other hand, the normal and spanwise components of velocity fluctuations are more likely to be associated with the sort of erupting eddies that effect most of the turbulent mixing. Thus, a monotonic decrease in u'/v' and u'/w' implies a monotonic increase in the general efficiency of turbulent mixing, and vice versa. As expected, the present results confirm an increase in turbulent mixing through transition and beyond.

4 CONCLUDING REMARKS

The main conclusions of this paper are summarised here.
1. A turbulent free-stream has a first order influence on the location of the start of transition and on the boundary layer state upstream, during and downstream of the transition zone.
2. The present data indicate that with low Tuo the main mechanism responsible for transition is the Tollmien-Schlichting instability. This instability seems to disappear at higher Tuo, being bypassed by some other mechanism. It is here noted that this switch-over in preferred mechanism could present severe problems for prediction methods.
All of the present work is analysed in detail in ref. 1.

5 ACKNOWLEDGEMENTS

The authors would like to thank Rolls-Royce plc for permission to publish this paper. This work has been carried out with the support of the Procurement Executive, Ministry of Defence.

6 REFERENCES

1. Roach, P.E. (1990)
   The influence of a turbulent free-stream on zero pressure
   gradient transitional boundary layer development.
   Part II : An analysis of the data.
   ERCOFTAC Workshop, Lausanne, 26-28 March 1990.

2. Roach, P.E. (1988)
   A new boundary layer wind tunnel.
   RAeS Aeron. Journ., Vol.92(916), 224-229.

3. Roach, P.E. (1987)
   The generation of nearly isotropic turbulence by means of
   grids.
   Int. J. Heat Fluid Flow, 8(2), 82-92.

4. Gostelow, J.P. (1989)
   Adverse pressure gradient effects on boundary layer
   transition in a turbulent free-stream.
   9th ISABE, Athens, Sept. 1989.

5. Abdel-Kareem, M.S.E. (1978)
   Effect of free-stream turbulence on boundary layer
   transition.
   PhD, Imperial College, University of London.

6. Charters, A.C. (1943)
   Transition between laminar and turbulent flow by
   transverse contamination.
   NACA TN 891.

7. Blair, M.F. and Werle, M.J. (1980)
   Influence of free-stream turbulence on the zero-pressure
   gradient fully turbulent boundary layer.
   AFOSR-TR 81-0514 and AD/A101094.

8. Schlichting, H. (1968)
   Boundary Layer Theory.
   McGraw-Hill, 6th ed.

9. Hart, M. (1985)
   Boundary layers on turbine blades.
   PhD, University of Cambridge.

10. Obremski, H.J., Morkovin, M.V. and Landahl, M. (1969)
    A portfolio of stability characteristics of
    incompressible boundary layers.
    AGARDograph 134.

11. Klebanoff, P.S. (1955)
    Characteristics of turbulence in a boundary layer with
    zero pressure gradient.
    NACA Rpt. 1247.

12. Murlis, J., Tsai, H.M. and Bradshaw, P. (1982)
    The structure of turbulent boundary layers at low
    Reynolds numbers.
    J.Fluid Mech., Vol.122, 13-56.

TABLE 1   THE PRESENT TEST CONDITIONS

Grid Type	d (mm)	M (mm)	File Number	Uo (m/s)	Rx ($\times 10^{-4}$)	Tuo (%)	Code
None	-	-	00101- 00201	6.0- 19.8	3.7- 190	0.16- 0.27	NOGRID
SMR	0.914	4.23	00202- 00307	14.9	9.3- 152	0.91- 0.48	SMRLO
SMR	"	"	00908- 01103	19.8	12.3- 202	0.87- 0.47	SMRHI
PR	6.35	25.4	00309- 00503	3.0	1.83- 29.5	2.8- 1.2	PRLO
PR	"	"	01104- 01209	5.3	1.52- 52.7	3.0- 1.1	PRHI (Case A)
PS	9.53	25.4	00505- 00610	3.0	0.84- 29.1	6.6- 3.3	PSLO
PS	"	"	00705- 00806 +00903 +00904 +00907	9.6	1.51- 95.7	6.0- 2.4	PSHI (Case B)

Notes: Grid type - None....Wind tunnel residual turbulence.
                  - SMR.....Square mesh, round wires.
                  - PR......Parallel array, round rods.
                  - PS......Parallel array, square rods.
                  - d = diameter and M = mesh of grids.
       File number - denotes file number in 1W?????.DAT
                     NB, final two digits range from 01 to
                     10 only
       Code   - denotes code name used for integral
                    parameter data sets in the title
                    ???????.INT

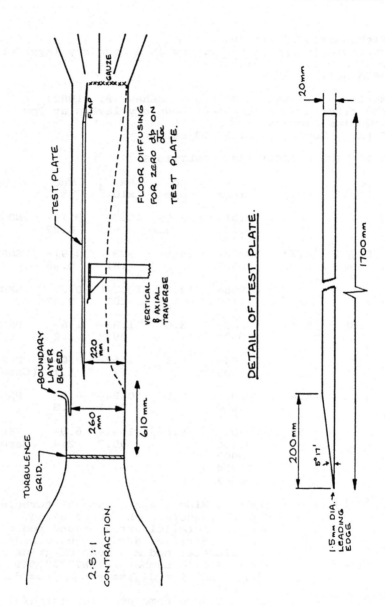

DETAIL OF TEST PLATE.

FIG. 1. A.R.L. WIND TUNNEL WORKING SECTION

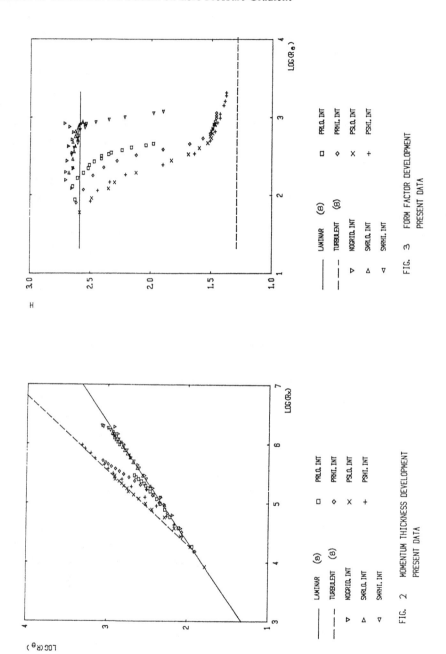

FIG. 3 FORM FACTOR DEVELOPMENT
PRESENT DATA

FIG. 2 MOMENTUM THICKNESS DEVELOPMENT
PRESENT DATA

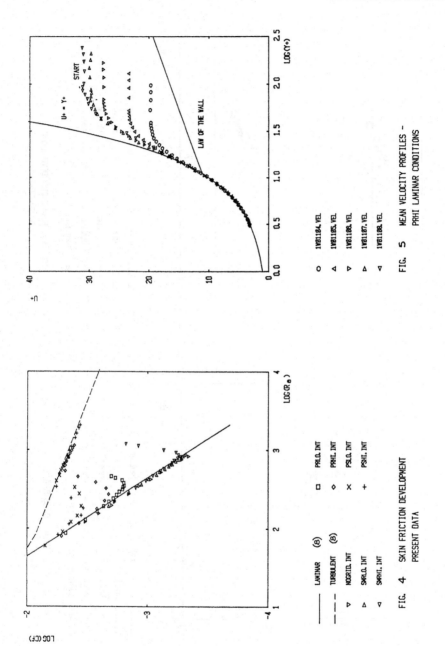

FIG. 5   MEAN VELOCITY PROFILES -
          PRHI LAMINAR CONDITIONS

FIG. 4   SKIN FRICTION DEVELOPMENT
          PRESENT DATA

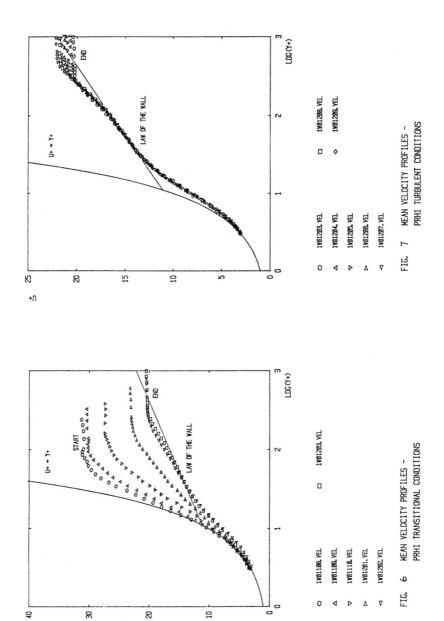

FIG. 7   MEAN VELOCITY PROFILES –
         PRHI TURBULENT CONDITIONS

FIG. 6   MEAN VELOCITY PROFILES –
         PRHI TRANSITIONAL CONDITIONS

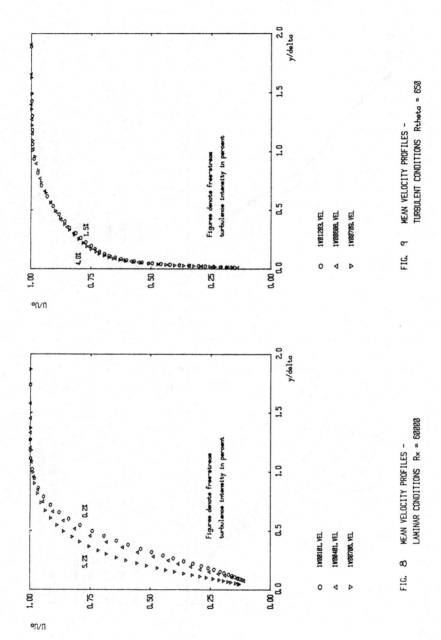

FIG. 8   MEAN VELOCITY PROFILES –
          LAMINAR CONDITIONS Rx = 60000

FIG. 9   MEAN VELOCITY PROFILES –
          TURBULENT CONDITIONS Rtheta = 650

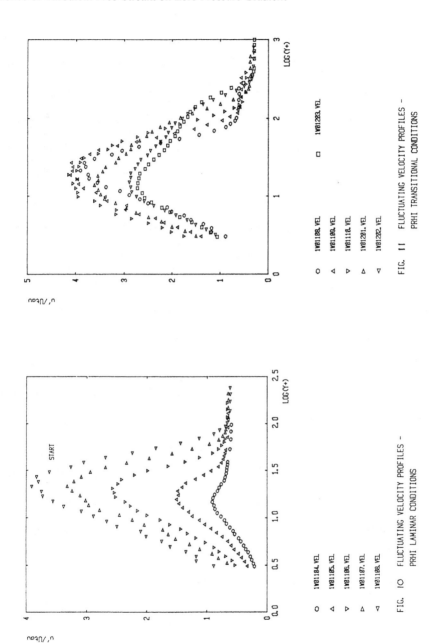

FIG. 11   FLUCTUATING VELOCITY PROFILES –
PRHI TRANSITIONAL CONDITIONS

FIG. 10   FLUCTUATING VELOCITY PROFILES –
PRHI LAMINAR CONDITIONS

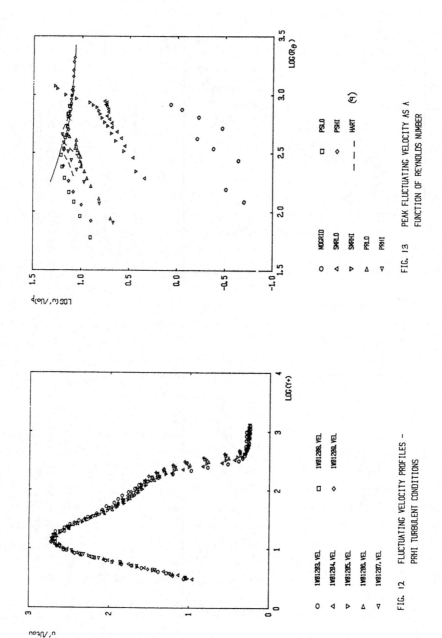

FIG. 12   FLUCTUATING VELOCITY PROFILES –
           PRHI TURBULENT CONDITIONS

FIG. 13   PEAK FLUCTUATING VELOCITY AS A
           FUNCTION OF REYNOLDS NUMBER

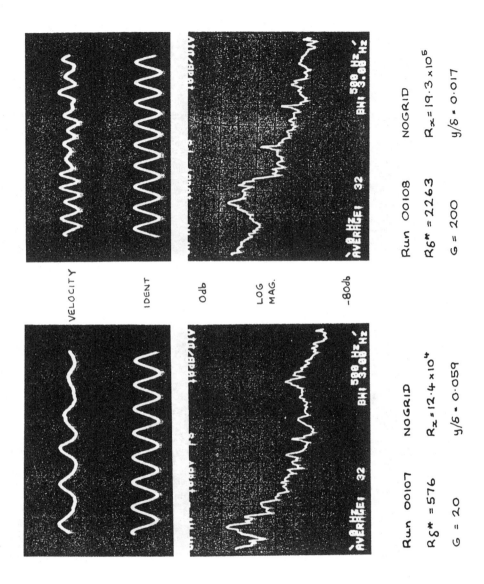

FIG. 14a.   TIME RECORDS AND SPECTRA WITHIN
THE BOUNDARY LAYER.

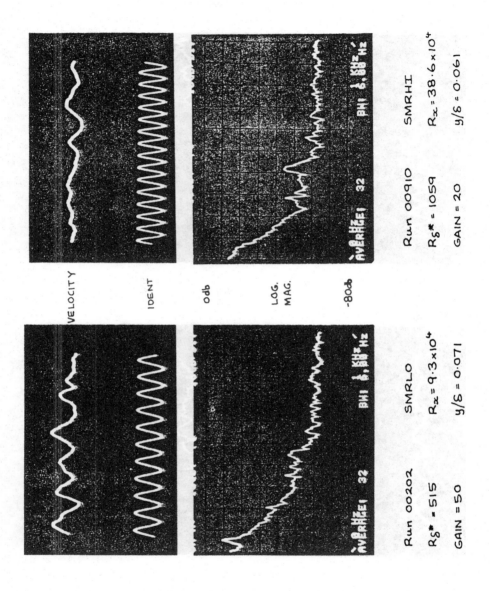

FIG. 14b.   TIME RECORDS AND SPECTRA WITHIN
THE BOUNDARY LAYER.

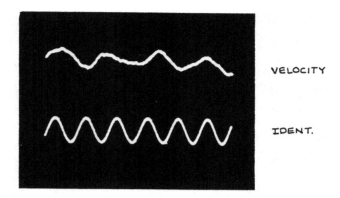

VELOCITY

IDENT.

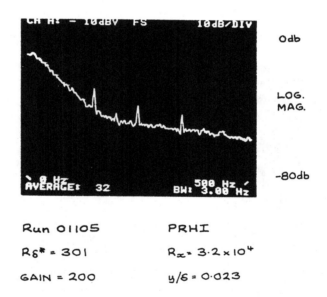

0db

LOG.
MAG.

-80db

Run 01105   PRHI

$R_{\delta^*} = 301$ $R_x = 3.2 \times 10^4$

GAIN = 200 $y/\delta = 0.023$

FIG. 14c. TIME RECORDS AND SPECTRA WITHIN

THE BOUNDARY LAYER

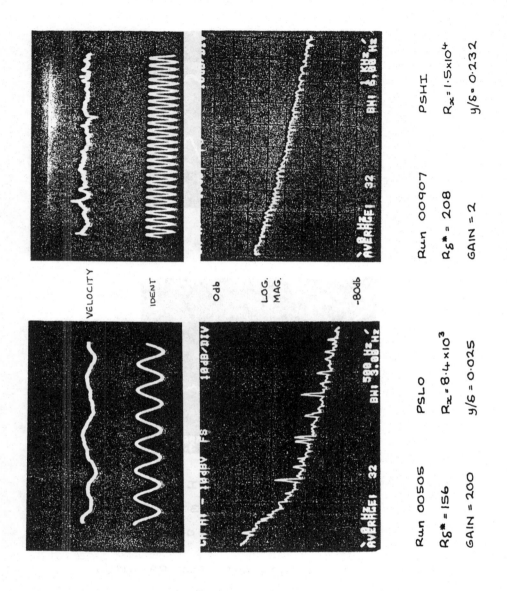

FIG. 14d.   TIME RECORDS AND SPECTRA WITHIN
             THE BOUNDARY LAYER.

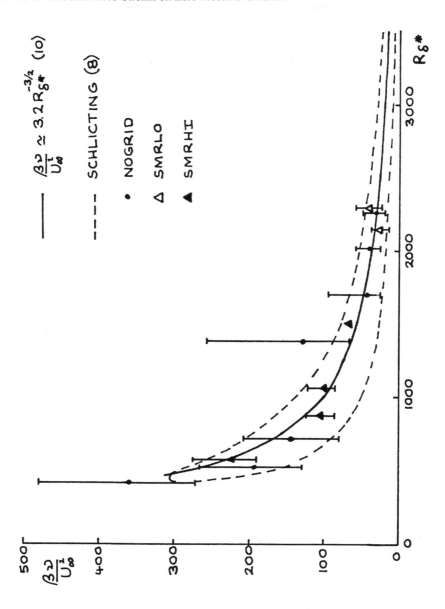

FIG. 15. PREDICTED AND OBSERVED MAXIMUM
TOLLMIEN - SCHLICHTING FREQUENCIES.

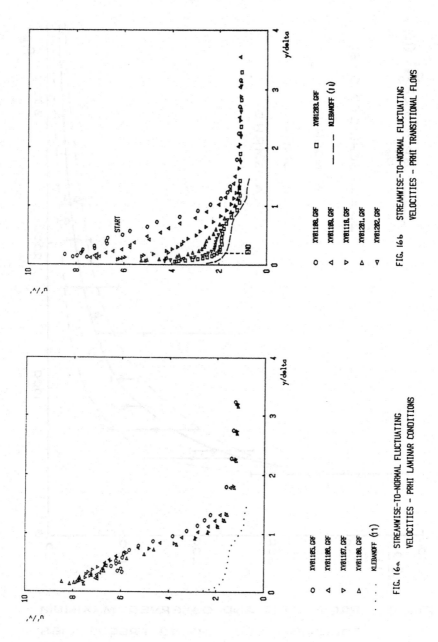

FIG. 16a   STREAMWISE-TO-NORMAL FLUCTUATING
            VELOCITIES - PRHI LAMINAR CONDITIONS

FIG. 16b   STREAMWISE-TO-NORMAL FLUCTUATING
            VELOCITIES - PRHI TRANSITIONAL FLOWS

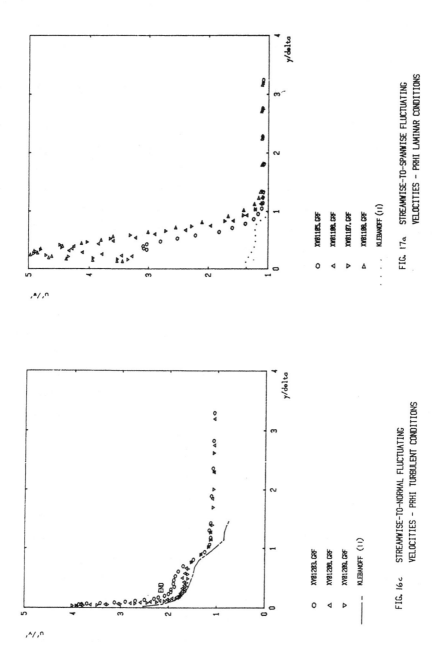

FIG. 17a STREAMWISE-TO-SPANWISE FLUCTUATING
VELOCITIES - PRHI LAMINAR CONDITIONS

o   XY01185.GRF
△   XY01186.GRF
▽   XY01187.GRF
△   XY01188.GRF
. . . . KLEBANOFF (11)

FIG. 16c STREAMWISE-TO-NORMAL FLUCTUATING
VELOCITIES - PRHI TURBULENT CONDITIONS

o   XY01203.GRF
△   XY01208.GRF
▽   XY01209.GRF
—   KLEBANOFF (11)

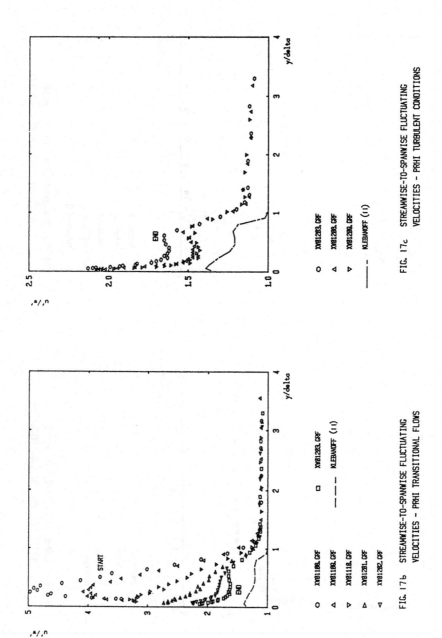

FIG. 17c STREAMWISE-TO-SPANWISE FLUCTUATING
          VELOCITIES - PRHI TURBULENT CONDITIONS

FIG. 17b STREAMWISE-TO-SPANWISE FLUCTUATING
          VELOCITIES - PRHI TRANSITIONAL FLOWS

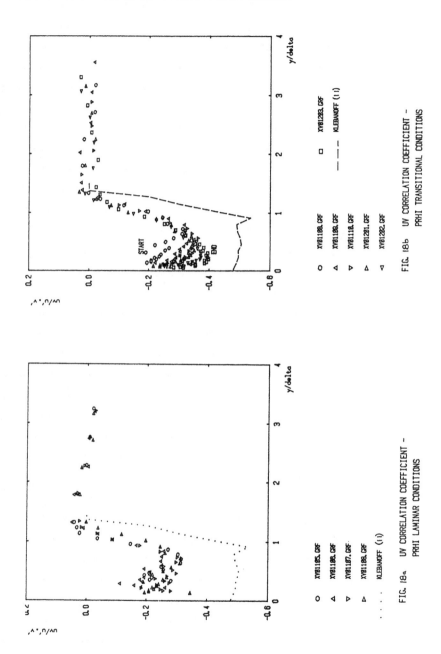

FIG. 18a UV CORRELATION COEFFICIENT –
PRHI LAMINAR CONDITIONS

FIG. 18b UV CORRELATION COEFFICIENT –
PRHI TRANSITIONAL CONDITIONS

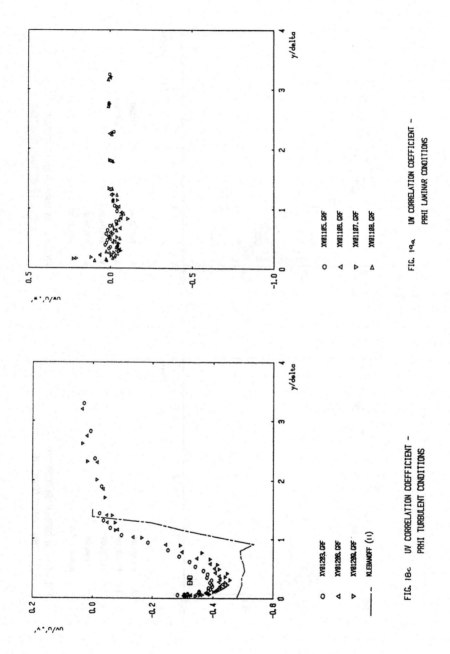

FIG. 19a   UV CORRELATION COEFFICIENT -
            PRHI LAMINAR CONDITIONS

FIG. 18c   UV CORRELATION COEFFICIENT -
            PRHI TURBULENT CONDITIONS

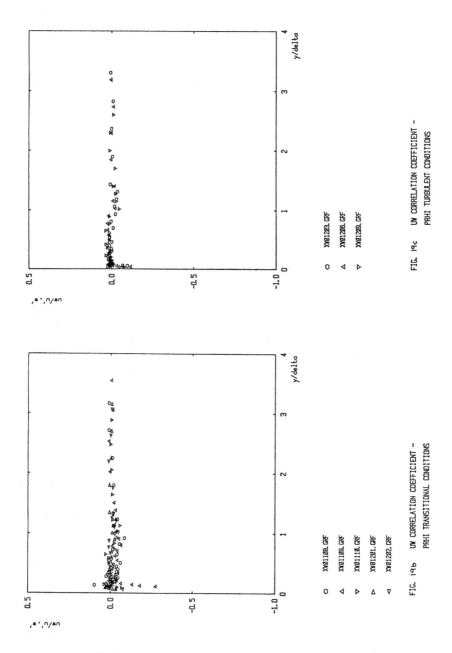

FIG. 19b UW CORRELATION COEFFICIENT -
PRHI TRANSITIONAL CONDITIONS

FIG. 19c UW CORRELATION COEFFICIENT -
PRHI TURBULENT CONDITIONS

A Correlation Analysis Approach to the T3 Test Case

P.E. ROACH

Advanced Research Laboratory, Rolls-Royce plc

1 INTRODUCTION

There are many models developed over the years which, together with a wide experimental database, have been used to in an effort to predict transitional boundary layer flows. This paper provides a correlation analysis of integral quantity data for the T3A & B Test Cases, and other zero pressure gradient data presented by Roach & Brierley (1990), together with earlier data, which should aid future modelling efforts as well as general understanding of the transition process. The influences of free-stream turbulence intensity, length scale, and leading edge geometry are addressed in turn and a number of conclusions drawn regarding the best way to correlate observed responses in each of the laminar, transitional, and fully turbulent flow regimes.

2 THE TRANSITION ZONE

2.1 Influence of free-stream turbulence

A detailed examination of all available experimental data reveals certain trends which may be used to untangle the complex phenomenon of boundary layer transition. Indeed, by restricting consideration to zero pressure gradient flat plate studies, as in the present Test Cases, it seems possible to account for the influence of free-stream turbulence intensity and length scale. In particular Taylor's (1936) proposal appears to offer a correlation of both the present and previous data superior to any similar models; such as that of van Driest & Blumer (1963). Taylor concluded that the 'critical Reynolds number ',Rec, of a sphere is a function of: $Tu(D/Lu)^{1/5}$ (where Tu & Lu are tintensity and length scale of the external turbulence and D the sphere diameter)

Using this relationship Taylor was able to demonstrate a good correlation for the critical Re of spheres obtained by several workers (similarly good correlations were also later obtained by Dryden et al.(1937), Scubauer (1939), and Evans (1971) for the start of transition on a sphere, on an elliptic cylinder, and in a 2-D compressor cascade, respectively).

348

For flat plate studies, numerous correlations exist which provide a relationship between the start of transition Reynolds number, Rexs, and the intensity of free-stream turbulence, for example those of van Driest & Blumer (1963) and of Abu-Ghannam & Shaw (1980). However the degree of scatter found for each of these correlations leaves much to be desired, see Figure 1, and it is clear that a simple correlation of the start of transition with external turbulence intensity alone is inadequate. However it can be shown that Taylor's proposal, which accounts also for the additional influence of free-stream turbulence length scale, can be used to correlate all of the earlier data obtained for different turbulence-generating grids by Hislop (1940). Although he was not able to measure length scales, and made the incorrect assumption about these that Lu was proportional to, the grid mesh size, Roach (1987) has recently documented the flow development behind such grids and it is clear that Taylor's parameter does accurately account for the systematic influence of length scale. At the same time it would appear that the end of transition can be correlated quite well with its start without introducing any additional parameters. In particular the correlation of Dhawan & Narasimha (1985) was found to be in reasonably good agreement with the present Test Case and other data, over a wide range of Reynolds number (Figure 2).

2.2 Influence of Leading Edge Geometry

The RDT analysis of Hunt (1973) showed that turbulence approaching a solid body undergoes changes in its component turbulence intensities near the leading edge which depend on the ratio of the length scale Lu of the eddies to the thickness, t, of the body. It has generally been assumed that such influences cannot be detected at distances 'sufficiently downstream of the nose. However Abdel-Kareem (1978) has shown that the pressure distribution around the nose of a flat test plate has a large effect on the position of transition, such that as the fineness ratio increases the laminar boundary layer becomes more stable and the position of transition onset moves downstream. Figure 3 presents a comparison of the Test Case configuration compared to results for other elliptic-nose geometries and it is clear that Rexs varies systematically also with nose fineness ratio. Indeed a simple power law dependence on this is suggested:

$$Rexs = F \{ Ta \ (Xn/t)^p \}$$

(where p=-3/7 for elliptic noses)   (1)

A similar trend is obtained for sharper leading edges (usually defined by two circular arcs) although the power law index p=-1/5 for these. The data also suggest that such sharper leading edge geometries induce earlier transition, perhaps because the sharper nose may be less susceptible to small, localised flow separation bubbles induced by the flow incidence variations associated with the eddying free-stream. Alternatively, or in addition, an elliptic nose will impose stronger favourable pressure gradients on the developing laminar boundary layer, thus possibly effecting a later transition.

3 BOUNDARY LAYER DEVELOPMENT

3.1 Laminar Flow

The influence of free-stream turbulence on the development of the initial laminar boundary layer is examined in Figure 4. Here the Test Case and other data are compared with the results of Dyban et al. (1976). The latter show a definite tendency for the wall shear stress to increase and the shape factor H to decrease with increasing free-stream turbulence intensity. Although the present data are broadly in agreement with this finding, the scatter is large and they even suggest that Cf might increase with decreasing turbulence activity in each of the individual flow cases. This is clearly contrary to expectations and requires the following explanation.

Starting with Polhausen's fourth order polynomial representation of the laminar velocity profile and following Taylor (1936) it can be shown that the increments in each of the int pgral quantities due to the influence of external turbulence will be functions Fd^*, $F\theta$, FCf & FH of Rx and the Taylor parameter. Analysis reveals that:

$$Fd^* = F\theta = Rx/Rxs^{6/5} \qquad FCf = Rx/Rxs^{1.9} \qquad FH = Rx/Rxs^{7/5}$$

Thus, although the changes in these quantities do not correlate with local external properties they all correlate with the free-stream turbulence properties at the start of transition, via Rxs which is itself a function of the Taylor parameter. This is a very important finding which suggests that bursts of turbulence at the onset of transition influence the laminar boundary layer region, perhaps by means of pressure waves propogating upstream which modify the pre-turbulent activity and hence the mean flow properties. In fact the present data can also be reasonably well correlated with the turbulence properties at the leading edge of the test plate. This finding provides a link between the onset of transition and the interaction of the free-stream turbulence with the test plate leading edge geometry.

3.2 Turbulent Flow

A good fit to general data on the development of boundary layer momentum in the absence of free-stream turbulence is provided by the relationship:

$$Re\theta = 162 + 0.032 (Rx - Rxs)^{4/5} \qquad\qquad (2)$$

The minimum value of 162 corresponds to the upper limit of stability of the laminar boundary layer to linear disturbances (see Schlicting (1968)). A study of the present data for 1%<Tu<6% suggests that the momentum thickness growth is unaffected by this. Other data sets (eg Hancock (1980)) do not support this contention, but it is interesting to note that all of these involved the use of additional heated wires located close to trip wires at the leading edge of test plates, and it would appear that the addition of a turbulent free-stream significantly increases the momentum loss due to these wires (an observation which Hancock himself alluded to).

The growth of the boundary layer thickness, δ_0, in the absence of free-stream turbulence shows a simple dependency on $Re\theta$ of: $\qquad R\delta_0 = 9Re\theta \qquad$ over a reasonable Re range.

This is similar to that derived from the 1/7 power law velocity profile (also discussed by Schlicting (1968)). With the addition of external turbulence the boundary layer thickness also seems to develop at the same rate as with a non-turbulent free-stream.

The decay of the skin friction coefficient Cf_0 in undisturbed boundary layers is well represented by the correlation: $Cf_0 = 0.200 \, (\log R\theta + 1.45)^{-2.584}$ (3)

essentially that given by Schultz-Grunow (1940) [See Schlichting (1968)], but with the constants slightly adjusted. However the alternative empirical relationship of Ludwieg & Tillman (1950), which relates Cf to $Re\theta$ and H by:

$$Cf_0 = 0.246 \times 10^{-0.678H} \, Re\theta^{-0.268} \equiv 0.246 \, LT \qquad (4)$$

is more useful since this applies even for the addition of free-stream turbulence (and for variable pressure gradients), although only for $Re\theta > 2000$. This needs extending to lower $Re\theta$. However a large selection of data, covering the range $400 < Re\theta < 10000$, is found to be very well correlated by: $Cf/LT = 0.240 \, A \, [- \log_e(H-1)]^B$ (5)

where: $A = 1 + \exp[-0.226(Re\theta)^{3/8}]$ & $B = 2.27 \times 10^{-4} \, Re\theta - 0.382$

Furthermore: $\Delta Cf/Cf_0 = (Cf - Cf_0)/Cf_0 = (K \, Tu^2)/Cf_0(Lu/\delta_0)$ (6)

where the subscript o denotes non-turbulent free-stream and the constant K is of order 2. The turbulence dissipation length scale Lu is the same as that adopted by Hancock (1980) and is readily evaluated from a knowledge of the free-stream turbulence intensity decay for both Test Case conditions - see Roach (1987).

Now it is to be expected that K will actually be a function of Re, again due to the essentially developing/non-equilibrium nature of such low-Re boundary layers. Castro (1984) has argued that a low-Re correction should be represented by a complex relationship between the free-stream turbulence properties and Re. Whilst not disagreeing with his argument an adequate correction factor Rcf for the present data is:

$$Rcf = 1.04 \, \{ \, 1 - \exp [-3.14 \times 10^{-4} \, (Re\theta - 162)] \} \qquad (7)$$

such that: $\Delta Cf/Cf_0 = Rcf \, Tu^2 / (Cf_0 \, Lu / \delta_0)$ (8)

This correlates the experimental data well, see Figure 5, and the asymptotic, high Re value of Rcf is quite close to the expected value of 2.

The undisturbed shape factor development is well represented by the solution to equations (3) and (5). With the addition of free-stream turbulence H is found to decrease and both Hancock (1980) and Castro (1984) have found that for high Re:

$$\Delta H/H_0 = -0.47 \, \Delta Cf/Cf_0 \qquad (9)$$

which is to be expected because of the universality of the Ludwieg-Tillman relationship (4). To include the effect of $Re\theta < 2000$ H can be deduced by solving equations (5) & (8). Another shape parameter of importance is the wake strength n, but it is possible to relate this quantity to Cf by solving the 'universal' wall-wake formula for a turbulent boundary layer (see Dean (1976)), which may be written:

$$U^+ = (\log_e y^+)/\kappa \; + \; C \; + \; g \tag{10}$$

where: $g = (\; (1+6\pi) \; (y/\delta) - (1+4\pi) \; (y/\delta)^3)/\kappa$ and $\kappa = 0.41, \; C = 5.0$

Solving this equation at $y = \delta$:

$$0.99\kappa/CF = \log_e (R\delta) + \log_e (CF) + \kappa C + 2\pi \tag{11}$$

where $CF = (Cf/2)$

Now equation (10) may be written for both a turbulent and a non-turbulent free-stream and these two equations solved for the same Rd, whence:

$$\Delta\pi / \pi_0 = (\pi - \pi_0)/\pi_0 = (-1/4 \; \pi_0) \; \Delta Cf/Cf_0 \; [1 + \kappa \; (2\sqrt{Cf_0})]$$

or $\Delta\pi / \pi_0 = - F\pi \; \Delta Cf/Cf_0$ $\tag{12}$

where $F\pi$ is a function of $Re\theta$ only and the available data suggests that $F\pi = 354 \; R\theta$ which provides a good correlation to the test case results.

3.3 Transitional Flow

The present experiments appear to be consistent with the correlations proposed by Abu-Ghannam & Shaw (1980) and by Narasimha (1985) for the development of the various integral parameters through transition. However the test case data shows a similarly high degree of scatter about the Abu-Ghannam & Shaw curves as exhibited by other data. Closer study of the present data reveals some interesting trends. In section 3.2 it has been shown that there exists a universal relationship between $Re\theta$ and $Rx-Rxs$ for turbulent boundary layers, both with and without free-stream turbulence. This was also shown to be the case for Rd versus $R\theta$. Surprisingly it appears that these relationships also represent the transition region data reasonably well, but only when the free-stream turbulence is sufficiently high (>0.5 - 1% as for both Test Cases T3A & B). Above this limit the flow appears to be dominated by vorticity-related disturbances, whereas below this limit the influences of acoustic and vibrational disturbances dominate. Low turbulence data do however appear to assymptote towards the turbulent boundary layer curves at sufficiently high Re.

In addition the skin friction coefficient Cf also seems to be represented by a modified Ludweig-Tillman relationship (although no such agreement is found for the individual Cf or H versus $Re\theta$ variations - these parameters gradually changing from laminar to turbulent levels). This good agreement of transition zone with fully turbulent data is rather remarkable. In the analysis of the turbulent flow region it was stated that the primary effect of free-stream turbulence is to increase Cf and reduce H, by redistributing momentum within the boundary layer. It appears the same is true for the transitional layer also. This suggest a strong similarity between the structure of transitional and turbulent boundary layers; a finding which tends to confirm earlier observations made by Blackwelder (1983).

Acknowledgements
Support provided for this work by DTI is gratefully acknowledged.

References

Abdel-Kareem, M.S.E. (1978) Effect of free-stream turbulence on boundary layer transittion. Ph.D thesis, University of London.

Abu-Ghannam, B.J. & Shaw, R. (1980) Natural transition of boundary layers - the effects of turbulence, pressure gradient and flow history. J. Mech. Eng. Sci. 22 (5), 213-228.

Blackwelder, R.F. (1983) Analogies between transitional and turbulent boundary layers. Phys. Fluids 26 (10), 2807-2815.

Castro, I.P. (1984) Effects of free-stream turbulence on low-Reynolds number boundary layers. Trans ASME JFE 106, 298-306.

Dean, R.B. (1976) A single formula for the complete velocity profile in a turbulent boundary layer. Trans ASME JFE 106, 723-727.

Dryden, H.l., Scubauer, G.B., Mock, W.C. & Skramstad, H.K. (1937) Measurement of intensity and scale of wind tunnel turbulence and their relation to the critical Reynolds number of spheres. NACA Rep.581.

Dyban, V.P., Epik, E.V. & Suprun, T.T. (1976) Characteristics of the laminar boundary layer in the presence of elavated free-stream turbulence. Fluid. Mech.-Soviet. Res. 5 (4), 30-36.

Evans, B.J. (1971) Effects of free-stream turbulence on blade performance in a compressor cascade. Ph.D thesis, University of Cambridge.

Hancock, P.E. (1980) The effect of free-stream turbulence on turbulent boundary layers. Ph.D. thesis, Imperial College.

Hislop, G.S. (1940) The transition of a laminar boundary layer in a wind tunnel. Ph.D. thesis, University of Cambridge.

Hunt, J.C.R. (1973) A theory of turbulent flow round two-dimensional bluff bodies. J.Fluid Mech. 61, 625.

Ludweig, H. & Tillman, W. (1950) Investigations of the wall shearing stress in turbulent boundary layers. NACA TM 1285.

Narasimha, R. (1985) The laminar-turbulent transition zone in the boundary layer. Prog. Aerospace. Sci. 22, 29-80.

Roach, P.E. (1987) The generation of nearly isotropic turbulence by means of grids. Int. J. Heat & Fluid Flow 8 (2), 82-92.

Roach, P.E. & Brierley, D.H. (1990) Transition under the influence of free-stream turbulence - T3A & B Test Cases. See this Proceedings Volume.

Schlicting, H. (1968) Boundary Layer Theory, McGraw-Hill.

Schubauer, G.B. (1939) Air flow in the boundary layer of an elliptic cylinder. NACA Rep.652.

Taylor, G.I. (1936) Statistical theory of turbulence. V - Effect of turbulence on boundary layers. Proc. Roy. Soc. 156(A), 307-317.

van Driest, E.R. & Blumer, C.B. (1963) Boundary layer transition: Free-stream turbulence and pressure gradient effects. AIAA J. 1 (6), 1303-1306.

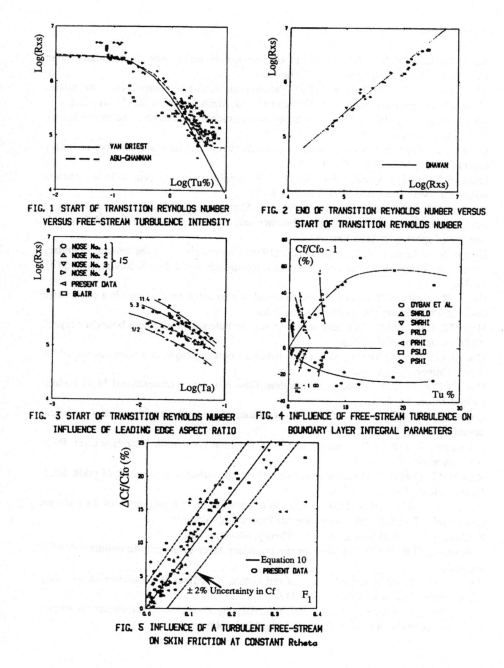

FIG. 1 START OF TRANSITION REYNOLDS NUMBER
VERSUS FREE-STREAM TURBULENCE INTENSITY

FIG. 2 END OF TRANSITION REYNOLDS NUMBER VERSUS
START OF TRANSITION REYNOLDS NUMBER

FIG. 3 START OF TRANSITION REYNOLDS NUMBER
INFLUENCE OF LEADING EDGE ASPECT RATIO

FIG. 4 INFLUENCE OF FREE-STREAM TURBULENCE ON
BOUNDARY LAYER INTEGRAL PARAMETERS

FIG. 5 INFLUENCE OF A TURBULENT FREE-STREAM
ON SKIN FRICTION AT CONSTANT R_{theta}

Calculations for the T3 Transition Test Cases

N.T. BIRCH, Y.K. HO & P. STOW

Theoretical Sciences Group, Rolls-Royce plc

1 COMPUTATIONAL METHOD

The standard Reynolds averaged Navier-Stokes equations were adopted with a one-equation turbulence/transition model. The equations were solved using a 3D fully elliptic, finite volume, pressure-correction technique due to J.G.Moore (see Moore & Moore (1984), Moore (1985)). A non-staggered grid system was adopted with all the flow variables stored at cell vertices. High order discretisation accuracy was achieved by using a bi-linear distribution of flow properties in two-dimensions over the cells. Stability of the scheme was ensured by adopting upwind control volumes; this ensured accurate representation of the convection terms and manitained stability. A SIMPLER-type pressure-correction approach was adopted.

2 TURBULENCE/TRANSITION MODEL

A one equation k-l model was adopted (k being the turbulence kinetic energy and lε a dissipation length scale). The transport equation employed for k was essentially the same as that used in the standard Hassid-Poreh (1975) k-l model, but with the constants adopted by Birch (1987). The dissipation length scale used was a modified Baldwin-Lomax three-layer formulation (for the inner layer, outer layer and free-stream) also proposed by Birch (1987).

3 APPLICATION TO THE T3 TEST CASES

For these flow problems the code was run in two-dimensional mode. The outer boundary was treated as inviscid and the inner wall viscous i.e. a no-slip boundary condition was applied. The computations were run as for a channel flow, but with an expanding channel width to allow for the boundary layer growth. This was assumed to be laminar and an appropriate expanding grid system was employed to take account of this boundary layer development. A typical grid mesh is illustrated in Figure 1. For the T3A case 116 x 51 grid points were used and for T3B 196 x 51.

The initial conditions, as given in the test case specification (for method 2), were implemented at the second axial station shown in Figure 1; the first axial grid line being a dummy station. A uniform free-stream velcity was therefore assumed together with a Blasius velocity profile for the initial laminar boundary layer. The initial turbulence kinetic energy distribution was the fourth power law one prescribed in the specification, and the external length scale lε was calculated from measured turbulence data (see Roach (1987)), assuming isotropic isotropic turbulence conditions in the free-stream, and:

$$\rho U \ dk/dx = - C_D \ k^{3/2} /l\varepsilon \qquad (1)$$

so that: $l\varepsilon \ = \ - C_D \ k^{3/2} / (\rho U \ dk/dx)$ where: $C = 0.164$ (2)

4 RESULTS

Predictions of C_f, H, and θ, for both T3A and T3B Test Cases, are presented in Figures 2,3,4 where they are compared with two correlation methods: An Integral approach based on Green's Lag Entrainment Equation which switches from laminar to turbulent in line with the Liverpool correlation of Abu-Ghannam & Shaw (1980), and a mixing length scheme based on the Cebeci-Smith model, but coupled to a slightly modified form of the Abu-Ghannam & Shaw correlation, with transition initiated when the local momentum thickness Reynolds number exceeded their critical Reynolds number.

It can be observed from Figures 2a&b that for the milder free-stream turbulence T3A case the 1-equation model predicts transition a little too late compared to the ARL data and the mixing length model, which is close to this (The onset of transition is equally well predicted by the integral method, but the overall Cf distribution is not well reproduced). However for the higher free-stream tubulence T3B case the k-l model predicts transition at approximately the same location as for T3A and hence much too late. The problem appears to be that the model is not diffusion controlled. Other tests have established that it is insensitive to variations in free-stream tubulence intensity in the range 1/4 to 1% (as one might expect), but is also relatively insensitive to progressively higher intensities in the range of 2-4% (approximately the levels at the transition location in the T3A & B cases). As a result it predicts transition too early for very low free-stream turbulence levels (< 1%) and too late for higher. In addition the fully turbulent Cf level is not recovered for either T3A or T3B case. Both the integral and mixing length models do at least achieve this and both do indicate an earlier transition onset with increasing free-stream turbulence level, but as indicated by Figure 2b this is insufficient to give as good predictions for T3B as for T3A.

The Shape factor results shown in Figures 3a&b reflect the Cf findings above. Thus H is better predicted for the T3A case than T3B and the k-l model performs progressively worse than the correlation methods. The fully turbulent flow level is attained by all of the models, but none of these correctly predict the reduction in H prior to transition onset (as indicated by departures from the laminar Cf curve) shown by the data. By comparison the momentum thickness distribution is quite well predicted by all three methods for T3A (see Figure 4a), but only the mixing length model comes close to the data for T3B (figure 4b).

The mean velocity profile development (Figure 5a&b) is somewhat underpredicted by the k-l model in both cases, the predicted profiles being rather less full than those measured, in keeping with the delayed transition onset. The k profiles (not shown) indicate the existence of a small peak within the initial 'laminar' boundary layer which grows outwards prior to transition and the development of a fuller turbulence energy profile.

5 REFERENCES

Abu-Ghannam, B.J. & Shaw, R. (1980) Natural Transition of Boundary Layers - the Effects of Turbulence, Pressure Gradient and Flow History. J. Mech. Eng. Sci. 22 (5).

Birch, N.T. (1987) Navier-Stokes predictions of transition, loss and heat transfer in a turbine cascade. ASME Paper No. 87-GT-22.

Hassid, S & Poreh, M. (1975) A Turbulent Energy Model For Flows With Drag Reduction. Trans ASME J.Fluids Eng. 97, 234-241.

Moore, J.G. (1985) An elliptic computational procedure for 3D viscous flow. AGARD LS 140 on 3D Computational Techniques Applied to Internal Flows in Propulsion Systems.

Moore, J.G & Moore, J. (1984) Calculation of horseshoe vortex flow without numerical mixing. ASME Paper No. 84-GT-141.

Roach, P.E. (1987) The generation of nearly isotropic turbulence by means of grids. Int. J. Heat & Fluid Flow 8 (2), 82-92.

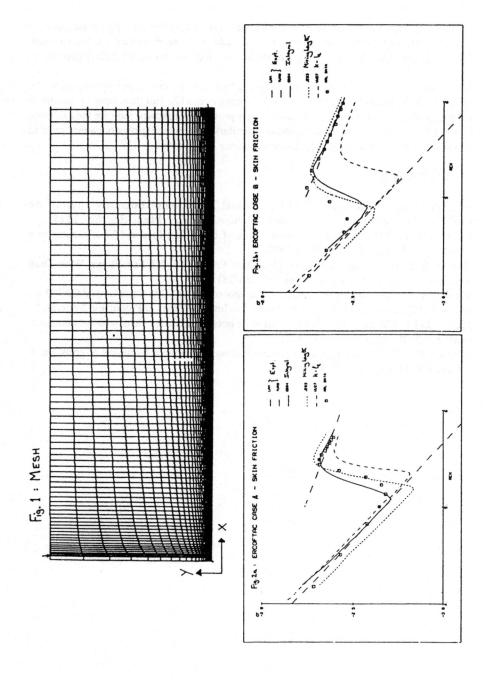

Fig. 1 : MESH

Fig. 2a : ERCOFTAC CASE A - SKIN FRICTION

Fig. 2b : ERCOFTAC CASE B - SKIN FRICTION

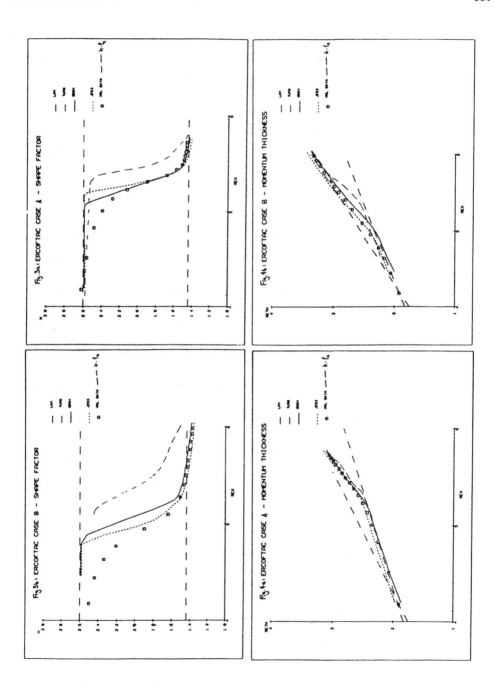

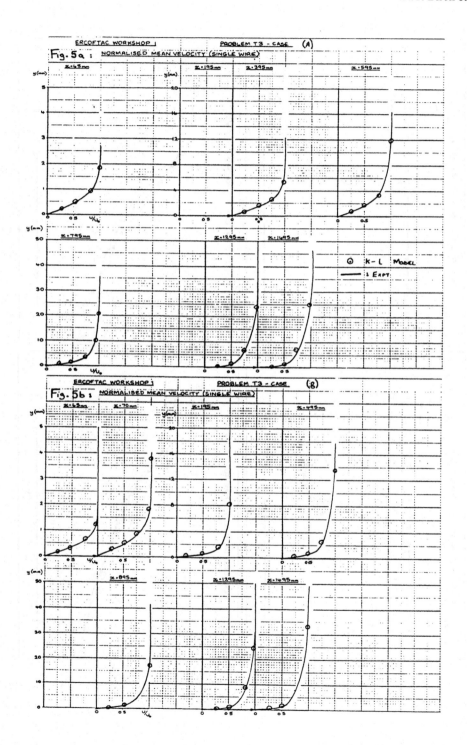

Fig. 5a : NORMALISED MEAN VELOCITY (SINGLE WIRE)

Fig. 5b : NORMALISED MEAN VELOCITY (SINGLE WIRE)

PREDICTION OF THE EFFECT OF FREE STREAM TURBULENCE ON A LAMINAR BOUNDARY LAYER AND TRANSITION

J.   BERNARD

J.M. FOUGERES

J.L. MANDERLIER

SNECMA

Accurate prediction of heat transfer on gas turbine blades in severe operating conditions requires a detailed knowledge of the boundary layer behaviour, in particular concerning transition to turbulence.

The present contribution deals with the prediction of the effects of free-stream turbulence on laminar and transitional boundary layers, using a mixing-length turbulence model. From an industrial point of view, this model provides a good compromise between computing time and accuracy.

The boundary layer code is based on Patankar and Spalding's algorithm and discretization scheme. The initialization of the boundary layer profiles is given by Polhausen's method. The turbulence model, based on Mac Donald and Fish's formulation, is designed to take into account the effect of free-stream turbulence and to determine the transition zone. Various improvements to the basic turbulence model resulted in increased numerical stability of the method and better detection of the onset of transition.

The code was applied to test cases T_{3A} and T_{3B}. The method is presented in the following.

1 - PRINCIPLES OF THE METHOD :

The method solves the boundary layer equations discretized using the finite difference technique of Patankar and Spalding (1).

361

Special attention was given to the following points in the development of the code (4) :

- The initial velocity and temperature profiles are given by Polhausen and the initial boundary layer thickness is determined iteratively.

- Close to the wall, the wall function suggested in (1) was used.

- The entrainment of inviscid fluid by turbulent eddies at the edge of the boundary layer was modelled as in (1).

- The mixing-length model, presented below, was carefully calibrated against a wide range of experimental cases.

2 - TRANSITION AND TURBULENCE MODELING :

The model used is that of Mac Donald and Fish (2) improved by Blair and Edwards in 1982 (3). It is a mixing length model with the eddy viscosity expressed as :

$$\mu t = \rho \, l^2 \left| \frac{\partial u}{\partial y} \right|$$

The mixing length, l, is given by :

$$l = D \, l_\infty \left\{ th(ky/l_\infty) + \tfrac{1}{2} \, (1-th(k\delta/l_\infty)) \, (1-\cos(\pi y/\delta_\tau)) \right\}$$

where : - k = 0.43 (Von Karman constant)

- l_∞ is the mixing length at the boundary layer edge

- δ is the boundary layer thickness

- δ_τ is the boundary layer "stress thickness", value of y where the shear stress reaches 2% of the maximum stress, starting from the outer edge

- D is the sublayer damping factor :

$$D = \sqrt{\tfrac{1}{2} \, (1 + erf \, ((y^+ - 23)/8))} \; ; \; y^+ = y \cdot \sqrt{\tau_w \rho}/\mu$$

An additional relation is necessary for the <u>calculation of</u>
<u>l∞</u> . The integral form of the Turbulence Kinetic Energy
equation together with an algebraic expression relating l∞
to a parameter Φ1 characterizing the integral T.K.E. are
used. The coupled set of equations is solved iteratively.

The <u>integral form of the T.K.E.</u> equation is defined as :

$$\frac{d}{dx}\left\{\frac{\rho_e \cdot u_e^3}{2a1}\ \Phi1\right\} = \rho_e\ u_e^3\ (\ \Phi2 - \Phi3) + E$$

Where Φ1, called the <u>convection integral</u>, is defined as :

$$\Phi1 = \int_0^{\delta\tau} \frac{\rho \cdot u}{\rho_e \cdot u_e}\left\{1\ \frac{\partial(u/u_e)}{\partial y} + a1\ f(y/\delta_\tau)\ \frac{q_e^2}{u_e^2}\right\}^2 dy$$

$$q_e = \sqrt{\overline{u'^2} + \overline{v'^2} + \overline{w'^2}}$$

and characterizes the integral T.K.E.

Φ2, the <u>net production integral</u>, is defined as :

$$\Phi2 = \int_0^{\delta\tau} \rho/\rho_e \cdot 1^2 \cdot \left\{\frac{\partial(u/u_e)}{\partial y}\right\}^3 \cdot (1 - 1/L)\ dy$$

L = 0,11. δ. D. th (ky/0,11.δ)

Φ3, the <u>normal stress production integral</u> is defined as

$$\Phi3 = \int_0^{\delta\tau} \rho/\rho_e \cdot \frac{a2-a3}{a1} \cdot \left\{(1\ \frac{\partial(u/u_e)}{\partial y}\ 2) + a1\ f(y/\delta_\tau)\ q_e^2/u_e^2\right\}$$
$$\cdot \frac{1}{u_e}\ \frac{du}{dx}\ dy$$

and E, the turbulence kinetic energy source term :

$$E = q_e^2/2 \ (\ \rho_e \ u_e \ d\delta_\tau/dx - (\ \rho v)_e \ )$$

In these integrals the function f is :

$$f \ (y/\delta_\tau) = \tfrac{1}{2} \ (\ 1-\cos(\pi y/\delta_\tau) \ )$$

Coefficients a1, a2, a3 relate the Reynolds shear stress and turbulence intensity components to the turbulence kinetic energy :

$$- \ \overline{u'v'} = a1 \ (\overline{q^2} - f \ (y/\delta_\tau) \ \overline{q_e^2})$$

$$\overline{u'^2} = a2 \ \overline{q^2}$$

$$\overline{v'^2} = a3 \ \overline{q^2}$$

$$\overline{w'^2} = (1 - a2 - a3) \ \overline{q^2}$$

Mac Donald and Fish suggested the values a2 = 0,5 and a3 = 0,2 . Coefficient a1 is determined by :

$$a1 = \frac{ao \ R_\theta/100}{1+ \ ao/0,15 \ (R_\theta/100-1)} \quad ; \ ao = 0,0115$$

where R_θ, a Reynolds number based on momentum thickness, is empirically related to a turbulent Reynolds number R_τ by :

$$R_\tau > 40 \qquad R_\theta = 68,1 \ R_\tau + 614,3$$

$$R_\tau < 0,2 \qquad R_\theta = 100 \ R_\tau^{\ 0,22}$$

This inferior bound of R_τ was proposed in (4). It gives excellent results on Consigny's test (5).

Between the bounds, a cubic polynomial is used to connect the two curves.

R_τ is defined as :

$$R\tau = \frac{1/\delta_\tau \int_0^{\delta\tau} \mathcal{V}_t \, dy}{1/\delta_s \int_0^{\delta s} \mathcal{V} \, dy}$$,where δ_s is the viscous sublayer thickness

To close this system of equations, $\Phi 1$ and l_e are related in the following fashion :

$$\Phi 1 = A \, l_e^2 + B \, l_e + C$$

where A,B and C are three empirical constants.

3 - COMPUTATION :

The computation was performed, starting from uniform inlet conditions at x = 0 (plate leading edge).

The inlet conditions are :
- total temperature : 293 K
- total pressure   : 1,013 10^5 Pa

The cpu time to obtain these results for one case is about 10 seconds on an IBM 3090.

The output values are skin-friction coefficient and shape factor against the local Reynolds number.

We observe that the results match with the laminar and turbulent correlations for the low and high Reynolds numbers. The prediction of the transition is not perfect for these cases but this industrial code provides a good compromise between computing time and accuracy.

REFERENCES :

1- Patankar SV and Spalding DB (1970)
   "Heat and Mass Transfer in Boundary Layers, a General
   Calculation Procedure"
   2nd Ed. Intertext Books London

2- Mac Donald H and Fish RW (1973)
   "Practical Calculation of Transitional Boundary Layers"
   International Journal of Heat and Mass Transfer, Vol 16
   pp 1729-1744

3- Blair MF and Edwards ED (1982)
   "The Effects of Free-stream Turbulence on the Turbulence
   Structure and Heat Transfer in Zero Pressure Gradient
   Boundary Layers"
   UTRC Technical Report, East Hartford, Connecticut

4- Guyon B and Arts T (1989)
   "Boundary Layer Calculation including the Prediction of
   Transition and Curvature Effects"
   ASME paper 89-GT-43.

5- Consigny H (1980)
   "Etude expérimentale des échanges thermiques convectifs
   à la surface des aubes de turbine"
   Ph D Thesis, Université de Bruxelles, Belgium

GRAPHICAL RESULTS

CASE A

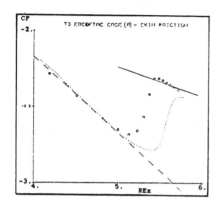

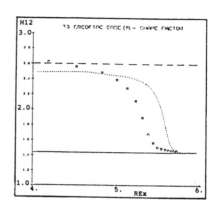

CASE B

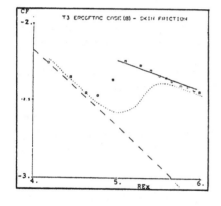

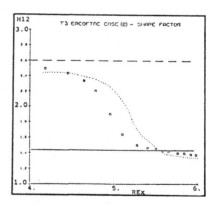

```
_ _ _   laminar
_____  turbulent
  □     experimental data
......  present calculation
```

Test Case T3 - Free Stream Turbulence.

THEODORIDIS G., PRINOS P., GOULAS A.
Lab. of Fluid Mechanics, Dept. of Mechanical Eng.
Univ. of Thessaloniki, Thessaloniki 540 06, GREECE

1. INTRODUCTION

The effect of free stream turbulence on the laminar boundary layer and on transition to turbulent has been studied using two low Reynolds number turbulence models : (a) the model of Nagano and Hishida (1987) and (b) the model of Launder and Sharma (1974). The effect of the initial conditions on predicting the onset of transition has been also studied using two starting positions : (a) Inlet plane at x=10mm downstream of the leading edge of the flat plate, (b) Inlet plane at the leading edge (x=0.0mm).

Comparisons with experimental data indicate that the Launder-Sharma model predicts better all the characteristics of transition than the Nagano-Hishida model.

As far as the effect of initial conditions is concerned, it is shown that computations with either model and inlet conditions at the leading edge of the flat plate predict earlier transition. Computation from a third inlet plane (x=-160mm) was also tried using an elliptic code, giving much earlier transition than the two previous cases and hence the upstream inlet condition was abandoned in the subsequent calculations.

For case B, due to the failure of both models with inlet conditions at the plate leading edge, predictions are shown using only the results of calculations starting downstream of the leading edge. In this case again the Launder-Sharma model is closer to the experimental data than the Nagano-Hishida model.

2. GOVERNING EQUATIONS

The two-dimensional, steady, incompressible flow over a flat plate is governed by the following equations:

Continuity equ.:
$$\frac{\partial U_j}{\partial x_j} - 0 \qquad (1)$$

Momentum equ.:
$$U_j \frac{\partial U_i}{\partial x_j} - -\frac{1}{\varrho}\frac{\partial P}{\partial x_i} + \frac{\partial}{\partial x_j}(v\frac{\partial U_i}{\partial x_j} - \overline{u_i u_j}) \qquad (2)$$

368

where U_i, U_j mean velocity components in the i- and j- direction respectively; ϱ = fluid density; P = pressure; v = kinematic viscosity, $\overline{u_i u_j}$ = Reynolds stresses.

The Reynolds stresses are calculated using the eddy viscosity concept:

$$-\overline{u_i u_j} - v_t (\frac{\partial U_i}{\partial x_j} + \frac{\partial U_j}{\partial x_i}) - \frac{2}{3} \delta_{ij} k \tag{3}$$

where v_t = eddy viscosity; δ_{ij} = Kronecker delta and k = turbulence kinetic energy.

The eddy viscosity v_t is related to k and its rate of dissipation ε through the Kolmogorov-Prandtl relationship:

$$v_t - c_\mu f_\mu \frac{k^2}{\varepsilon} \tag{4}$$

The turbulence characteristics k and ε are calculated using the two Low-Reynolds k-ε models proposed by Nagano-Hishida and Launder-Sharma. These models use the following transport equations:

k-equation:
$$U_j \frac{\partial k}{\partial x_j} - \frac{\partial}{\partial x_j} ((v + \frac{v_t}{\sigma_k}) \frac{\partial k}{\partial x_j}) - \overline{u_i u_j} \frac{\partial U_i}{\partial x_j} - (\varepsilon + D) \tag{5}$$

ε-equation:
$$U_j \frac{\partial \varepsilon}{\partial x_j} - \frac{\partial}{\partial x_j} ((v + \frac{v_t}{\sigma_\varepsilon}) \frac{\partial \varepsilon}{\partial x_j}) - \frac{\varepsilon}{k} (c_{\varepsilon 1} f_1 \overline{u_i u_j} \frac{\partial U_i}{\partial x_j} + c_{\varepsilon 2} f_2 \varepsilon) + E \tag{6}$$

The values of the constants c_μ, $c_{\varepsilon 1}$, $c_{\varepsilon 2}$, σ_k, σ_ε, the model functions f_μ, f_1, f_2 which account for the low Reynolds number and wall proximity effects as well as the extra terms D, E can be found in table 1. In this table $R_t = k^2/(v\varepsilon)$. A detailed review of the low Reynolds models can be found in Patel et al (1985).

3. SOLUTION PROCEDURE
The equations (1) - (6) are solved using a modified iterative (implicit) forward marching scheme, originally proposed by Patankar and Spalding (1971) in cartesian coordinates.

In this case 100 grid lines were used in the y-direction (of which 50 in the region of

$y^+ < 50$) and the Δx step used was of the order of 10^{-4} m. Runs with finer mesh in both directions produced no changes in the required results. The CPU time for a typical run was approximately 1/2 hour in a HP 9000/825

4.   BOUNDARY CONDITIONS
The following boundary conditions were used:

Free stream: Gradients of all variables with respect to the vertical axis (y) were set to zero.

Flat plate: Non-slip conditions were imposed.

Inlet (x = 0.0mm):   Uniform values of all variables were used according to problem suggestions.

Inlet (x = 10mm):   For the velocity U the Pohlhausen profile was prescribed (Schlichting (1979)) together with the following empirical profiles for k and ε:

$$k - k_e (\frac{U}{U_e})^2 \quad \varepsilon - \alpha_1 k \frac{\partial U}{\partial y} \quad \varepsilon \geq \varepsilon_e \tag{7}$$

proposed by Rodi (1985). The subscript e denotes values at the free stream which in turn for the inlet station are calculated from the following relationships:

$$k_e - 1.5 \overline{u^2} \quad \varepsilon_e - - U_e \frac{\partial k}{\partial x} \tag{8}$$

where the normal stress $\overline{u^2}$, is calculated using the proposed relationships for the decay of grid turbulence.

5.   COMPUTATIONAL RESULTS
For case A, comparison of the two turbulence models used against experimental data is shown in figures 1,2,3 with computations starting from x = 10mm and x = 0.0mm. The figures show experimental and computed results for the skin friction c_f the shape factor H and the momentum thickness Reynolds number Re_θ respectively for various Re_x.

Similar comparisons for case B are shown in figures 4,5,6 with inlet conditions at x = 10.mm.

For both cases it is shown that the Launder-Sharma model predicts better the characteristics of transition. The model predicts fairly well the onset of transition, however underestimates the length of transition. The failure of the Nagano-Hishida model is most probably due to the modelling of f_μ as a function of y^+ instead of R_t which is used in the model of Launder-Sharma.

The effect of the inlet conditions on transition is shown also in figures 1,2,3 for both models. It is shown that both models are affected by the inlet conditions. They predict earlier transition when the calculations start at the leading edge. The results are getting worse (not shown here) as the inlet plane of computations is moved upstream.

Finally computed velocity distributions by both models are compared with experimental data at selected stations in figures 7,8 for case A and B respectively. In both cases the computations were started at $x = 10.0mm$. In general computed velocity distributions by the Launder-Sharma model are closer to the experimental values, especially in the laminar and transitional regimes, while both models predict fairly well the turbulent velocity profiles.

6. REFERENCES

LAUNDER, B.E. and SHARMA, B.I. (1974) "Application of the Energy-Dissipation Model of Turbulence to the Calculation of Flow Near a Spinning Disk", Letters in heat and mass transfer 1, pp 131-138.

NAGANO, Y. and HISHIDA, M. (1987) "Improved Form of the k-ε Model for Wall Turbulent Shear Flows", ASME, Journal of Fluids Eng., vol 109, pp 156-160.

PATANKAR, S.V. and SPALDING, D.B. (1971) "A Calculation Procedure for Heat, Mass and Momentum Transfer in Three-dimensional Parabolic Flows", J. Heat Mass Transfer, vol. 15, pp 1787-1806.

PATEL, V.C., RODI, W. and SCHEUERER, G. (1985) "Turbulence Models for Near-wall and Low Reynolds Number Flows : A Review", AIAA Journal, vol. 23, No 9 , pp 1306-1319.

RODI, W. and SCHEUERER, G. (1985) "Calculation of Heat Transfer to Convection-cooled Gas Turbine Blades", transactions of ASME, vol 107, pp 620-627.

SCHLICHTING, H. (1979) "Boundary Layer Theory", 7th edition, McGraw-Hill, NY.

TABLE 1.

Model	c_μ	$c_{\varepsilon 1}$	$c_{\varepsilon 2}$	σ_k	σ_ε	f_μ	f_1	f_2	D	E
N-H	0.09	1.45	1.9	1.0	1.3	$\{1-\exp(-\frac{y^+}{26.5})\}^2$	1.0	$1-0.3\exp(-R_t^2)$	$2\nu(\frac{\partial\sqrt{k}}{\partial y})^2$	$\nu\nu_t(1-f_\mu)(\frac{\partial^2 U}{\partial y^2})^2$
L-S	0.09	1.44	1.92	1.0	1.3	$\exp\{-\frac{3.4}{(1+R_t/50)^2}\}$	1.0	$1-0.3\exp(-R_t^2)$	$2\nu(\frac{\partial\sqrt{k}}{\partial y})^2$	$2\nu\nu_t(\frac{\partial^2 U}{\partial y^2})$

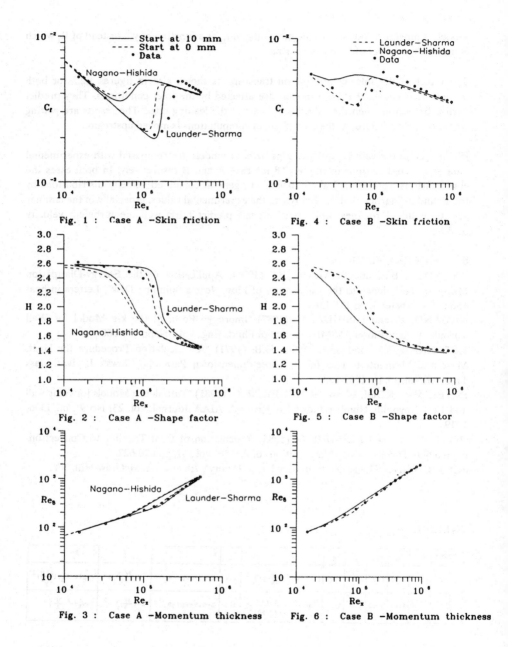

Fig. 1 : Case A –Skin friction

Fig. 4 : Case B –Skin friction

Fig. 2 : Case A –Shape factor

Fig. 5 : Case B –Shape factor

Fig. 3 : Case A –Momentum thickness

Fig. 6 : Case B –Momentum thickness

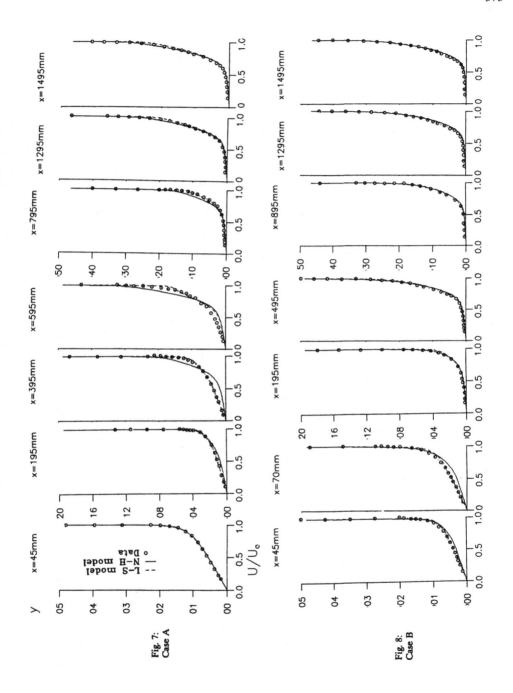

Fig. 7:
Case A

Fig. 8:
Case B

A New Version of the $k - \epsilon$ Model of Turbulence Applied to Boundary-Layer Transition

F. Tarada

Mott MacDonald Consultants Ltd, Croydon, England

1   Summary

Predictions of the transition from laminar to turbulent flow of boundary-layers disturbed by free-stream turbulence are presented in this contribution, using a modified $k - \epsilon$ model of turbulence. The model has been derived to guarantee balancing of the terms in the turbulent transport equations near the wall.

The resulting predictions give earlier transition locations than those indicated by the two flat-plate test cases. The primary cause of this discrepancy is thought to be the inadequate modelling of the flow and turbulence structure around the leading edge of the plate.

2   The Transport Models

The normal continuity and momentum equations were solved, along with the turbulent transport equations listed below (Tarada (1987)) :

Kinetic Energy of Turbulence :

$$\rho U \frac{\partial k}{\partial x} + \rho V \frac{\partial k}{\partial y} = \frac{\partial}{\partial y}((\mu + \frac{\mu_t}{\sigma_k})\frac{\partial k}{\partial y})$$
$$+\mu_t(\frac{\partial U}{\partial y})^2 - \rho \tilde{\epsilon} - 2\mu(\frac{\partial k^{\frac{1}{2}}}{\partial y})^2$$

374

'Isotropic' Dissipation Rate:

$$\rho U \frac{\partial \tilde{\epsilon}}{\partial x} + \rho V \frac{\partial \tilde{\epsilon}}{\partial y} = \frac{\partial}{\partial y}((\mu + \frac{\mu_t}{\sigma_\epsilon})\frac{\partial \tilde{\epsilon}}{\partial y})$$

$$+\frac{\tilde{\epsilon}}{k}[C_{\epsilon 1}\mu_t(\frac{\partial U}{\partial y})^2 - \rho C_{\epsilon 2}\tilde{\epsilon}] + 2\rho\nu\nu_t[\frac{\partial^2 U}{\partial y^2}]^2$$

$$-2\mu(\frac{\partial \tilde{\epsilon}^{\frac{1}{2}}}{\partial y})^2$$

The above equations were developed to suit the near-wall behaviour of both k and $\tilde{\epsilon}$, both of which tend to a quadratic behaviour near a wall (Hanjalić and Launder (1976)). In addition, a (correct) cubic near-wall variation of turbulent viscosity $\nu_t(= f_\mu k^2/\tilde{\epsilon})$ was obtained using Chien's (1982) recommendations that

$f_\mu = 1 - exp(-0.0115y^+),$
$(C_\mu, C_{\epsilon 1}, C_{\epsilon 20}, \sigma_\epsilon, \sigma_k) = (0.09, 1.35, 1.8, 1.3, 1.0)$
with $C_{\epsilon 2} = C_{\epsilon 20}(1 - 0.22\exp(-(R_T/6)^2)).$

Fluctuation profiles of streamwise velocity, u'/U_e, were calculated via an Algebraic Stress Model :

$$\frac{\overline{u^2}}{k} = \frac{2}{3}[1 + \frac{P}{\epsilon}\phi_4] + \phi_5\frac{\overline{v^2}}{k}$$

where

$$\phi_4 = (2 - 2c_2 + c_2 c_2' f)/\delta,$$
$$\phi_5 = c_1' f/\delta,$$
$$\delta = P/\epsilon - 1 + c_1$$

and

$$(c_1, c_2, c_1', c_2') = (3.0, , 0.3, 0.75, 0.5)$$

following Gibson and Younis (1986).

The Algebraic Stress Model was used for the purposes of data comparison only and did not feature in the marching calculation of the flow.

3   Method of Solution

The above transport equations were transformed via a Falkner-Skan technique to ensure economy of computation. A Keller-Box Method was employed to solve the parabolic equations, with around 70 points across the boundary-layer to satisfy grid-independence. A geometric spacing factor of 1.1 was applied between successive nodes in the cross-stream direction.

To accurately compute the decay of free-stream turbulence, it was found necessary to extend the computational domain well into the 'free-stream'. This was to account for the long 'tail' of the dissipation variable. The transport equations were therefore solved up to $y^+ = 100$ in the initial (laminar) regions and up to $y^+ = 2000$ in the turbulent regions.

Typical computing times were of the order of 60 seconds on a VAX 11/780 for a full aerodynamic run.

4   Initial Profiles

The initial momentum thickness was computed from Thwaites's integral equation, and the initial (laminar) velocity profile was described by a Pohlhausen Fourth-Degree type equation. The initial turbulent kinetic energy profile was given by

$$k = k_\infty (U/U_\infty)^2$$

after Rodi and Scheuerer (1985).

The initial dissipation profile was given by

$$\bar{\epsilon} = \frac{k^{3/2}}{2.5y}[1 + (-6 + 3\Pi)(y/\delta)^2 + (8 - 3\Pi)(y/\delta)^3 + (\Pi - 3)(y/\delta)^4],$$

where $\Pi = 2.5\bar{\epsilon}_\infty \delta / k_\infty^{3/2}$.

The above profile ensured a smooth transition between the inner boundary-layer 'equilibrium' values of $\bar{\epsilon}$ (found by equating the production to dissipation of k), and the free-stream value of $\bar{\epsilon}$, namely $\bar{\epsilon}_\infty$.

The free-stream values of k and $\bar{\epsilon}$ were found from the given measurements of the decay of turbulence behind grids. These can be expressed as

$$Tu = \frac{u'}{U_\infty} = a(\frac{x}{b})^{-n}$$

where 'b' is the effective bar size ($b = 3.5$ mm for Case 1 and $b = 9.5$ mm for Case 2) and (a, n) equal $(1.12, 5/7)$.

Assuming an isotropic free-stream, we may estimate

$$k_\infty = \frac{3}{2}Tu^2 U_\infty^2$$

and

$$\bar{\epsilon}_{\infty} = 3\frac{n}{x}U_{\infty}^3 Tu^2.$$

The above initial profiles were applied at 5 mm and 10 mm from the leading edge to check for sensitivity with respect to starting location. No significant influence on predicted transition locations was found.

5 Discussion of Results

Figures 1 to 8 show the results for the low free-stream turbulence Test Case 1, and figures 9 to 16 show the results for the high free-stream Test Case 2. Turbulence and mean-velocity profiles in the fully-turbulent flow regions are seen to be adequately predicted for both Test Cases. However, the predicted transition locations are seen to be too early, particularily for Test Case 2. In addition, the transitional zone between laminar and turbulent flow is also too short when compared to the experimental results.

The calculated transition process was found to be sensitive to the initial turbulent kinetic energy and dissipation profiles. It may therefore be possible to improve upon these results by choosing better initial profiles. In particular, the assumption of turbulence isotropy employed in the setting up of the initial turbulence profiles is questionable. The early development of the flat-plate boundary-layer is governed by the nature of the flow at the leading-edge stagnation point. This is an area where no detailed experimental data is available from the current Test Cases.

In order to correctly model the anisotropy of the leading-edge stagnation region, it may be preferable to employ a full Reynolds Stress turbulence model for all non-zero second order stresses. Such an approach would ideally involve the solution of the full elliptic set of Navier-Stokes equations rather than the parabolic approximations presented here.

However, even the current model shows that it *is* possible to predict transition from laminar to turbulent flow using simple "phenomenological" models of turbulence such as the $k - \epsilon$ model. This is valid only for situations where the normal transition mechanisms are "bypassed" due to external disturbances, such as free-stream turbulence.

6 References

Chien K. Y.,1982,'Predictions of Channel and Boundary-Layer Flows with a Low-Reynolds-Number Turbulence Model', AIAA J. Vol. 20, No. 1, pp. 33–38.

Gibson, M. M. and Younis, B. A.,1986,'Calculation of Swirling Jets with a Reynolds Stress Closure', Phys. Fluids, 29(1), pp. 38–48.

Hanjalić, K. and Launder, B. E., 1976, 'Contribution towards a Reynolds-Stress Closure for Low-Reynolds-Number Turbulence', J. Fluid Mech., Vol. 74, Part 4, pp. 593–610.

Rodi,W. and Scheuerer,G., 1985,'Calculation of Heat Transfer to Convection-Cooled Gas Turbine Blades', J. Eng. for Gas Turbines and Power, Vol. 107, pp. 620–627.

Tarada, F. H. A., 1987, 'Heat Transfer to Rough Turbine Blading', D. Phil. thesis, University of Sussex, England.

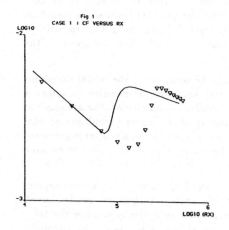

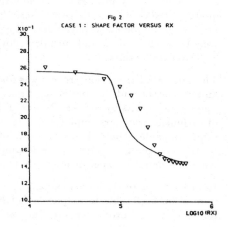

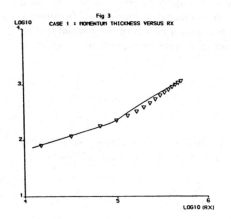

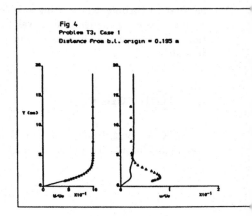

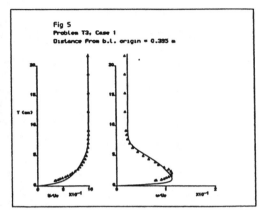

Fig 5
Problem T3, Case 1
Distance from b.l. origin = 0.395 m

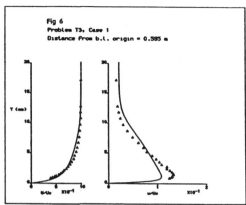

Fig 6
Problem T3, Case 1
Distance from b.l. origin = 0.595 m

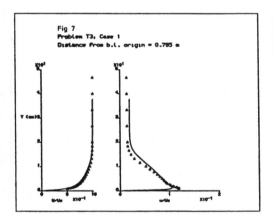

Fig 7
Problem T3, Case 1
Distance from b.l. origin = 0.795 m

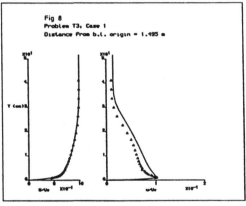

Fig 8
Problem T3, Case 1
Distance from b.l. origin = 1.495 m

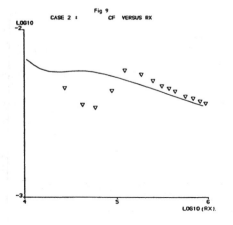

Fig 9
CASE 2 : CF VERSUS RX

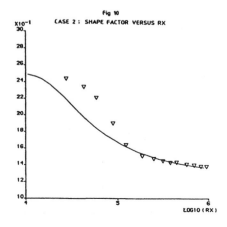

Fig 10
CASE 2 : SHAPE FACTOR VERSUS RX

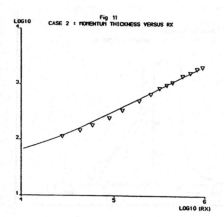

Fig 11
CASE 2 : MOMENTUM THICKNESS VERSUS RX

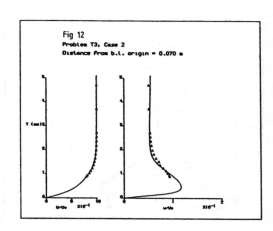

Fig 12
Problem T3, Case 2
Distance from b.l. origin = 0.070 m

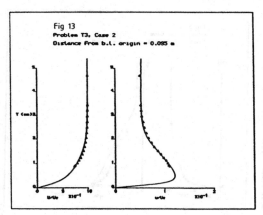

Fig 13
Problem T3, Case 2
Distance from b.l. origin = 0.095 m

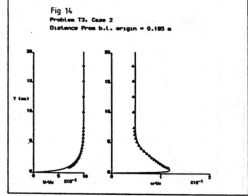

Fig 14
Problem T3, Case 2
Distance from b.l. origin = 0.195 m

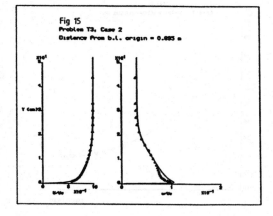

Fig 15
Problem T3, Case 2
Distance from b.l. origin = 0.895 m

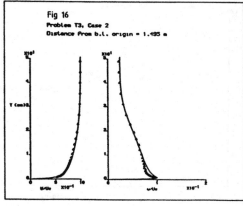

Fig 16
Problem T3, Case 2
Distance from b.l. origin = 1.195 m

Calculation of Transitional Boundary Layers under the Influence of Free-Stream Turbulence

W. Rodi, N. Fujisawa*, B. Schönung**

Institute for Hydromechanics, University of Karlsruhe,
D 7500 Karlsruhe, F.R. Germany

1. INTRODUCTION

This paper describes the calculation procedure used by the team coordinated by W. Rodi for the test case T3 - transitional boundary layers at 2 different free-stream turbulence levels. In contrast to the calculations performed by this team for the other test cases (Rodi and Zhu) with an elliptic procedure, the calculations for T3 were carried out with a 2D boundary-layer procedure solving 2D parabolic equations. Further, as opposed to the test cases T1 and T4, wall functions could not be used for the transitional flows of T3. Rather, the viscosity-affected near-wall layer was resolved with various low-Reynolds number turbulence models.

2. MEAN-FLOW EQUATIONS

The mean flow in the boundary layer along a flat plate (Fig.1) was determined by solving the usual boundary layer form of the two-dimensional x-momentum equation and the 2D continuity equation for incompressible flows. These equations were solved right up to the wall, where the velocity components U and V were set to zero.

3. TURBULENCE MODELS

The turbulent shear stress $\overline{uv}$ appearing in the x-momentum equation was determined with the aid of three different turbulence models, namely two low-Reynolds-number versions of the k-ε model and one two-layer model using the k-ε model only outside the viscosity-affected near-wall layer and employing a one-equation model in this layer.

---

*   present address: Gunma University, Japan
**   present address: ABB Turbo Systems AG, Baden, Switzerland

a)   Low-Reynolds-Number k-ε Models

The versions due to Launder and Sharma (1974) and Lam and Bremhorst (1981) as reviewed by Patel et al. (1985) were used. Both models are described by the common equations in Table 1. The differences between the two models lie in the choice of the variables D and E, the boundary conditions for the dissipation quantity $\tilde{\varepsilon}$ and the functions f_μ, f_1 and f_2. The choices for the two models are also given in Table 1. The boundary condition for k at the wall is k = 0.

For higher free-stream turbulence levels (say above 1%), these models lead by themselves to transition due to the mechanism of diffusion of free-stream turbulence into the boundary layer. However, the transition performance depends on the initial distribution specified for k and ε in the laminar boundary layer. The following initial profile was used for k:

$$(k/k_e) = (U/U_e)^2 \tag{1}$$

where "e" indicates the value in the free stream. This relation is in contrast to the test case specification. However, this quadratic relation seems more reasonable than the specified relation with power 4 and it had to be used because, during the model development, the initial profile for ε was tuned on this distribution. The initial distribution of ε is calculated from

$$\varepsilon = a_1 k \partial U/\partial y \tag{2}$$

but with ε set equal to ε_e whenever $\varepsilon < \varepsilon_e$ would result from the above relation. The parameter a_1 was set equal to a constant value of 0.3 for the Launder-Sharma model but was obtained from an empirical relation due to Rodi and Scheuerer (1985) given in Fig.2. For momentum thickness Reynolds numbers Re_θ below 100 this correlation was modified by

$$a_1 = a_{1,100} (0.0027\, Re_\theta + 0.73) \tag{3}$$

where $a_{1,100}$ is the value of a_1 from Fig. 2.

b)   Two-Layer Model

The equations describing the two-layer model are compiled in Table 1. The damping function f_μ involves the parameter A^+ which is given a large value (here 300) when the boundary layer is laminar and a value of 25 when the boundary layer is fully turbulent (at least in zero pressure gradient conditions). In the transition region, A^+ is assumed to vary between the two limiting values according to the following formula (after Crawford and Kays 1976):

$$A^+ = A_t^+(300 - A_t^+) \left(1 - \sin(\frac{\pi}{2} \frac{Re_\theta - Re_{tr}}{Re_{tr}})\right)^2 \tag{4}$$

This is effective when the momentum thickness Reynolds number Re_θ is larger than the critical transition Reynolds number Re_{tr} and smaller than twice Re_{tr}. The critical Reynolds number is obtained from a correlation of Abu Ghannam (1979) which for boundary layers with zero pressure gradient reads:

$$Re_{tr} = 163 + \exp (6.91 - Tu) \qquad (5)$$

where Tu is the turbulence intensity in the free stream.

The initial k-distribution was again determined from equation (1). In the outer region the initial ε was set equal to $\varepsilon_e = k_e^{3/2}/L_e$. In the inner region the length scale L was obtained from a ramp function

$$
\begin{aligned}
L_1 &= \min (c_L y, L_e) \\
L_2 &= \min (c_L y, 0.517\delta) \qquad (6) \\
L &= \max (L_1, L_2)
\end{aligned}
$$

where δ is the boundary layer thickness.

4.   NUMERICAL SOLUTION PROCEDURE

The mean-flow equations and the k- and ε-equations were solved with a modified version of the GRAFTUS boundary layer program (Schönung 1986). This is a 2D forward-marching finite-volume method, using a numerical grid which stretches with the growth of the boundary layer. The stretching is obtained by non-dimensionalising the local wall distance with the boundary layer thickness. For discretization in the normal direction, central/upwind hybrid differencing is used. In the main flow direction, an implicit discretization is applied. The resulting tridiagonal matrix is solved with a Thomas algorithm, which is a simplified version of the Gauss elimination algorithm.

For the calculations, 80 grid points were used in the normal direction. The forward step was 25% of the local momentum thickness. This resulted in about 7000 forward steps necessary for the present calculations. A typical calculation took about 4 minutes CPU time on an IBM 3090 computer.

5.   CALCULATION OF TEST CASES A AND B

The calculations were started at a distance of $x = 10$ mm from the leading edge (for geometry see Fig. 1) with the Blasius velocity profiles for laminar boundary layers. The initial value of ε_e at this location was determined from

$$\varepsilon_e = - U_e dk_e/dx \qquad (7)$$

leading to $L_e = k_e^{3/2}/\varepsilon_e = 0.0301$ m for case A and of 0.418 m for case B.

Fig. 3 compares the calculation results for the shape factor H with the experimental data provided by the test case supervisor. For case A with the lower free-stream turbulence level, transition takes place in the range $Re_x \approx 4.6 - 5.6$. The Launder-Sharma model predicts this rather well, while both the Lam-Bremhorst and the 2-layer model predict transition somewhat late ($Re_x \approx 5 - 5.7$). For case B with the higher free-stream turbulence level, transition takes place earlier in the range $Re_x \approx 4 - 5.3$. This is predicted reasonably well by all three models, but the shape of the H-distribution in the transition region is not reproduced very accurately. Further results can be found in the synthesis paper on test case T3.

6. REFERENCES

Abu Ghannam, B.J., 1979, "Boundary-layer transition in relation to turbomachinery blades", Ph.D. Thesis, University of Liverpool, England.

Crawford, M.E. and Kays, W.M., 1976, "STAN5 - A program for numerical computation of two-dimensional internal and external boundary-layer flow, NASA CR 2742.

Lam, C.K.G. and Bremhorst, K.A., 1981, "Modified form of the k-ε model for predicting wall turbulence", J. Fluids Eng., Vol. 103, pp. 456-460.

Launder, B.E. and Sharma, B.I., 1974, "Application of the energy dissipation model of turbulence to the calculation of flow near a spinning disk", Letters in Heat and Mass Transfer, Vol. 1, pp. 131-138.

Norris, L.H. and Reynolds, W.C., 1975, "Turbulent channel flow with a moving wavy boundary", Report No. FM-10, Stanford Universit, Dept. of Mech. Eng..

Patel, V.C., Rodi, W., Scheuerer, G., 1985, "Turbulence models for near-wall and low-Reynolds-number flows: A review", AIAA J., Vol. 23, No. 9, pp. 1308-1319.

Scheuerer, G. and Rodi, W., 1985, "Calculation of laminar-turbulent boundary-layer transition on turbine blades", Proc. AGARD-PEP 65th Symposium, Bergen, Norway.

Schönung, B., 1986, "Untersuchung des Grenzschichtverhaltens an Turbinenschaufeln unter Berücksichtigung von Filmkühlung und von lokalen Ablöseblasen - Programmbeschreibung GRAFTUS", FVV-Forschungsberichte Heft 383-2.

Low-Re-k-ε-models

$$-\overline{uv} = \nu_t \frac{\partial U}{\partial y}, \qquad \nu_t = c_\mu \cdot f_\mu \cdot \frac{k^2}{\tilde{\varepsilon}}$$

$$U\frac{\partial k}{\partial x} + V\frac{\partial k}{\partial y} = \frac{\partial}{\partial y}\left[\left(\nu + \frac{\nu_t}{\sigma_k}\right)\frac{\partial k}{\partial y}\right] + \nu_t\left(\frac{\partial U}{\partial y}\right)^2 - \varepsilon$$

$$U\frac{\partial \tilde{\varepsilon}}{\partial x} + V\frac{\partial \tilde{\varepsilon}}{\partial y} = \frac{\partial}{\partial y}\left[\left(\nu + \frac{\nu_t}{\sigma_\varepsilon}\right)\frac{\partial \tilde{\varepsilon}}{\partial y}\right] + c_{\varepsilon 1}f_1\frac{\tilde{\varepsilon}}{k}\nu_t\left(\frac{\partial U}{\partial y}\right)^2 - c_{\varepsilon 2}f_2\frac{\tilde{\varepsilon}^2}{k} + E$$

$$\varepsilon = \tilde{\varepsilon} + D$$
$$R_T = \frac{k^2}{\nu \cdot \tilde{\varepsilon}}$$
$$R_y = \frac{\sqrt{k} \cdot y}{\nu}$$
$$y^+ = \frac{y \cdot u_\tau}{\nu}$$

MODEL		D	ε_w – B.C.	c_μ	$c_{\varepsilon 1}$	$c_{\varepsilon 2}$	σ_k	σ_ε	f_μ	f_1	f_2	E
LAUNDER-SHARMA	LS	$2\nu\left(\frac{\partial k^{1/2}}{\partial y}\right)^2$	0	0.09	1.44	1.92	1.0	1.3	$\exp\left[\frac{-3.4}{(1+R_T/50)^2}\right]$	1.0	$1-0.3\exp(-R_T^2)$	$2\nu \cdot \nu_t\left(\frac{\partial^2 U}{\partial y^2}\right)^2$
LAM-BREMHORST	LB	0	$\frac{\partial \varepsilon}{\partial y} = 0$	0.09	1.44	1.92	1.0	1.3	$[1-\exp(-0.0160\,R_y)]^2 \cdot \left(1+\frac{19.5}{R_T}\right)$	$1+(0.06/f_\mu)^3$	$1-\exp(-R_T^2)$	0

Two-layer model:

Away from wall: standard k-ε model (above model with D = E = 0, $f_\mu = f_1 = f_2 = 1$, ν neglected in k and ε equations compared with $\nu_t/\sigma_{k,\varepsilon}$)
Near wall: One-equation model due to Norris and Reynolds (1975):

$$\nu_t = c_\mu \cdot f_\mu \cdot \frac{k^2}{\varepsilon}, \qquad f_\mu = 1 - \exp(-0.495\,R_y/A^+), \qquad R_y = \frac{\sqrt{k}y}{\nu}$$

$$\varepsilon = \frac{k^{3/2}}{L}\left(1 + \frac{c_\varepsilon}{\sqrt{kL/\nu}}\right), \qquad L = c_L y, \qquad c_\mu = 0.09, \qquad c_L = 2.5, \qquad c_\varepsilon = 13.2$$

The two models are matched where $f_\mu = 0.95$ (effectively at $y^+ \approx 80$)

Table 1: Turbulence models used for T3 calculations

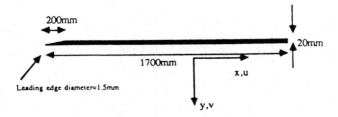

Fig. 1: Flow configuration for Test case T3

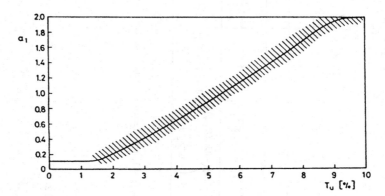

Fig. 2: Dependence of a_1 factor in equation (2)
on turbulence level Tu

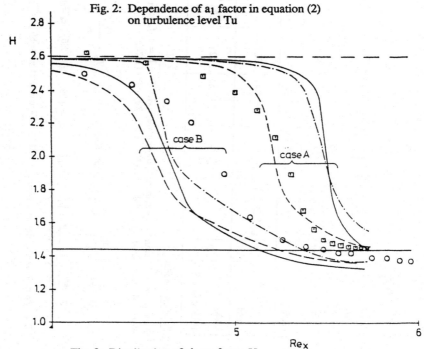

Fig. 3: Distribution of shape factor H,
—— Lam-Bremhorst model, - - - Launder-
Sharma model, —.— 2 layer model,
-.. D.O. experiments

Application of a low-Reynolds number Reynolds Stress Transport model to the prediction of transition under the influence of free-stream turbulence.

A.M. SAVILL

Rolls-Royce Senior Research Associate, University of Cambridge, England.

1 INTRODUCTION

This paper presents a submission to the T3 Test Case problem of the Lausanne ERCOFTAC Workshop on Numerical Simulation of Unsteady Flows, Transition to Turbulence and Combustion. The boundary layer model used for the present calculations is a low-Re Reynolds Stress Transport (RST) scheme, originated by Younis [1] from the earlier proposals of Hanjalic & Launder [2], which is now undergoing futher development and has been applied to the T3 Test Case by the author. To his knowledge there have been only two other attempts to develop a similar high level closure low-Re model (one in which the computations are extended right down to the wall surface instead of employing wall functions for the near-wall region), by So & Yoo [3], and Shima [4], and these have only been applied to fully-developed turbulent flows. (An alternative RST model for transition, incorporating low Reynolds number damping terms, but omitting wall pressure reflection terms, was proposed earlier by Finson [5]). The present model has elements in common with these other low-Re RST approaches, and also draws on experience gained in devising similar low-Re model approximations for lower level k-ε closure schemes. More than ten different versions of the latter have now been proposed. The majority of these have been reviewed in detail by Patel, Rodi & Scheuerer [6], but see also Jones & Launder [7] and Nagano & Hishida [8] in particular. These last two models have both been applied to re-laminarising (or re-transitional) flows with some success, as have some of the others [see 6]. The Jones & Launder model has also been applied successfully to some transitional flow cases, and more recently the Lam & Bremhorst low-Re k-ε model [see 6] has also been used by Liu [9] to predict transition. In addition several previous attempts have been made to model the effect of free-stream turbulence on a fully turbulent boundary layer using high-Re versions of k-ε [10], Algebraic Stress Model (ASM) and RST schemes [11], but so far it has proved impossible to exactly reproduce the influence of both free-stream turbulence intensity and length scale on Cf (eg. as in Stanford Test Case 0210).

2 THE REYNOLDS STRESS TRANSPORT MODEL

The present scheme is based on the modelled versions of transport equations for the Reynolds stresses $\overline{u_i u_j}$ and Dissipation ε.

$$\frac{D\,\overline{u_i u_j}}{Dt} = P_{ij} + D_{ij} + \Phi_{ij} - \frac{2}{3}\delta_{ij}\,\varepsilon\,(1 - f_s D) \tag{1}$$

$$\quad\quad\quad\quad\text{[Production] [Diffusion] \quad [Pressure-Strain] \quad [Anisotropic]}$$
$$\quad\quad\quad\quad\quad\quad\quad\quad\quad\quad\quad\quad\text{Redistribution \quad\quad Dissipation}$$

and: $\quad\dfrac{D\varepsilon}{Dt} = f_1\,c_{\varepsilon_1}\left(\dfrac{\varepsilon}{k}\right)P_k + D_\varepsilon - f_2\,c_{\varepsilon_2}\left(\dfrac{\varepsilon}{k}\right)\varepsilon + E \tag{2}$

which differ from their high-Re counterparts by the addition of the f (damping) factors and the additional terms D & E, and where:

$\quad\quad$ P_{ij} and P_k which are derived from P_{ij} by summation over indices, are explicit;

$$D_{ij} = C_s\frac{\partial}{\partial y}\left[\left(\frac{k}{\varepsilon}\right)\overline{u_m u_l}\,\frac{\partial\,\overline{u_i u_j}}{\partial x_m}\right]\quad\quad D_\varepsilon = C_\varepsilon\frac{\partial}{\partial y}\left[\left(\frac{k}{\varepsilon}\right)\overline{v^2}\,\frac{\partial\varepsilon}{\partial y}\right] \tag{3}$$

and $\quad \Phi_{ij} = \Phi_{ij1} + \Phi_{ij2} + \Phi_{ij\omega 1} + \Phi_{ij\omega 2} \tag{4}$
$$\quad\quad\quad\quad\text{[Return-to-isotropy] \quad [Rapid] \quad [Wall-Pressure Reflection]}$$

$$\Phi_{ij1} = -C_1\left(\frac{\varepsilon}{k}\right)\left(\overline{u_i u_j} - \frac{2}{3}\delta_{ijk}\right) \quad\text{(Note alternative from [4]: } C_1^* = C_1[1 - (1 - \frac{1}{c_1})\,f_\omega]\tag{5}$$

$$\Phi_{ij2} = -C_2\left(P_{ij} - \frac{2}{3}\delta_{ijk}\,P_k\right)\quad\quad\text{(LRR Model 2 [12])} \tag{6}$$

$$\Phi_{ij\omega 1} = -C_1'\left(\frac{\varepsilon}{k}\right)\left(\overline{u_n^2}\,\delta_{ij} - \frac{3}{2}\overline{u_n u_i}\,\delta_{nj} - \frac{3}{2}\overline{u_n u_j}\,\delta_{ni}\right)f \tag{7}$$

$$\Phi_{ij\omega 2} = -C_2'\left(\Phi_{kn_2}\,\delta_{ij} - \frac{3}{2}\Phi_{ni_2}\,\delta_{nj} - \frac{3}{2}\Phi_{nj_2}\,\delta_{ni}\right)f \tag{8}$$

where: $f = (k^{3/2}/\varepsilon_y)$ [compared to [4]:

$$\Phi_{ij\omega} = \left\{\alpha(P_{ij} - \frac{2}{3}\delta_{ij}\,P_k) + \beta(D_{ij} - \frac{2}{3}\delta_{ij}\,D_k) + \gamma k\left(\frac{\partial U_i}{\partial x_j} + \frac{\partial U_j}{\partial x_i}\right)\right\}f_\omega \tag{9}$$

with $\alpha = 0.45$, $\beta = -0.03$, $\gamma = 0.08$, $f_\omega = \exp[-\{0.015 R_y\}^4]$ and the model constants used are the same as the 'standard' high Re RST model values (eg as in [12], except where other frequently used values have been indicated in brackets) :

$C_\mu = 0.09$, $\quad C_s = 0.22\,(0.25)\quad\quad C_\varepsilon = 0.15\,(0.18, 0.17)\quad\quad C_1 = 1.8$
$C_2 = 0.6\quad\quad C_1' = 0.5\quad\quad\quad\quad C_2' = 0.3\quad\quad\quad\quad\quad\quad C_{\varepsilon_1} = 1.275\,(1.44\,,\,1.35)$
$C_{\varepsilon_2} = 1.8\,(1.92)$

(where C_{ε_1} is consistent with $\kappa = 0.41$ and equal to $(C_{\varepsilon_2} - 3.5\,C_\varepsilon)$.

The following low-Reynolds number model approximations are employed:

$$\varepsilon_{ij} = \frac{2}{3}\delta_{ij}\,(1 - f_s)\,\varepsilon + f_s D \tag{10}$$

where from [2]: $f_s = \dfrac{1}{(1 + R_t/10)}$ (compare $f_s = \exp\left(-\dfrac{R_t}{40}\right)$ [3]) : $R_t = \dfrac{k^2}{\varepsilon v}$

and:
$$D = F\left(\frac{\varepsilon}{k}\right)\left(\overline{u_iu_j} + \overline{u_iu_k}\,n_kn_j + \overline{u_ju_k}\,n_kn_i + \delta_{ij}\,\overline{u_ku_L}\,n_kn_L\right) \tag{11}$$

with: $F = (1 + 2.5\,(\,\overline{v^2}\,/k\,))$ (as in [3])

In addition: $f_1 = \text{MAX}\,[\,\frac{1}{C_{\varepsilon 1}}(\,2 - \frac{0.725}{65}\,R_y\,)\,,\,1\,]$ \hfill (12)

and: $f_2 = 1 - (\frac{0.4}{C_{\varepsilon 2}})\exp\{-[\,\text{MIN}\,(\frac{R_t}{6}\,,\,20\,)\,]\,\}$ \hfill (13)

(also as in [3], and in the Chien low-Re k-ε model, see [6])

And one of two proposed alternative E terms:

The first based on the Dutoya-Michard low-Re k-ε model - see [6] :

$$E = 2C_{\varepsilon 2}\,\nu\left(\frac{\varepsilon}{k}\right)\left(\frac{\partial\sqrt{k}}{\partial y}\right)^2 \tag{14}$$

and the second based on the Jones & Launder low-Re k-ε model [7]:

$$E = f_\mu\,C_{\varepsilon 3}\,\nu\left(\frac{\overline{v^2}}{k}\right)\left(\frac{k^2}{\varepsilon}\right)\left(\frac{\partial^2 U}{\partial y^2}\right)^2 \tag{15}$$

with either: $f_\mu = 1$, $C_{\varepsilon 3} = 0.25$ or $f_\mu = \exp\left[\frac{-3.4}{(1 + (R_t/50))^2}\right]$, $C_{\varepsilon 3} = 1.12$

where E is then a generalised form of that in the Launder-Sharma low-Re k-ε model [6].

For both these latter model approximations the value of the additional constant $C_{\varepsilon 3}$ was optimised from a series of fully turbulent flow computations, conducted by the author prior to the present study, to ensure best agreement with a correlation of experimental data on Cf versus Reθ (Fig.1), compiled by Agoropoulos and used previously for validating the performance of high-Re k-ε, ASM and RST models [see 11,13]. A limited evaluation of an alternative Lam & Bremhorst f_μ factor (again see [6]) suggested that implementing this instead of the Launder-Sharma version above was less beneficial.

3 IMPLEMENTATION

The parabolic thin-shear layer equations were solved using a finite-difference, forward-marching computational scheme on an orthogonal grid incorporating an allowance for (turbulent) boundary layer growth. A conditional central/upwind differencing scheme was employed in the direction normal to the wall (y) while the equations were implicit in the stream direction (x).

For each of the Test Case configurations A & B, computations were performed starting at x =10mm downstream of the test plate leading edge (620mm from the free-stream turbulence generating grids), with an initial Blasius mean velocity profile computed using the method of Haj-Ibrahim [14] and all other conditions exactly as given in the Test Case specification:

i.e. $(u'/u'_0) = (U/U_0)^2$, $v' = 0$, $w' = 0.2u'$, with $\overline{uv} = 0$

so that: $\overline{u^2} = u'^2$, $\overline{v^2} = v'^2 = 0$, $\overline{w^2} = 0.04u'^2$ and $k = (\overline{u^2} + \overline{v^2} + \overline{w^2})/2$

The necessary initial dissipation profile was derived assuming the normal ramp function mixing length for turbulent boundary layer flow to be valid in the laminar region, whence:

$$\varepsilon(y) = k(y)^{3/2} / l\varepsilon \qquad \text{with:} \qquad l\varepsilon = C\mu^{-3/4} \{ \text{Max} [\kappa y , 0.085\delta] \}$$

At the wall the following boundary conditions were applied:

$$U = 0 , \quad \overline{u^2} = \overline{v^2} = \overline{w^2} = 0 , \quad \overline{uv} = \overline{uv} (y_2) , \varepsilon = \varepsilon (y_2)$$

At the external edge of the boundary layer the free-stream turbulence was assumed to be isotropic, since Roach [15] has shown this is very nearly the case, and accordingly the following boundary conditions were assumed: $U = U_0,$ $\overline{u^2} = \overline{v^2} = \overline{w^2} = [u'_{oi}(x/x_i)^{-5/7}]^2$ (16)

in order to allow for the grid-turbulence decay as specified.

In additon the free-stream turbulence dissipation length scale L was estimated from the measured rate of decay of longitudinal velocity fluctuations behind such grids [15], using the formula adopted by Rodi & Sheuerer [9]: $L_u^e = (u'_{oi})^{3/2} / (U_{od} \overline{u^2} /dx)$ (17)

for their own k-ε modelling, and also for subsequent ASM [11], RST computations of free-stream turbulence effects on turbulent boundary layers. The external dissipation boundary value was then computed as: $\varepsilon = k_0^{3/2} /L_u^e (x/x_i)^{1/2}$ (18)

to accomodate the expected rate of growth of the length scale [see 15].

The computations were performed on an expanding grid, derived from the initial Blasius velocity profile computations, with 80 points spanning the boundary layer. The resolution in the near wall region was considerably greater than for the outer part of the flow with 30 points below $y^+=10$, with $y_2^+=0.04$, compared to 28 points above $y=0.1\delta$ (Tests showed that doubling or halving the grid resolution, in the inner and outer layers respectively, only produced changes of order $\pm 1\%$ in the computed results). An x-step increment of 0.025δ was adopted to avoid the need for additional in-step iterations, so that some 4500 time or x-steps were needed to cover the whole flow development up to x=1.495m. The code was run on the Cambridge University IBM 3084Q mainframe and typically computations to generate all of the required output at the 7 specified Stations consumed 135s Cpu. This corresponded to 0.03s/time step, as compared to an average figure of 0.015s/time step recorded for previous high-Re RST computations of fully turbulent boundary layers [11].

4 RESULTS

Surprisingly no transition was predicted for the lower free-stream turbulence intensity T3A Test Case. Instead starting from laminar conditions an essentially Blasius velocity profile was maintained, and the turbulence intensity within the boundary layer decayed rapidly from its initially specified level. The computed skin friction coefficient Cf therefore remained close to the expected laminar flow correlation with Rex, as shown by Fig. 2, and the same was true of the calculated momentum thickness θ and shape factor H distributions. It appeared that one reason for this failure might be the fact that the initial $\delta+$ was only about 30, so that the whole of the turbulence profile was within the heavily damped region of the low-Re model. In order to test whether transition would be predicted with a higher initial level of turbulent activity, further computations were performed with alternative starting conditions. First assuming isotropic turbulence,

with: $\overline{u^2} = \overline{v^2} = \overline{w^2} = 2k/3$

(either sheared or un-sheared: $\overline{uv} = -0.3k$ or $\overline{uv} = 0$)

and/or: $k(y) = u'^2 \, (y/\delta), \ \varepsilon(y) = k^{3/2} / l\varepsilon + \varepsilon(y/\delta)$

as used by Liu [9] to successfully predict transition using the Lam & Bremhorst k-ε model. However the results were virtually unaffected by these changes, although subsequent computations conducted from fully turbulent starting conditions did satisfactorily reproduce the Cf versus Rex correlation for turbulent flow, again as indicated on Fig.2. Since leading edge effects were not accounted for in the model or in the specified starting conditions, it must be assumed that these had an influence on the subsequent flow development.

Transition was successfully predicted for the higher free-stream T3B Test Case as illustrated by the computed Cf, H and θ variations with Rex presented in Fig.3-5. Results are shown for all three of the suggested E factors. It can be seen that there was little difference between the results obtained with the two $C_{\varepsilon3} = 0.25$ & 1.12 versions, although the first, without any f_μ, influence predicted transition slightly earlier than the latter incorporating a [Launder-Sharma] f_μ factor, while use of the $C_{\varepsilon2}$ [Dutoya-Michard] approximation led to a delayed transition. Comparative results with $f_\mu=1$, $C_{\varepsilon3}=0.3$, instead of 0.25, also showed that the results were rather insensitive to small changes in this parameter. In fact the predictions obtained with the latter value were closest of all four sets of results to the experimental data for Cf, H and θ, with which they proved in reasonable agreement although transition actually occured a little earlier than the model indicated (see discussion in [16]). The mean velocity profile development was also predicted with reasonable accuracy, although the computed profile was initially slightly less full than the measured one and later the reverse was true. By comparison the overall magnitude and the peak of the turbulence intensity profile was considerably under-predicted at all stations, although the discrepancy diminished rapidly by the end of the measured development section. This finding is consistent with the delay between the measured and predicted transition locations and, since the results were virtually the same with the alternative starting conditions tested for the T3A Case, again probably reflects the missing influence of leading edge effects.

Acknowledgements:
This work has been funded by Rolls-Royce plc through University Research brochure PVA3-100D. It would not have been possible to undertake it without considerable assistance from Dr.B.A.Younis who not only made his newly-developed low-Re RST code freely available, but also assisted with the implementation of this on the Cambridge main-

frame. Additional travel and subsistence funds provided by the EPFL ERCOFTAC Pilot Centre are also gratefully acknowledged.

References:

[1] B.A.Younis (1987) Unpublished work provided as a private communication.

[2] K.Hanjalic & B.E.Launder (1972) A Reynolds-Stress Model of Turbulence and Its Application to Thin Shear Flows. JFM 52, pp.609.

[3] R.M.C.So & G.J.Yoo (1987) Low-Reynolds Number Modelling of Turbulent Flows With and Without Wall Transpiration. AIAA J. 25, pp.1556.

[4] N.Shima (1988) A Reynolds-Stress Model for Near-Wall and Low-Reynolds Number Regions. Trans. ASME JFE 110, pp.38.

[5] M.L.Finson (1975) A Reynolds Stress Model For Boundary Layer Transition With Application to Rough Surfaces. Physical Sciences Inc. Rep PSI TR-34.

[6] V.C.Patel, W.Rodi & G.Scheuerer (1985) Turbulence Models for Near-Wall and Low-Reynolds Number Flows: A Review. AIAA J. 23, pp.1308.

[7] W.P.Jones & B.E.Launder (1972) The Prediction of Laminarization with a Two-Equation Model of Turbulence. Int. J. Heat & Mass Transfr. 15, pp.301 (see also 16, pp.1119, 1973).

[8] Y.Nagano & M.Hishida (1987) Improved Form of the k-ε Model for Wall Turbulent Shear Flows. Trans. ASME JFE 109, pp.156.

[19] S.L.Liu (1989) The Prediction of Boundary Layer Transition Using low-Reynolds Number k-ε Turbulence Model. Proc. 4th Asian Congr. of Fluid Mech., 1. pp.A9, Hong Kong University.

[10] W.Rodi & G.Scheuerer (1983) Calculation of turbulent boundary layers under the influence of free-stream turbulence. Proc 4th Turbulent Shear Flows Symp. pp.2.19, see also TSF4 Springer-Verlag.

[11] A.M.Savill (1987) Algebraic and Reynolds Stress Modelling of Manipulated Boundary Layers Including Effects of Free-Stream Turbulence. Proc. Royal Aero. Soc. Int. Conf. on Turbulent Drag Reduction By Passive Means. 1, pp.89, RAeS.

[12] B.E.Launder, G.J.Reece & W.Rodi (1975) Progress in the development of a Reynolds Stress turbulence closure. JFM 68, pp.537.

[13] D.Agoropoulos (1986) Interactions between wakes and boundary layers. Ph.D thesis, University of Cambridge.

[14] Haj-Ibrahim Widah (1985) Programme de Calcul de Solutions de Similitude Couche Limite Laminaire, Incompressible, Gaz Parfait, Methode de Resolution Pseudo-Instationnaire. ONERA/CERT, Toulouse.

[15] P.E.Roach (1987) The generation of nearly isotropic turbulence by means of grids. Int. J. Heat & Fluid Flow 8, pp.82.

[16] A.M.Savill (1991) Synthesis of T3 Test Case Submissions, in this Proceedings Volume, CUP

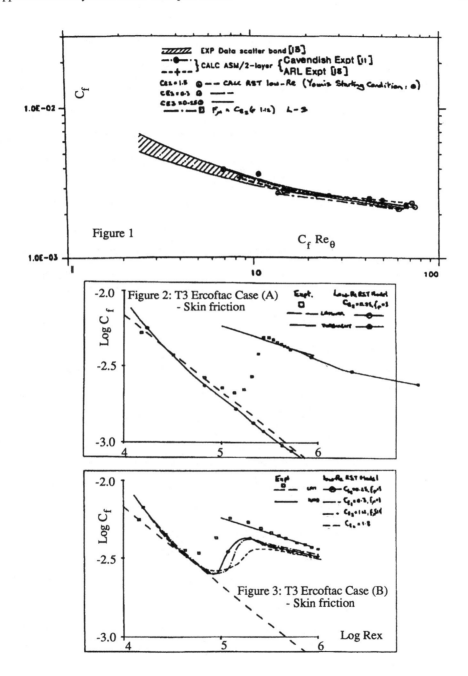

Figure 1

Figure 2: T3 Ercoftac Case (A) - Skin friction

Figure 3: T3 Ercoftac Case (B) - Skin friction

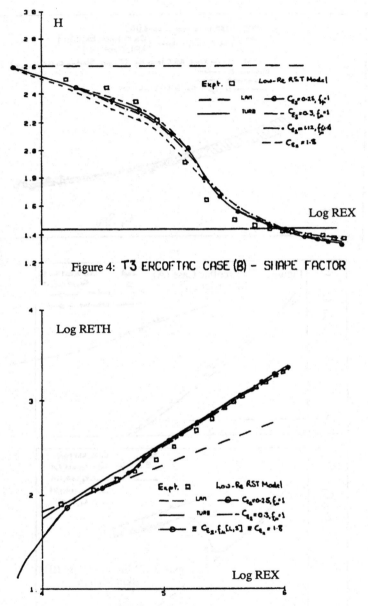

Figure 4: **T3 ERCOFTAC CASE (B) - SHAPE FACTOR**

Figure 5: **T3 ERCOFTAC CASE (B) - MOMENTUM THICKNESS**

Contribution to the ERCOFTAC Workshop at EPFL, March 26-28, 1990
FLAT PLATE TRANSITION TESTCASE T3

Hans E. Mårtensson
Lars-Erik Eriksson
Per J. Albråten
CFD group
VOLVO Flygmotor AB
S-461 81 Trollhättan
Sweden

SUMMARY AND CONCLUSIONS
A study is undertaken of the simulation of transition to turbulence on a flat plate with two
levels of freestream turbulence. An explicit Finite Volume scheme is used with an optional
Smagorinsky like model for Subgrid scale (SGS) turbulence. In this model the grid spacing
effectively defines a mixing length. Four different attempts were made:

1.   2D mesh with an area upstream of the leading edge and SGS model.
2.   2D mesh including an area upstream of the leading edge and only a 4th difference
      artificial viscosity.
3.   3D mesh including an area upstream of the leading edge and SGS model
4.   2D mesh with a circular leading edge circumscribed by the mesh

No signs of transition were found. In case 4 an unsteady separation bubble, shedding long
lived vortices from the leading edge, occured on the flat side when no measures were taken
to remove this by controlling the circulation. The separation bubble was found to be highly
sensitive to the local angle of attack around the leading edge. In cases 1-3 the grid did not
circumscribe the leading edge so effects of leading edge could not be reproduced. Simulation
of incoming turbulence was simulated by the use of direct random number perturbations of
the boundary grid point values. The simulations involve a number of uncertainties, some of
which are related to the use of the SGS model through the laminar boundary layer and some
related to the inflow of turbulence. The SGS model generally raises the effective viscosity in
all sheared regions, regardless of whether there is any turbulence or not. This puts a demand
on the resolution if the laminar part of the boundary layer is to be well represented. General
objections can be raised to the use of random numbers to represent incoming turbulence on
the account of turbulence not being strictly random in nature, the more critical point for
this attempt is probably that when perturbations are specified on the grid the effects will
scale on the computational grid and not on the physical devices used in the experiments.

Turbulence in the freestream is rapidly reduced to low levels by the action of the artificial
dissipation which also produces slight overshoots of the streamwise velocity on the edge of
the boundary layer. Using the SGS model more freestream perturbations are retained but the
boundary layer grows to rapidly and does not represent the laminar profile very well due to
too large grid distances through the boundary layer. Far from being conclusive it is felt that

a higher order method with less numerical diffusion is generally needed to suppress the damping effects of artificial viscosity and if the the SGS model is to be used at all it should use grid distances in the direction normal to the boundary layer.

NUMERICAL METHOD

Spatial discretization is done with a $O(h^2)$ Finite Volume technique for the compressible Navier-Stokes equations. Compact difference molecules are used, centered around the cell walls. A 4th difference artificial viscosity may be used to ensure damping of sawtooth waves at high cell Peclet numbers. Time marching is done using a 3-stage $O(t^3)$ Runge-Kutta scheme with a global time step. The stability limit of this scheme is CFL=1.8. All codes used are general purpose solvers based on multi-block structured grids.

Using compressible codes at very low Mach numbers where the time step is determined mostly by the acoustic propagation velocities is clearly not optimal. The primary reason for doing so is the convenience of working with existing in house codes. In order to get around the problem of stiffness a freestream Mach number is chosen that is considered low enough not to add effects of compressibility but still high enough to give convection a fair chance. In this study M = 0.3 is used consistently.

SGS MODELLING

In the L.E.S. concept the flow field is split into resolved and sub grid scale (SGS) quantities. What is resolved and what is SGS is determined by the filter width which is the smallest length scale that can be accurately represented by the discretization. Depending on the choice of numerical scheme the filter width is related to grid spacing by some ratio Δ_f/Δ_m. In the present application this is usually taken to be 3.

After filtering the Navier-Stokes equations a model is needed to account for the action of small scale turbulence. Here a compressible counterpart of the Smagorinsky SGS. model is used for this purpose. For the compressible model the resulting SGS stress tensor is split and modelled as an isotropic part and a deviatoric part wheras in the incompressible case the isotropic part can be lumped with pressure. For a more extensive discussion on the derivation of this model the reader is referred to [1].

The S.G.S. stress tensor is modelled as follows:

$$\tau_{ij} = <\overline{-\rho U_i U_j}> = \tau_{ij}^D + \tau_{ij}^I \qquad\qquad \text{where}$$

$$\tau_{ij}^D = 2C_D \rho \Delta_f^2 |S| (S_{ij} - \frac{1}{3} S_{kk}\delta_{ij}) \qquad - \text{ Deviatoric}$$

$$\tau_{ij}^I = - \frac{2}{3} C_I \rho \Delta_f^2 |S|^2 \delta_{ij} \qquad\qquad - \text{ Isotropic}$$

$$S_{ij} = \frac{\partial u_i}{\partial x_j} + \frac{\partial u_i}{\partial x_i}$$

with the model constants C_D & C_I as recommended in [1]
$C_D=0.012$; $C_I=0.0066$

f is the local filter width and $_m$. Here this is taken to be 3 times the maximum grid distance in the three directions.
Using the maximum distance seems to be the most conservative choice in some sense, but is a disadvantage for the resolution of the laminar boundary layer. In principle the model value still vanishes as $O(\Delta_f)$ so that a point can always be reached where the the grid is fine enough to resolve everything given a large enough computer.

GEOMETRY
Some different grid sizes were tried, typically the boxes were length*height = 1*0.16 with a maximum lengthwise Reynolds number of 200000 from the leading edge to the end of the plate. In the 2d case grids of size 801*71 cells and in the 3d case 501*41*8 cells were used.

REFERENCES
[1]   Erlebacher G., Hussaini M.Y., Speziale C.G., Zang T.A.
        Toward the Large Eddy Simulation of Compressible Turbulent Flows
        ICASE-87-20. NASA-CR-178273.

Numerical Simulation of Boundary Layer Transition in the Presence of Free Stream Turbulence

Z Y YANG and P R VOKE[*]

The Turbulence Unit, School of Engineering, Queen Mary and Westfield College
Mile End Road, London E1 4NS U.K.
[*] Present address: ERCOFTAC Pilot Centre, EPFL, CH1015 Lausanne.

SUMMARY

Transition of the zero-pressure-gradient boundary layer on a flat plate has been numerically simulated using a finite volume code, with two levels of free-stream turbulence. In spite of the relatively coarse resolution used, the predictions compare well with experiments.

1 INTRODUCTION

Transition from laminar to turbulent flow is a fundamental flow phenomenon of fluid dynamics. It is only partly understood, particularly in the case where free-stream turbulence affects the character of the laminar and transitional flow [1,2]. In the work reported here we concentrated on the study of the effects of free stream turbulence on the laminar boundary layer and transition on a flat plate at zero incidence.

It is very difficult to apply conventional approaches to cope with transition problems as available turbulence closure models deal only with time-averaged quantities. With the development of large-memory high-speed supercomputers it is now possible to simulate some turbulent flows directly. In the present study numerical simulation is adopted instead of more conventional approaches. No subgrid model is used because of the very low Reynolds numbers involved.

2 COMPUTATION

The three-dimensional time dependent Navier-Stokes equations are integrated forward in time numerically by the explicit second order Adams-Bashforth method. The spatial dicretisation is by the second-order finite volume method which exactly maintains local and global conservation of mass, momentum and convected kinetic energy, while approximating the values of certain fluxes to second order. A Poisson equation

398

is solved to obtain pressure at each time step in such a way that continuity is enforced to machine accuracy.

The viscous terms as well as the advection terms are treated explictly, and the time step is chosen to keep well within both the viscous and convection Courant limits. The maximum values of the viscous and Courant numbers in the simulations reported were less than 0.03 and 0.22 respectively. While expensive compared to viscous-implicit methods when the viscosity is small, the explicit treatment of the viscous terms is prefered in problems where molecular viscosity plays a crucial role and the viscous limit is less pressing, such as the disturbed laminar and transitional boundary layers reported here.

A no-slip boundary condition is applied on the solid wall (natural boundary conditions) and a free slip boundary condition for the upper surface. No flow is allowed through the upper boundary, so there is a small streamwise pressure gradient present in both simulations, largest at the leading edge but dropping to below 0.09 beyond x=43mm. In the spanwise direction, periodic boundary conditions are used. At the outlet advective boundary conditions are used for all the velocities, based on an advecting velocity equal to the mean streamwise velocity. The inlet conditions imposed are described in detail below.

Mesh resolution is very important for the simulations as no subgrid model is used. In this pilot phase of a three year study, less than one quarter of the possible utilisable memory of a Cray XMP/48 computer was employed, to reduce the computational costs and ensure that preliminary results were obtained in a short time. The resolution used to obtain the results reported here was 255 (streamwise) x 16 (spanwise) x 32 (cross-stream). In spite of this very limited resolution, we consider the results to be surprisingly good.

In our study the streamwise length of the computational box was taken to be 600mm instead of 1700mm as specified in the T3 test problem to improve the resolution that could be obtained in the streamwise direction. We are mainly interested in the prediction of transition and the preceding disturbed laminar boundary layer, so it is reasonable to cut short the streamwise length as the boundary layer flow will become fully turbulent flow beyond 600mm from the leading edge. The computational volume was not shortened upstream of the transition (the simulation was started from the leading edge) since it was found that any attempt to move the upstream boundary close to the transition point severely affected the quality of the predictions.

It is well known that the Tollmien-Schlichting (TS) wave plays an important role in unforced boundary layer transition. We assumed that sufficient streamwise resolution would be obtained if there were several grid points within one TS wavelength. Orszag [3] has suggested that as the TS wavelength is of order 10δ (δ is the laminar boundary layer thickness), if the body has length scale L then L/δ grid points would be necessary along the body; in other words a streamwise spacing of δ is sufficient. In our case the average value for δ is about 3mm so that 200 grid points would be necessary. In fact 255 cells were used in the streamwise direction.

The height of the computational box in the cross stream direction was fixed at 30mm as the laminar boundary layer thickness is only about 6mm just before transition. 32 grid points were used in an exponentially stretched mesh to make sure that there were about 20 grid points within the boundary layer. The stretching ratio from one grid to the next above was 1.14. In the homogeneous (spanwise) direction the length was taken to be 20mm and 16 grid points were used, adequate to resolve 1/15 of a TS wavelength. The geometry of the computation is illustrated in Figure 1. The grid resolution employed is summarised in Table 1; the viscous units are based on the viscous length scale v/u_τ at transition, and the resolution in terms of the momentum thickness θ and the approximate ($u = 0.99\ U$) boundary layer thickness δ at transition are also given.

TABLE 1

	stream Δx	span Δz	cross Δymin	Δymax
(mm)	2.353	1.25	0.072	3.7
(δ)	0.87	0.463	0.027	1.37
(θ)	7.35	3.91	0.225	11.56
(viscous)	58.8	31.25	1.8	92.5

Computations have been performed for both cases (a) and (b) starting from uniform inlet conditions at x=0, the plate leading edge. The free-stream turbulence was created by superimposing pseudorandom disturbances on the flat inlet profile. In later phases of the project the inflowing free-stream turbulence will be conditioned by a separate simulation and introduced through the inflow plane using the "slicing" technique [4,5]. In this pilot study it was found that the pseudorandom "turbulence" superimposed on the flat profile dropped very quickly to about 10% of the initial intensity, before adopting a more physical decay rate. The intensity at the inlet was therefore raised by the appropriate factor in order to obtain an approximately correct streamwise profile of the decay of the free-stream turbulence over most of the plate. The decay of the imposed free-stream turbulence in the simulations (averaged over z and time) compared with the functions specified in the test data, is shown in Figure 2. It is important to note that the integral length scale of the disturbances is not independent of the mesh and the box dimensions. At the inlet, all wavelengths from the box dimensions down to the mesh size are present, but the higher wavenumber motions are rapidly depleted by viscous action, leaving a more realistic spectrum dominated by larger eddies.

3 RESULTS

All results presented in the summary paper [6] are spanwise (z) averages. The mean velocity profile was also averaged over time between t=0.372 and 0.393s (steps 8000 to 8500), ten samples being taken. The remaining results are based on the instantaneous situation at t=0.37s.

It can be seen from the simulation results presented in the summary paper [6] that transition is clearly indicated by the strong increase of the skin friction coefficients; it occurs earlier in case (b) than in case (a). This can only be attributed to the higher free stream turbulence for case (b). The transition region is also indicated by the higher rate of increase of θ although this is not so obvious as the increase of the skin friction coefficients. There is a sharp decrease of shape factor at one point in each case which implies again that transition occurs there.

For case (a), up to x=395mm the mean velocity clearly has a laminar boundary layer profile, while at x=595mm it is evident that the velocity has changed to a turbulent profile. On the other hand for case (b) even at x=195mm the velocity already has a typical turbulent boundary layer profile, again confirming that transition takes place much earlier.

It can be seen from the intensity profiles that within the boundary layer there is a level where the turbulence intensity reaches its peak value. This indicates that the turbulence inside the layer may be generated within the boundary layer itself from the high shear stresses there rather than diffused from the free stream. Nevertheless, diffusion of the free stream turbulence must have an important influence on the position of transition, since this occurs so much earlier for case (b) than for case (a).

4 CONCLUSIONS AND FUTURE WORK

It is encouraging to have found that transition from laminar to turbulent boundary layer flow can be predicted by a finite-volume numerical simulation even with a rather low resolution. Considering the preliminary nature of the present study, the agreement with the available experimental results is surprisingly good. The transition region is clearly indicated by the strong increase of the skin friction coefficients for both cases. We have also predicted that transition takes place earlier in the presence of higher intensity free stream turbulence, in agreement with the experiments. The correct prediction of the peaks in the intensity of the disturbances in the laminar layers prior to transition strongly suggests that the simulation is reproducing the important features of the dynamics in the disturbed laminar flow.

We have recently found that the use of pseudorandom disturbances is suspect in simulations at higher resolutions, producing earlier transition. This may be due to triggering of transition by the higher wavenumber disturbances present initially on the finer mesh, propagating acoustically. Current research aims to replace the unphysical pseudorandom inflow disturbances by properly conditioned grid turbulence. We shall be seeking answers to many of the outstanding problems highlighted by the present study: how the free stream turbulence influences the laminar boundary layer, and the origin of the disturbances seen in the laminar boundary layer prior to transition; the nature of the transition process in the presence of free stream turbulence as compared to natural transition; and the physical mechanisms by which the free stream turbulence influences transition.

REFERENCES

1. Goldstein, M. E. (1983) Generation of Tollmien-Schlichting waves by free stream disturbances at low Mach number, NASA TM-83026.

2. Goldstein, M. E. and Hultgren, L. S. (1989) Boundary-layer receptivity to long-wave free stream disturbances, Ann. Rev. Fluid Mech., Vol. 21, pp. 137-166.

3. Orszag, S. A. (1979) Numerical studies of transition in planar shear flows, IUTAM Symposium, Stuttgart, Germany. Laminar - Turbulent Transition, Editors R.Eppler and H. Fasel.

4. Tsai, H. M., Voke, P. R. and Leslie, D. C. (1987) Large-eddy simulation of turbulent free shear flows, Proc. 5th Conf. on Numerical Methods in Laminar and Turbulent Flow, Montreal, Canada (ed C. Taylor, W. G. Habashi and M. M. Hafez) Pineridge Press, Swansea, UK.

5. Tsai, H. M., Leslie, D. C. and Voke, P. R. (1989) Prediction of extremal events by numerical simulation, Proc. 7th Turbulent Shear Flow Conf., Stanford University, U.S.A.

6. Savill, A. M. (1991) Synthesis of T3 test case submissions, in this volume.

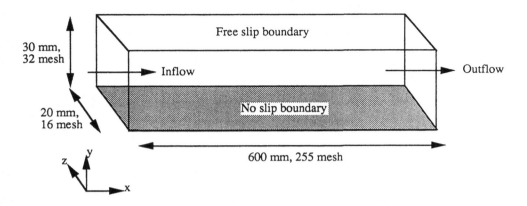

Figure 1. Geometry of the Simulation.

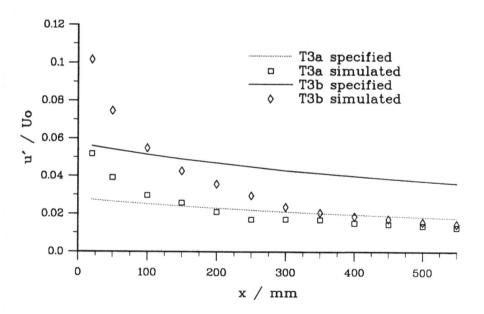

Figure 2. Decay of the Streamwise Turbulence.

A Synthesis of T3 Test Case Predictions

A.M. SAVILL

Test Case Supervisor
Rolls-Royce Senior Research Associate
Department of Engineering
University of Cambridge
Cambridge CB2 1PZ
England

1 THE T3 TEST CASE SPECIFICATIONS

The object of this Test Case problem, as originally suggested by Drs. P.Stow and N.T.Birch of the Theoretical Sciences Group, Rolls-Royce plc, was to predict the initial effect of isotropic free-stream turbulence on a laminar boundary layer, and the subsequent transition, in zero pressure gradient. The geometry to be considered consisted of a flat test plate with a grid mounted upstream to generate the external turbulence (see Figure1).

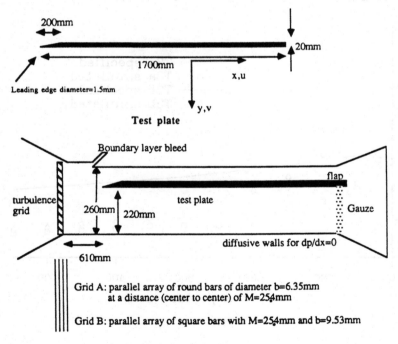

Figure 1: Flow Geometry under consideration

Two separate flow cases were in fact proposed:

Test Case T3A: in which the external mean flow velocity Uo=5.2m/s, and approximately 3% free-stream turbulence intensity was generated at the test plate leading edge by a grid composed of a parallel array of round bars b=6.35mm in diameter with a mesh separation M=25.4mm, mounted 610mm upstream;

Test Case T3B: with Uo=9.6m/s and approximately 6% external turbulence at the leading edge due to a parallel array of square bars, with b=9.53mm and M=25.4mm, mounted at the same upstream location.

Experimental data for these two flow conditions had been obtained by Drs.P.E.Roach & D.H.Brierley at the Advanced Research Laboratory, Rolls-Royce plc, as part of a wider study of the influence of a turbulent free-stream on transitional boundary layer development (T3A corresponds to case PRHI and T3B to case PSHI in their presentation of the Test Case database [1]). Considerable care was taken in both the experimental set-up and the measurements to ensure that the data were truly representative of fully-attached 2-D boundary layer flow on an accurately flat surface in the absence of any pressure gradient. Care was also taken to avoid any unsteadiness of the leading edge flow (This entailed inclining the test plate by half a degree to the mean flow - something which was not mentioned in the actual problem specification). Earlier studies [2] had revealed that the turbulence generated by such grids was indeed remarkably accurately isotropic over the downstream extent considered in the present cases, and obeyed the following decay law:

$$u'o\,/Uo = \{\ C\ (\ x + 610)^{-5/7}\ \}\ \%\qquad\qquad(1)$$

where: C=274 for the T3A grid; C=560 for T3B grid; x = distance from leading edge in mm; and $u'o\ =\ v'o\ =w'o$.

This being the case two possible types of computations were proposed:

1: Starting from uniform inlet conditions at or upstream the test plate leading edge x=0.
2: Starting at x=10mm downstream of the leading edge and assuming an initial Blasius boundary layer mean velocity profile with Reynolds stress intensity distributions of the form:

$$u'/u'o\ =\ (U/Uo)^2\qquad\qquad(2)$$
$$w'\ =\ 0.2\ u'\qquad\qquad(3)$$
$$v'\ =\ 0\qquad\qquad(4)$$
$$uv\ =\ 0\qquad\qquad(5)$$

However Rodi (private communication) pointed out that, on the basis of a wide range of other data, a more appropriate alternative to (2) should be:

$$k'/ko\ =\ (U/Uo)^2\qquad\qquad(6)$$

This was adopted by both Rodi et al.[7] and Tarada [8].

Most groups in fact assumed: $k\ =\ u'$ (7)
with the exception of Tarada [8], who employed Algebraic Stress relations for the normal stresses, and Savill [9] who used (2)-(5) to initialise his RST model.

A significant omission from the Test Case specification was that no free-stream dissipation length scale Lo was specified; although Roach [2] had previously reported that ,for the both the grids considered, this obeyed the following decay law:

$$Lo\ =\ Loi\ [\ (x + 610)/\ 610]^{0.5}\qquad\qquad(8)$$

TABLE 1: Classification of turbulence models

Computors	Code No.	Model
ROACH	0	Correlation Analysis
BIRCH+H0+STOW	1	Green's Lag Entrainment Method
" " "	2	Cebeci-Smith Mixing Length +Abu Ghannam & Shaw Correlation
" " "	3	Hassid-Poreh k equation + Birch lε formulation
BERNARD+ FOUGERES+ MANDELIER	4	MacDonald &Fish + Blair &Edwards Integrated k-ε
THEODORIDIS+ PRINOS+GOULAS	5	Nagano-Hishida low-Re k-ε
TARADA	6	Chien (or Launder-Sharma) low-Re k-ε + Gibson-Younis ASM
RODI+FUJISAWA	7	Launder-Sharma low-Re k-ε
+SHONUNG	8	Lam & Bremhorst low-Re k-ε
	9	Standard k-ε + Norris &Reynolds near-wall k-lε + Abu Ghannam & Shaw Correlation
YANG+VOKE	10	Fully Resolved Direct Numerical Simulation
MARTENSSON+ ERIKSSON+ ALBRATEN	11	Large Eddy Simulation+Smagorinsky Sub-Grid Scale Model or + 4th Order Difference Artificial Viscosity
SAVILL	12	Younis low-Re RST: based on Hanjalic-Launder & Chien k-ε
	13	or alternative Hanjalic-Launder & Launder-Sharma closure
	14	or alternative Hanjalic-Launder & Dutoya-Michard closure

(Details of all the methods are either given in the computors individual papers [3-11] or in the references there contained. For a general review of low-Re k-ε models see also the paper by Patel et al. [12] which discusses most of the closure schemes mentioned in the Table. Details of the more recent Nagano-Hishida approach can be found in their paper outlining this [13].)

All of the computors attempted both the T3A and T3B Test cases, but using different methods as indicated by the summary Table 2. This table is intended only to give a quick guide to the methods for comparison purposes. Once again further details of the numerical schemes used in each case can be found in their individual papers [3-11].

The initial value of Lϵi indicated by his earlier measurements, assuming:

$$Loi = \frac{u'^3}{Uo\ d(u'^2)/dx} \qquad (9)$$

was 13.4mm at x=10mm, for the T3A grid with a similar value being indicated for the T3B grid also.

However the corresponding values determined from the decay of the free-stream turbulence intensity using:

$$\epsilon_O = -Uo\ dko/dx \qquad (10)$$

were 30.1mm for T3A and 41.8mm for T3B, and these were the values adopted in practice by the majority of computors who used Method 2 above. Thodoridis et al.[6], who initially attempted computations using Method 1 above, obtained an estimate for $\epsilon_O = 0.903 m^2/s^3$ at x=-160mm in a similar manner.

Computors adopting Method 2 had also to estimate the initial dissipation profile. Most chose to do this by assuming that within the pseudo-laminar boundary layer:

$$L = lm/C\mu \qquad (11)$$

where lm is the fully turbulent mixing length, lm=min[κy, 0.09δ] and Cμ=0.09, although Tarada adopted the more complex expression:

$$\epsilon = k^{3/2} [\ 1\ (-6 + 3\pi)(y/\delta)^2 + (8-3\pi)(y/\delta)^3 + (\pi-3)(y/\delta)^4] \qquad (12)$$

where: $\pi = \alpha\ 2.5\ \delta\ \epsilon\ /\ ko^{3/2}$

and both Theodoridis et al.[6] and Rodi et al.[7] preferred to use the alternative prescription: $\epsilon = a_1\ k\ dU/dy \qquad (13)$

where a_1 (=uv/k) was assigned the value 0.3; again representative of fully turbulent conditons.

{Some computors also tested the effect of varying the initial and boundary conditions, as well as some model constants - see Discussion}

Those attempting Simulations [10,11] chose to generate the free-stream turbulence fluctuations pseudo-randomly with adjustments to get approximately the correct intensity level Tu and decay rate over the test plate. In practice the integral length scale of the resulting turbulence was close to that measured by Roach [2] eg. for the DNS [11] the integral length scales at x=10mm from the leading edge were approximately 12mm for T3A and 13,5mm for T3B compared to the experimental estimates of ~10mm.

2 CLASSIFICATION OF COMPUTATIONAL METHODS AND MODELS

Including the analysis of the data provided by Roach [3] a total of 16 different sets of results were presented by the 9 particpating research groups at the Workshop (see   [3-11] & Table 1). These covered a very wide range of turbulence models, from simple low-Reynolds number one-equation k-l schemes [4,5], through several low-Re k-ϵ approaches [6-8], and a so-called 'partial-low Re' two-layer k-ϵ/k-l model [7], to a full Reynolds Stress transport scheme [9], and even a Large Eddy Simulation [10] without any specific correction for transition. The results of a first Direct Simulation of the Test Case configuration were also presented [11]. This wide range of approaches allowed a comparison to be made between some typical design methods and more advanced research codes; including not only some of the more commonly used low-Re k-ϵ models (e.g.Chien[CH],Launder-Sharma[LS],Lam & Bremhorst[LB]), but also a more recent

Table 2: Classification of Computational Approaches Adopted.

Computors	No.	Numerical Scheme	Initialisation	Initial lε +	Development
BIRCH+HO+STOW	1-3	TIME-MARCHING or ELLIPTIC	Method 2	lε from k+	From decay law
BERNARD+ FOUGERES+ MANDELIER	4	ELLIPTIC	Method 1 at x=0mm	1 ε from k	+ Computed
THEODORIDIS+ PRINOS+GOULAS	5	ELLIPTIC	Method 1 * at x=-160mm	1 ε from k	+ Computed
TARADA	6	PARABOLIC	Method 2 at x=1mm	1 ε from k	+ Computed
RODI+FUJISAWA +SHONUNG	7-9	PARABOLIC	Method 2	1 ε from k	+ Computed
YANG+VOKE	10	TIME-STEPPING	Method 1 at x=0mm	1 ε from pseudo-random fst	+ Computed
MARTENSSON+ ERIKSSON+ ALBRATEN	11	TIME-STEPPING	Method 2	1 ε from randomly generated fst	+ Computed
SAVILL	12-14	PARABOLIC	Method 2	lε from k	+ From decay law

* NB: Refers to elliptic results originally presented at the Workshop [6] and not later parabolic predictions [18]

refined version (the Nagano-Hishida[NH] model which had already demonstrated superior performance on a wide range of fully turbulent flows), and a state-of-the-art low-Re stress transport model.

Although the Test Case specification specifically requested that correlations for the start and end of transition should not be applied, comparative results obtained using both an Integral method and a mixing length model, with correlations for start of transition of the Abu Ghannam & Shaw [14] type, were presented by the test case proponents themselves [4]. In addition Rodi et al. [7] also made use of correlations for the structure parameter a1 (as a function of Re and Tu due to Rodi & Sheuerer [15]), and for the near wall damping factor (as a function of a critical transition Reynolds number derived from the Abu Ghannam & Shaw correlation [14]). It is therefore important to note that Roach [3] found that the latter did in fact correlate the T3A and T3B data with a very wide range of other similar transition data sets.

The various numerical approaches adopted by the computors are detailed in Table 2. Most computors used only one numerical scheme so the range of these was insufficient to make any detailed comparison of various combinations of models and numerics. Indeed only three elliptic computations were attempted [4-6] (although with three different k-ε models), and only one of these started upstream of the leading edge [6]. Since the x-y grids adopted consisted of 85-200 x 50-75 grid points, and did not incorporate any local mesh refinement, severe resolution problems were encountered in the leading edge region for the case where computations were started upstream of this (only one point in the initial boundary layer), and only about one third of the points were contained within the boundary layer generally. It is therefore uncertain whether any of the results were completely grid independent; no detailed grid sensitivity tests wrere reported. By comparison the parabolic codes typically employed a similar number of points (~60-80) in the y-direction, but these were concentrated within the boundary thickness (about half below y+=30 although the initial δ+ was ~30-40), and several thousand x-steps of (~0.025δ) were used to cover the whole streamwise development. Tests conducted by several participants indicated that this was sufficient to generate grid-independent predictions.

The Simulations employed either 2D x-y grids of 801 x 71 points (LES only) or x-y-z 3D grids of 501 x 41 x 8 (LES) and 255 x 32 x 16 (DNS). In each case there were only about 20 points covering the width of the boundary layer through the transition region. For the DNS the streamwise grid spacing was set equal to the initial δ, while the spanwise extent was limited to one Tolmein-Schlicting wave length (This was chosen as the appropriate span-wise length scale despite the fact that there is conflicting evidence in the litterature as to whether such T-S waves exist in the by-pass transition induced by free-stream turbulence levels greater than ~1%).

Computation times varied from 1-4 minutes cpu on a dedictaed Vax or IBM mainframe for parabolic computations, up to several minutes or even hours on similar machines for elliptic computations, and many tens (LES) or hundreds (DNS) of hours on a CRAY XMP for the Simulations.

3 RESULTS WITH DISCUSSION

Participants were asked to provide results for the integral parameters Cf, H and θ , as a function of log Rex ,and also mean velocity profiles, normalised by the free-stream value, at x=45, 195(95), 395(195), 595(495), 795(895), 1295 & 1495mm for T3A (and T3B). In addition those that could provide such information had the option to supply profiles of the

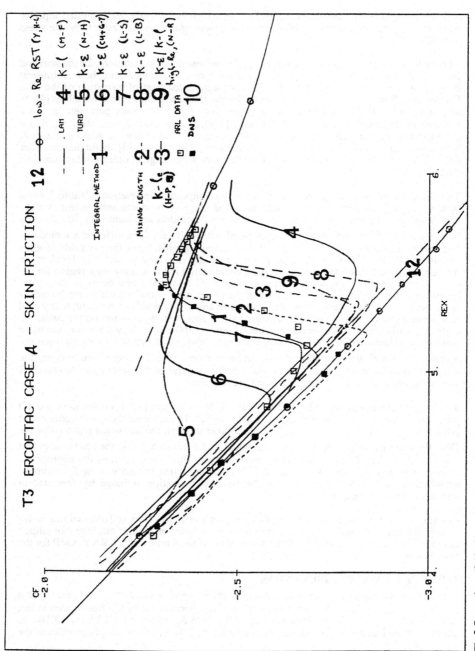

Fig.2 Comparison of skin friction predictions with experimental data for T3A

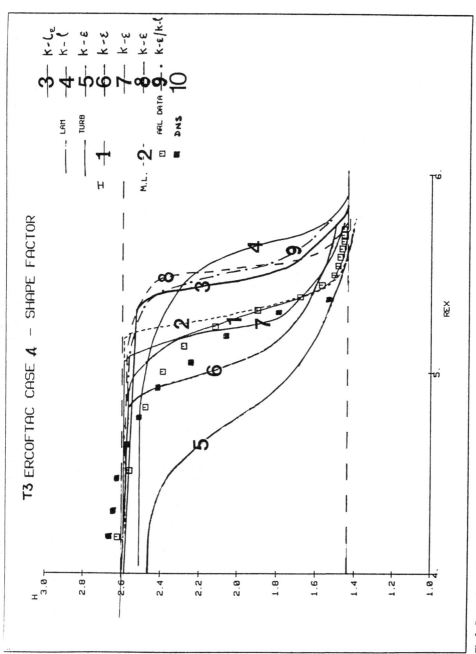

Fig.3 Comparison of shape factor predictions with experimental data for T3A

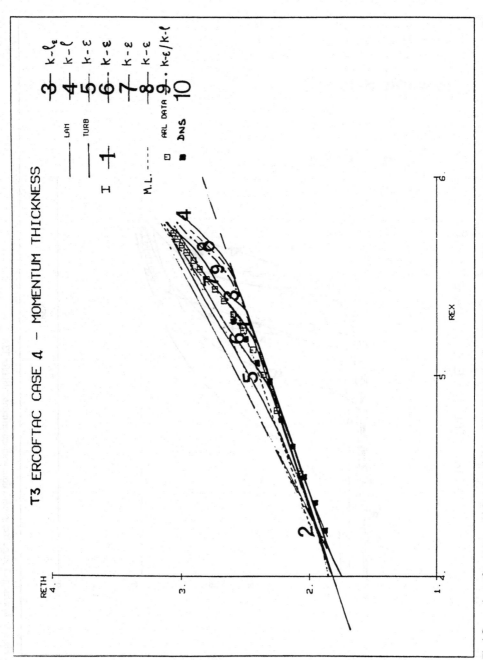

Fig.4 Comparison of momentum thickness predictions with experimental data for T3A

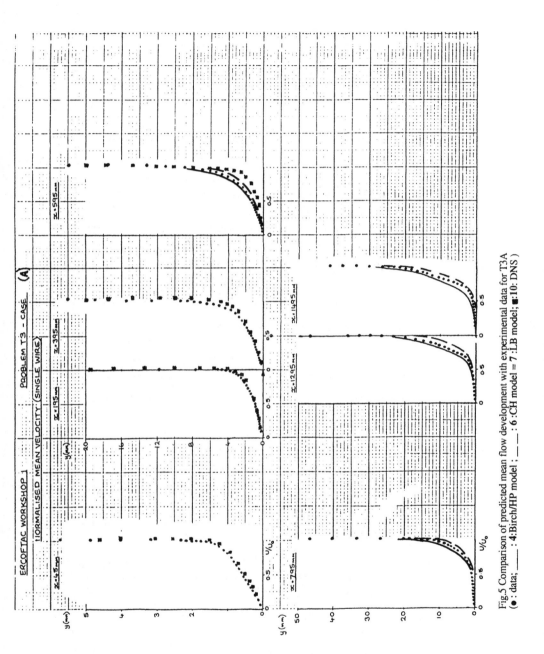

Fig.5 Comparison of predicted mean flow development with experimental data for T3A
(● : data; ———— : 4:Birch/HP model ; ——— : 6 :CH model = 7 :LB model; ■:10: DNS)

streamwise turbulence intensity distribution, normalised by either the local mean velocity and/or the free-stream value, at x=195(95), 595(195), 995 (895) & 1495mm for T3A(T3B). Standard plot formats and layout were predefined, and appropriate mats provided to the computors, so that comparisons could be made easily with transparent overlays of the experimental data (at least within a small photocopying scaling factor!). In practice the Cf, H and u' plots proved to be the most useful for discriminating between models.

3.1 T3A Predictions

Skin friction Cf: As can be seen from Figure 2 there was more than an order of magnitude variation in the predicted location of transition (onset defined as the first departure from laminar Cf curve which in fact occured at x=295mm downstream of the leading edge; equivalent to $Rex_{tr}=10^5$), although the models were at least about evenly spread about this, and, as one would have expected, the elliptic computations were more widely spread than the parabolic code predictions. In addition, as might also have been anticipated in view of the analysis performed by Roach [3], the integral and mixing length calculations which made use of Abu Ghannam & Shaw correlations agreed most closely with the experimental data. The k-ε[LS] scheme actually produced a better prediction for Rex_{tr}, but did not produce quite the correct variation of Cf through transition. In common with many of the other transport equation models the peak Cf (overshoot) was not captured; the predicted transition length was rather too short; and the fully turbulent Cf level was only approached assymptotically from below. Surprisingly the low-Re RST scheme did not show any evidence of transition within the Rex range covered by the Figure and the same was true of the Large Eddy Simulation, although both reproduced the correct turbulent region behaviour if turbulent initial conditions (and, in the case of the LES, a sub-grid scale model) were used.

The Direct (or Fully Resolved) Numerical Simulation, however, provided the best overall 'reproduction' (not strictly a 'prediction' since such a Simulation should ideally be directly equivalent to the experiment , and hence the use of alternative 'data' point symbols plotted on the Figures) of the experimental results; in particular capturing the Cf overshoot, although the necessarily limited extent of the computational box prevented any computation of the subsequent turbulent flow region development.

Shape factor H: The results presented in figure 3 reflect, and are entirely consistent with, the wide variation of Cf results. The predictions were again evenly spread over an order of magnitude in Rex about the transition location, with the elliptic scheme results once more oustside the range of the parabolic predictions. However all of the methods (with the exception of the RST and LES schemes) did show the correct changeover from laminar to fully turbulent limiting values. The correlated mixing length and integral methods, the k-ε[LS] model, and the DNS were again closest to the data, but only the DNS correctly reproduced the important reduction in H prior to the Cf-indicated Rex_{tr}.

Momentum thickness θ*:* Figure 4 shows clearly that the variation in θ is a less useful indicator of model performance although it is evident that the same four computation methods gave the best results. None of the other models could be regarded as close to the data, although, as with Cf and H, the k-l model of Hassid-Poreh/BirchHP-B] and the Chien k-ε[CH] provided the next closest results to experiment.

Mean & turbulent velocity profiles: Most methods in fact produced apparently reasonable predictions of the mean velcoity profile development, as indicated by Figure 5. The only

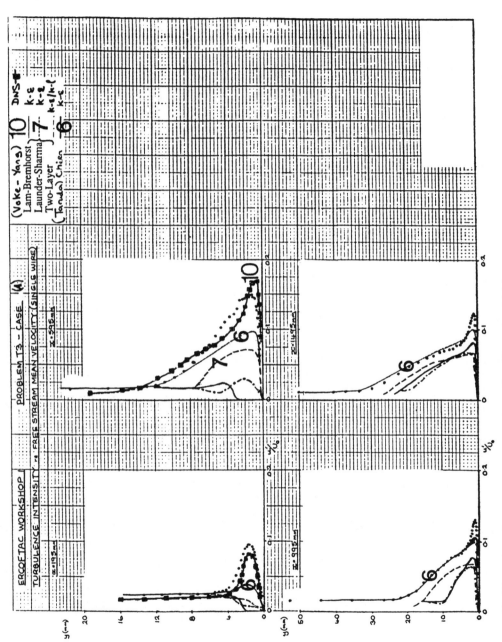

Fig.6 Comparison of predicted turbulence intensity profiles with experimental data
(a) normalised by Uo,

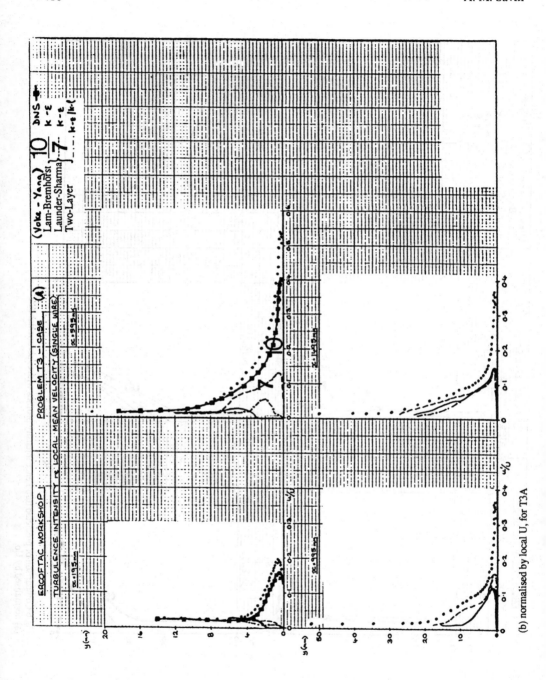

(b) normalised by local U, for T3A

exceptions were the Birch k-l[HP-B] model which underpredicted the fullness of the profiles rather more than might have been expected from the slightly late Cf transition indicated by this model, and the k-ε[LB] & [CH] models which both predicted profiles rather fuller than might have been anticipated from their correspondingly early predictions of Cf transition. Overall though it is clear that the integral parameters provide a more subtle indication of discrepancies in the precise form and development of the mean profiles.

It should be noted that the DNS also predicted much too full velocity profiles, but this was due to the fact that cpu limitations placed a restriction on the averaging time that could be employed to derive the mean values (as well as limiting the streamwise extent of the computational box), rather than due to any defficiency of the approach itself. This is clearly illustrated by the fact that the DNS also very accurately reproduced the streamise turbulence intensity profile development (see Figure 6). The only reported turbulence model results, for the k-ε [LS], [CH], [LB] and k-ε/k-l models, all lagged well behind the experimental measurements, and in particular none of the models picked up the initial growth of u' peak near the wall. Furthermore, although the various predictions did appear to catch up with experiment at the furthest downstream station, this was misleading. The apparently good performance of the k-ε [CH] model was due to the fact that it predicted transition so early - see Figure 2 (and also partly because the computations using this model were started closer to the leading edge than the other parabolic codes and consequently the turbulence had a little more time to develop). By comparison the k-ε [LS] model considerably underpredicted the fullness of the profiles even in the turbulent region.

Plotting the same results normalised by the local mean velocity, as in Figure 6(b), only served to highlight these same model defficiencies compared to the DNS scheme, although it is evident from the Figure that even this failed to reproduce the very near wall beahviour; probably due to the relatively coarse resolution in this region.

3.2 T3B Predictions
Skin friction: Perhaps not surprisingly Figure 7 shows an even wider spread (factor 20 in Re_x) of results for the higher free stream turbulence case, with again the elliptic predictions being more extreme than the parabolic. However the predicticted transition locations were once more evenly distributed either side of the experimentally measured position (transition onset located at about x=9mm; equivalent to $Re_{x_{tr}}$=7 x 10^4), and the transition length was generally rather better predicted by most models, with the exception of the Chien k-ε model. Despite predicting premature transition this model exhibited a damped response through the transition region (features which have also been noted by Stephens & Crawford [16] in separate comparative studies of the Chien, Launder-Sharma and Lam & Bremhorst k-ε models for transition with more than 1% free-stream turbulence). Otherwise the results were consistent with those for T3A ,in the sense that those models which predicted transition early (or late) for T3A also predicted transition early (or late) for T3B, and, more significantly, most models predicted approximately the correct shift forwards of transition (in terms of Re_x) with increasing Tu. The k-ε [LS] model, however, performed rather worse on this case, predicting transition too early, and it would seem that both this model and the k-ε [LB] model are a little too sensitive to increasing Tu; a finding supported by a recent independent study conducted by Fujisawa [17]. The same is also true to a lesser extent for the k-ε/k-l model; whereas the Birch k-l model proved to be very insensitive to changes in Tu, predicting virtually the same transition location for both T3A and T3B.

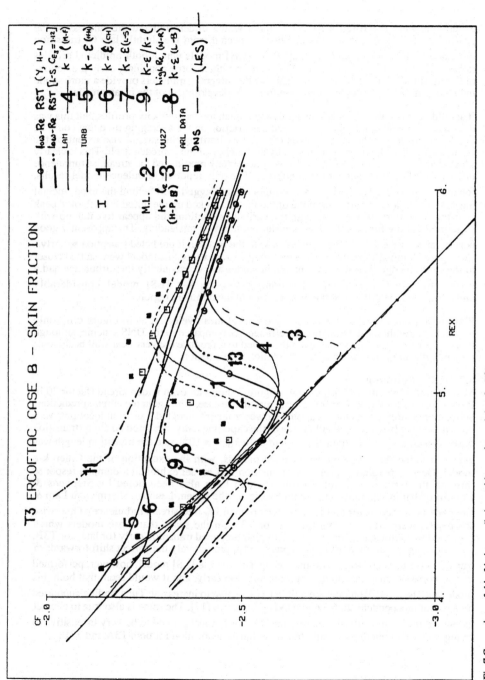

Fig.7 Comparison of skin friction predictions with experimental data for T3B

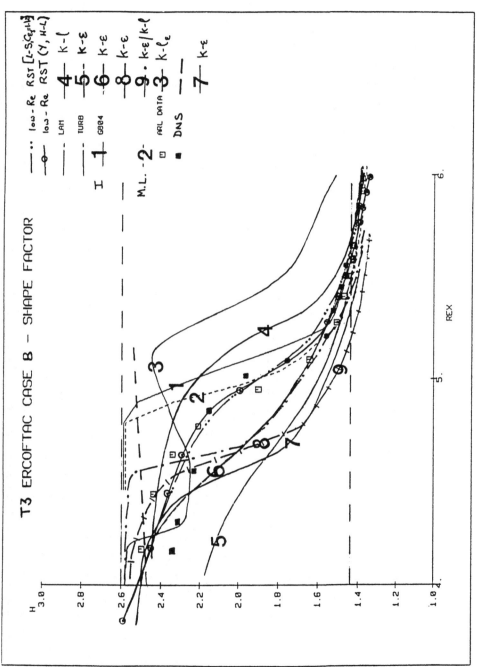

Fig.8 Comparison of shape factor predictions with experimental data for T3B

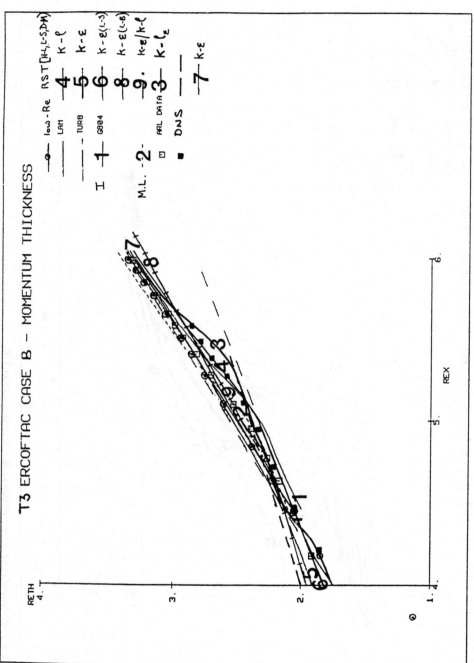

Fig.9 Comparison of momentum thickness predictions with experimental data for T3B

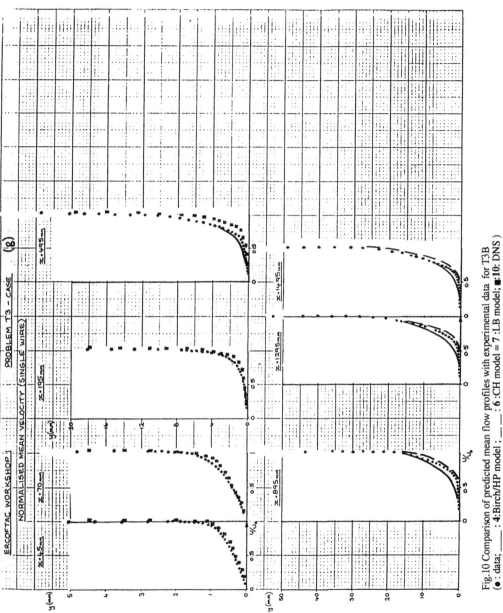

Fig.10 Comparison of predicted mean flow profiles with experimental data for T3B
(● : data; ——— : 4:Birch/HP model; — — : 6 :CH model = 7 :LB model; ■:10: DNS)

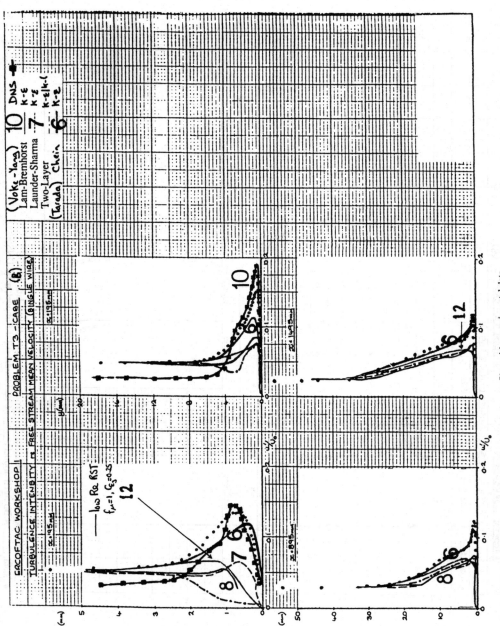

Fig.11 Comparison of predicted turbulence intensity profiles with experimental data

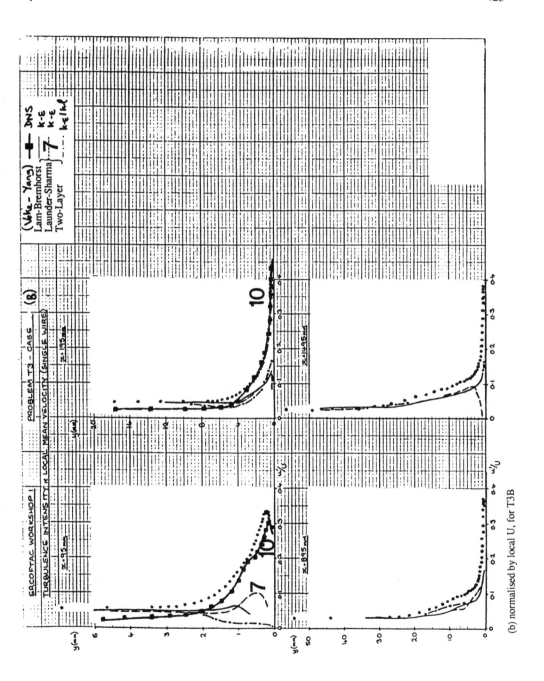

(b) normalised by local U, for T3B

The best improvement was seen from the low-Re RST model which generated the best turbulence model predictions (The H-L closure producing results almost equivalent to the Abu-Ghannam & Shaw correlated mixing length and integral methods, slightly better than the alternative LS and DM versions) for this higher free-stream turbulence test case, although with all three model variants transition was predicted late, and the Cf distribution through transition showed similar defficiencies to the best k-ε [LS] model predictions for T3A. The LES code again failed to predict transition when the sub-grid scale model was omitted, but reproduced the post-transition behaviour when this was included. Once more the Direct Simulation most closely reproduced the experimental data, and since transition occured earlier within the computational box employed, the DNS was able to follow the development into the turbulent region equally successfully in this case. All the turbulence models again tended to assymptote to the final Cf value slowly and from below, rather than via the small overshoot shown by the experimental and DNS 'data'.

Shape factor & momentum thickness: The shape factor and momentum thickness distributions shown in Figures 8 & 9 respectively are again consistent with the Cf results, and reflect the wider variation in predictions for T3B, although once more it is difficult to discriminate so easily between models in the latter case. Several of the turbulence models did predict a reduction in H prior to Re_{xtr}, which was something the integral method and mixing length approach failed to do, however in most cases this was largely due to the fact that they predicted Cf transition too early. The low-Re RST model, however, did correctly predict the inital drop in H and indeed predicted the variation of H right through transition remarkably accurately. The latter was not even so well reproduced by the DNS.

Mean & turbulent velocity profiles: The mean velocity profile development again appeared to be generally well-predicted, as illustrated by Figure 10, but once more this was simply because it was so difficult to distinquish the small discrepacies in profile shape which contributed to more obvious errors in the integral quantities. However the fullness of the profiles was again considerably underpredicted by the k-l [HP-B] model, and overpredicted by the DNS, for the same reasons as in case T3A. However the DNS again provided the best reproduction of the turbulence intensity profiles, whether normalised by the free-stream mean flow velocity, as in Figure 11(a), or the local velocity as in Figure 11(b); whereas the k-ε [LS], [LB]; k-ε/k-l turbulence model; and even the RST closure predictions once again lagged well behind the experimentally measured profil development. Again the Chien k-ε model appeared to provide excellent predictions solely because it predicted transition far too early.

4 DISCUSSION

The relative performance of the various computational approaches (excluding the correlated mixing length and integral methods) can best be summarised by Table 3. This attempts to rank the 'best' three sets of predictions for each of the requested flow parameters, irrespective of how satisfactory these actually were. (In most cases it was in fact possible to make a relatively clear distinction between the best three, or at most four, sets of predictions - those showing errors of <50% - and the remainder). Needless to say any such assessment is bound to be somewhat subjective, but, provided this is always borne in mind, such an analysis can nevertheless be an informative exercise, and in the present case some useful points do emerge.

Thus it is immediately apparent that all of the methods used predicted at least one flow quantity (very) approximately correct, but none produced anywhere near acceptable predictions for all of the parameters requested in both cases.

Table 3: Overall Summary of 'Three Best' Model Predictions for Different Flow Parameters

T3A

Cf	H	θ	U	u'
DNS	k-ε (LS)*,**	DNS=k-ε (LS)*,**	k-l (FM)=k-l/k-ε	DNS
k-ε (LS)*,**	DNS	-	k-ε (CH)*	k-ε (CH)*
k-l(HP-B)	k-ε (CH)*	k-ε (CH)*	k-ε (NH)**	-

T3B

Cf	H	θ	U	u'
DNS	RST *,**	k-ε (CH)*	k-ε (LS)*,**	DNS
-	DNS	RST *,**	RST *,**=k-l(FM)	k-ε (CH)*
RST *,**=k-ε (CH)*	k-l(FM)	DNS	-	-

*: Satisfies wall limiting condition for uv
**: Satisfies wall limiting condition for ε

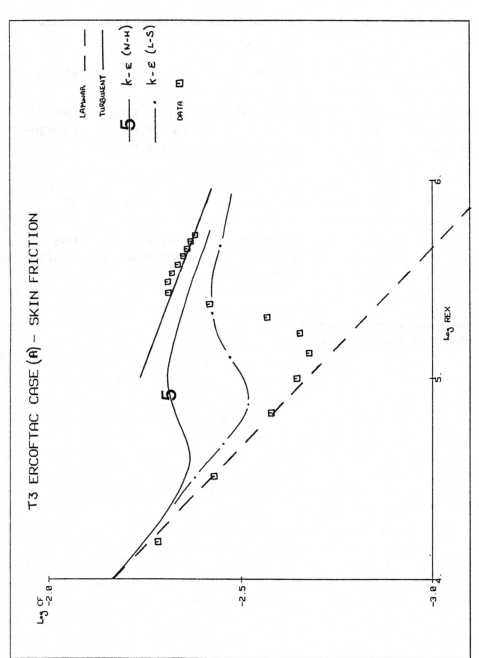

Fig. 12 Comparison of new elliptic Launder-Sharma k-ε model predictions [18] for T3A with previous Nagano-Hishida results [6] obtained using same numerical scheme

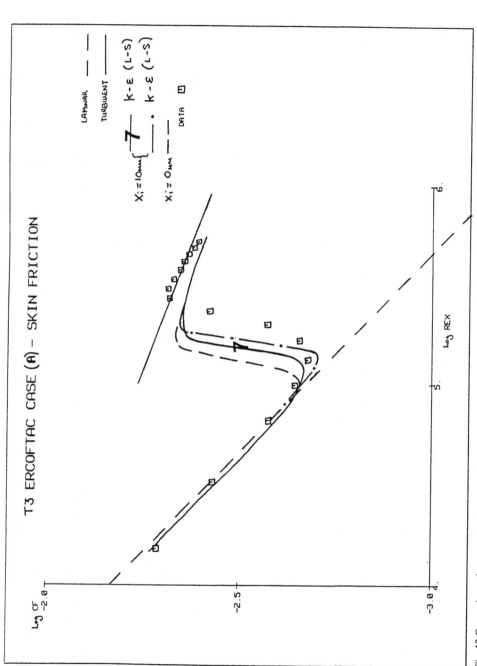

Fig. 13 Comparison of new parabolic Launder-Sharma k-ε model predictions [19] with Workshop predictions using the same model [7], but different numerical scheme: Cf

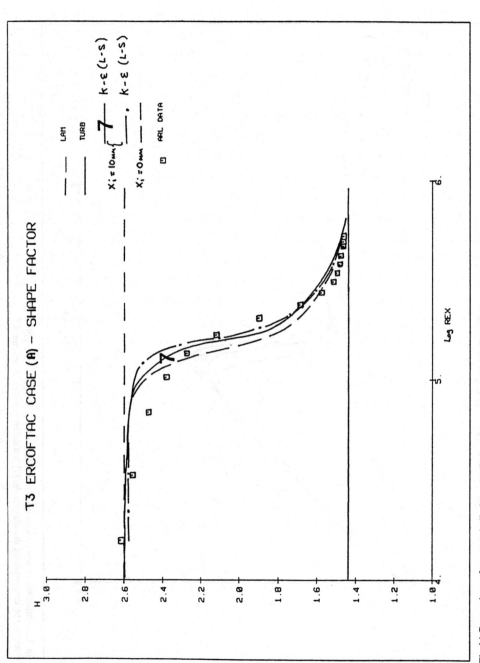

Fig.14 Comparison of new parabolic Launder-Sharma k-ε model predictions [19] with Workshop predictions using the same model [7], but different numerical scheme: H

The DNS clearly proved to be the 'best ' overall, particularly as the deficiencies in mean flow predictions could be at least partly attributed to cpu limitations rather than any defficiency in the method itself. (Indeed Yang & Voke (private communication) have subsequently obtained much better agreement with experiment, by averaging over much longer run times than could be achieved prior to the Workshop).

Of the turbulence models the Launder-Sharma, and to a lesser extent the Chien, k-ε models produced the most consistently satisfactory results overall, while the low-Re RST scheme (Hanjalic & Launder or Launder-Sharma closure variants) provided the best model predictions for T3B. The fact that these three models out-perfomed all the other k-ε variants, as well as the k-l models, is rather encouraging because they are the only ones which correctly satisfy the wall limiting behaviour for uv (uv+ α y+^{3}), while the k-ε[LS] model and the RST [H-L& LS] schemes in addition satisfy the correct wall limiting behaviour for ε (ε+α 2A+$+$ 4B+y+). These criteria were both identified by Patel et al.[12] as important requirements for any such low-Reynolds number model . The Nagano-Hishida k-ε model and the Dutoya-Michard k-ε model, on which the alternative RST [DM] closure is based, also satisfy the ε condition. However both of these, as well as the Lam & Bremhorst k-ε model and Hassid-Poreh (here used in modified k-l form) give uv+ α y+^{4} near the wall (as does the Van Driest damping scheme), which is only true if uv tends to zero there.

Tarada [8] in fact made some direct comparisons of the Launder-Sharma k-ε model with the alternative Chien closure, using the same parabolic numerical scheme, which also suggested that the former produced the better predictions, at least when starting computations from x=10mm as specified by the Test Case proponents.

In addition more recent comparative elliptic computations, performed by Prinos et al. [18] since the Workshop, have shown that implementing the Launder-Sharma k-ε model in place of their original Nagano-Hishida closure reduced the error in their Cf predictions for T3A by a factor of two in Rex, making these the equal fourth best results despite the grid refinement difficulties. Indeed, as Figure 12 illustrates, the predicted Rex$_{tr}$ was then very close to the experimentally determined value, although the level of Cf in the initial pre-transition flow was rather too high. Furthermore, when they switched to a parabolic approach [19] the predictions they obtained with the Launder-Sharma model (with the possible exception of the unreported u' results) were as good (for T3A) or better (for T3B) than those obtained by Rodi et al . [7]; the T3A results for Cf and H, starting at x+0mm and x=10mm bracketing the latter - see Figures 13 & 14 (although emphasising that the initial drop in H is not captured by this model).

Finally, Savill [20] has recently found that implementing the Launder-Sharma-type closure in the RST model did produce a predicted T3A transition, within the range of other parabolic models, if C$_{\varepsilon 3}$=0.25 and Loi=30.1mm (But transition was still not predicted when either the alternative H-L or DM closures were applied). Although the Cf transition was still too late (see Figure 15), the shape factor predictions (Figure 16) then become the fourth best. With the same closure, constant setting, and value for Loi, the Cf predictions for T3B were also improved (see Figure 17), but only at the expense of a rather poorer shape factor distribution (Figure 18); this being predicted to drop rather too rapidly in the pre-transitional boundary layer as Arnal [21] has anticipated might then be the case.

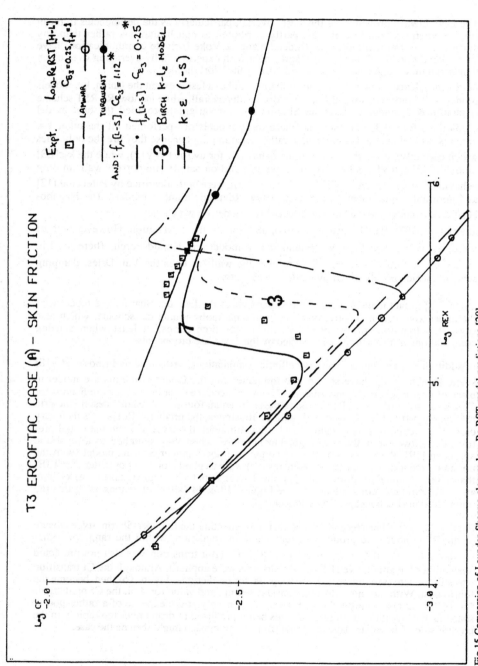

Fig.15 Comparsion of Launder-Sharma closure low-Re RST model predictions [20]
with Workshop predictions of Cf in Case T3A (*: Loi=30.1mm rather than 13.4mm)

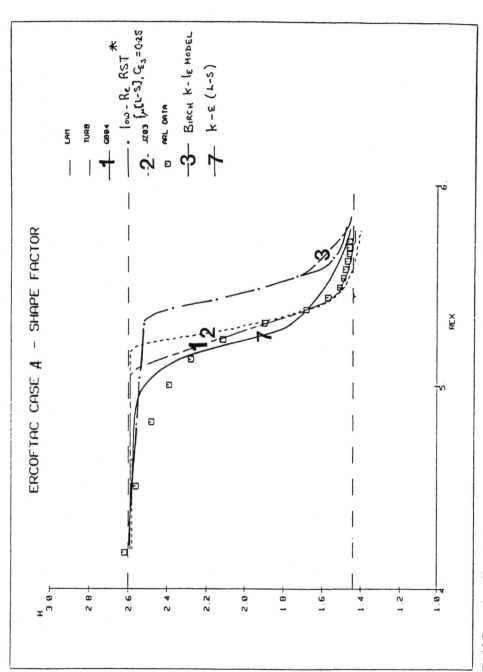

Fig.16 Comparsion of Launder-Sharma closure low-Re RST model predictions [20]
with Workshop predictions of H in Case T3A

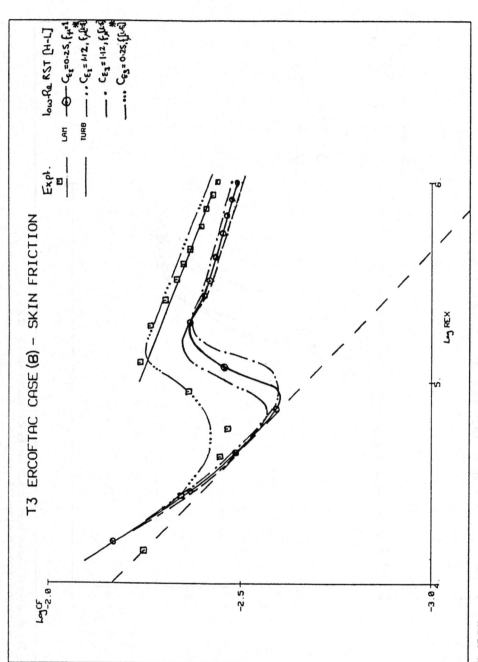

Fig.17 Effect of modified low-Re RST closure (and initial free-stream dissipation length-scale: *: Loi=41.8mm rather than 13.4mm) on Cf predictions [20] for Case T3B

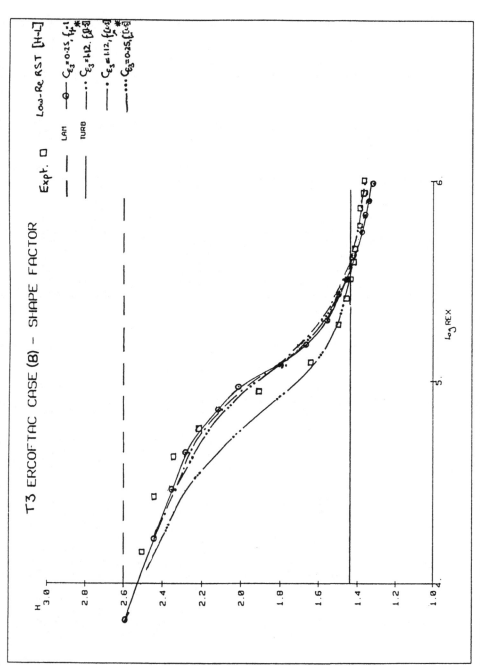

Fig.18 Effect of modifying low-Re RST closure (and initial free-stream dissipation length-scale: * : Loi=41.8mm rather than 13.4mm) on H predictions [20] for Case T3B

In Figures 19 & 20 the best T3B skin friction and shape factor predictions, obtained by employing Launder-Sharma-type damping factors within the k-ε [19] and RST [20] closures, are compared. The two sets of predictions generally bracket the experimental data points, except that Cf is underpredicted in the post-transition region at both levels of closure and, as noted above, the pre-transition drop in H is not captured by the lower-level closure scheme.

It is clear then that the Launder-Sharma near wall-treatment has some advantages over the other low-Re models evaluated at the Workshop whether implemented at the k-ε or full Reynolds stress level. The main reason for this appears to be that the LS treatment is assymptotically correct for uv and ε near the wall boundary, since other models which satisfy only one of these criteria or neither of them generally produced progressively worse predictions. However this may be a rather false conclusion for two reasons.

First, Chapman & Kuhn [22] have shown that any such assymptotically correct model should also have for consistency a principal eddy viscosity damping factor $f\mu \propto y+^{-1}$, and none of the models evaluated here, or those considered by Patel et al. [12], meet this condition. In fact Nagano & Tagawa [23] have already proposed a refined version for van Driest/Launder-Sharma damping functions which does satisfy this criteria, as well as the wall-limiting behaviour for uv and ε, and Myong & Kasagi [24] have proposed a similar extension for a van Driest/Chien closure. At the same time Speziale et al.[25] have suggested employing the Myong & Kasagi closure (with a tanh rather than exponential blending function) in an alternative k-τ model where the time scale τ ($=k/\varepsilon$) is used in place of ε itself in order to overcome some of the numerical stiffness associated with even the asymptotically correct form of D factor.

All of these models also ensure $f_2 \propto y+^2$ (which is not generally satisfied by earlier low-Re models - see Patel et al.[12]) and $f_1 \propto y+^0$ (which is generally satisfied). They have all been shown to provide excellent results for a range of fully turbulent flows (in some cases providing considerably better predictions than the Launder-Sharma model), however none of them has yet been tested on transitonal flows. One problem may be that they all make $f\mu$ a specific function of y+, which the Launder-Sharma model avoids because the LS damping factors are all functions of the turbulence Reynolds number (and hence the turbulence energy) instead, and hence $f\mu \propto y+^0$.

(NB. Since $k+ \propto y+^2$ near the wall a q-l model such as that used by Birch et al.[4], where $q=k^{1/2}$ ($\propto y+$) might be expected to have some advantages because of its greater linearity).

Secondly, one has to remember that in using any of the present low-Re models to predict transition one is implicitly assuming that approximations which have been devised to handle low-Reynolds number flow near walls are equally valid for low-Reynolds number transitional flows. There is of course no reason why this need be the case and indeed intuitively one might not expect it to be so. This is because the assumption requires that the vertical (y) damping functions, which (ideally) produce the correct blending between the (essentially laminar) viscous sub-layer and the fully turbulent. outer boundary layer region, also provide the correct streamwise (x) blending between the pre-transitional pseudo-laminar boundary layer and the post-transitional fully turbulent flow. Physically this is very unlikely to be the case, and one reason why the Launder-Sharma model gives anywhere

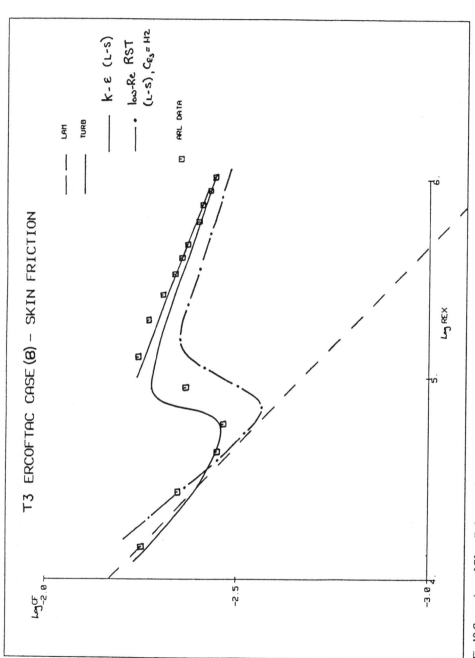

Fig.19 Comparison of Cf predictions obtained for T3B with the Launder-Sharma low-Re model applied at either the k-ε [19] or RST [20] closure level (Loi=41.8mm)

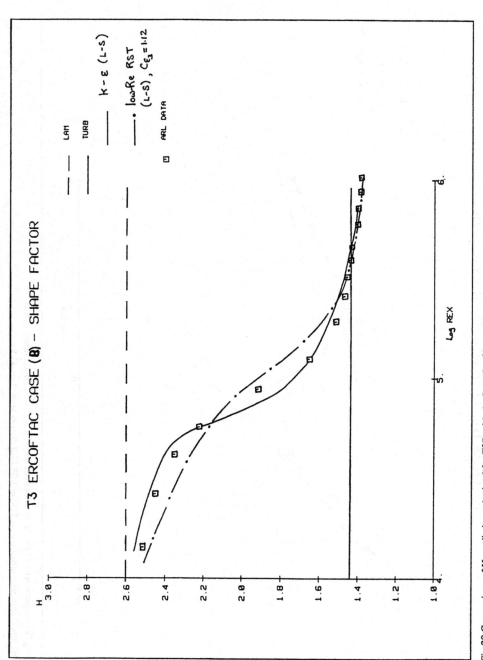

Fig.20 Comparison of H predictions obtained for T3B with the Launder-Sharma low-Re model applied at either the k-ε [19] or RST [20] closure level (Loi=41.8mm)

near correct predictions may well be that the associated damping factors are functions of a quantity (the turbulence Reynolds number) which is only a rather general indicator of the degree of turbulent activity at any x or y location in the flow, rather than a specific function of the location itself. Models which contain damping factors that are functions of Rey or y+ are clearly less likely to predict transition as well as they predict near-wall behaviour. This was certainly found to be the case in this evaluation study, and the same conclusion has been reached by both Fujisawa [16] and Theodoridis et al.[19].

In view of the above comment is perhaps not surprising that even the Launder-Sharma treatment failed to predict the turbulence intensity development correctly. However the observation that even models such as this, which predicted the location and duration of transition approximately correctly, considerably under-predicted the growth of the turbulence fluctuations can be attributed at least partly to inconsistencies in the specified initial conditions. It is clear from an examination of the u' data (see Figures 6a & 11a) that a small peak quickly developed in the inner part of the profile and that this continued to grow as the whole profile filled out towards its final turbulent form. In fact this peak could clearly be identified at only x=10mm from the leading edge (the starting location for parabolic computations), in both the experimental and Simulated data, but no allowance was made for this in the simple damped initial turbulence profiles. In fact some recent additional computation made with the HP-B k-1 model (J.Coupland - private communication) do show such a peak growing outwards, but rather too slowly. In other predictions the peak intensity appeared to develop later and from a point closer to the wall. The cross-wire data which is now available for both Test Cases also indicates that equation (7) generally over-estimates u', that the v' and w' profiles at x=10mm also exhibit significant features, and that uv is not equal to zero at this Station. (In addition it is important to note that the free-stream turbulence intensities quoted in the specification were those measured at approximately 3δ above the test plate, where the external turbulence was in practice actually very nearly exactly isotropic, but at δ_{995} Tu was higher in both cases (~3.4% for T3A & ~6.5% for T3B) and the normal stress distribution markedly anisotropic).

Some computors did in fact evaluate the effects of varying selected flow parameters about the specified values and these studies indicated that the predictions were surprisingly insensitive to the initial turbulence intensity or even, in the case of the RST model, the normal stress distribution (although the variations tested did not in any case approximate the actual experimental initial conditions), but they were particularly sensitive to the (assumed) initial dissipation profile, distribution.

For example, Tarada found that as ε was increased Rex_{tr} as predicted by the Chien k-ε model tended to a constant value of 10^5, whereas when ε was reduced the predicted transition region continually moved ever further forwards. Since the Workshop, Savill [20] has also found a similar effect with the RST model in that, when Lo was increased from the value deduced by Roach to that estimated by Rodi et al. (and hence ε_0 reduced so that the free-stream turbulence intensity did not decay do rapidly) the predicted T3B Cf transition moved forwards (see Figure 17), although the predicted shape factor distribution was only very slightly altered (Figure 18). On the other hand reducing Lo by a factor 2 from the Roach value (and hence increasing ε_0, and thereby more rapidly reducing the influence of the free-stream turbulence intensity) delayed transition for T3A by a similar amount.

In contrast Savill [9] earlier found that the alternative choices for the initial dissipation (equations (11) & (13)), and turbulence intensity profiles (equations (2) & (6)), produced equivalent predictions; provided the fully turbulent values for κ, λ and a_1 were adopted. It is far from obvious however that these are the values that should be used, and the results obtained are undoubtedly sensitive to them (for example Fujisawa [17] has shown that, although the Launder-Sharma k-ε model produces good transition predictions with a_1=0.3, a value of 0.07 is required to obtain comparable results with the Lam & Bremhorst model). Nevertheless within these limitations the assessments of relative model performance made in Table 3 are probably valid.

Generally it would appear that the predictions were not particularly sensitive to the values adopted for the various model constants, except where these were introduced specifically to handle transition and therefore required some calibration, and provided they remained within the accepted limits for fully turbulent flows.

The failure of the Large Eddy Simulation to predict transition in the absence of a sub-grid scale model highlights the importance of capturing the influence of the smallest scale instabilities. The Direct Simulation was performed assuming that it would be necessary to resolve Tolmein-Schlicting waves (hence the spanwise grid spacing of ~1/15 λ_{TS}) even though there was no a priori evidence that such waves played a role in the transition process at the relatively high free-stream turbulence intensities of the Test Case problems. There is, however, some evidence in the litterature that T-S waves may form, between turbulent spots, irrespective of the incoming turbulence level, and certainly the DNS 'data' are very satisfactory; especially bearing in mind their relatively low-resolution and averaging time. The LES with a turbulent sub-grid scale model quite naturally produced turbulent results, so again transition could not be predicted. The same problem has been encountered by Piomelli et al.[26] when attempting to Simulate transition in a plane channel flow. By comparing LES and DNS results they have already established that the residual stress and sub-grid scale dissipation in such flows are very different to those in turbulent flow, and are considerably over-estimated by Smagorinsky-type sub-grid scale models. They have also found that an appropriately re-scaled Smagorinsky model can be used to obtained predictions for boundary layer transition which compare favourably to DNS results (and require only a fraction of the computational effort needed to aquire the latter), although a new non-dissipative model which they have proposed may really be required.

5 CONCLUSIONS & RECOMMENDATIONS
The Launder-Sharma model emerged from this albeit limited Test Case study as the best immediate choice of low-Reynolds number treatment for engineering computations of (by-pass) transitional flows, at least at the k-ε closure level of many industrial codes, and possibly at a more advanced stress transport level also. However a number of caveats must be attached to this conclusion. Foremost among these is the fact that the initial conditions supplied as part of the Test Case specification were incomplete and did not accord with the experimental measurements. It is therefore very important that a comparative evaluation is made for parabolic computations starting from experimentally measured conditions, with a fixed value for Loi and specified initial dissipation profile. In addition the present evaluation was restricted to a, possibly flow geometry dependent, narrow range of free-stream turbulence intensities under zero pressure gradient conditions. The Launder-Sharma model has been successfully tested by other researchers over a wider range of experimental conditions and Tu>1% [e.g.16,17], but has not so far been linked into an intermittency model which might permit predictions of spot-dominated 'natural' transition. In addition, although this model has been shown to quite acurately predict the effect of strong pressure gradients on fully turbulent flows (e.g. see [12]), Fujisawa [17] has found some defficiencies in predicting transition in non-zero pressure gradients (which are not entirely

overcome by his suggested corrections). It is clear then that this model, and perhaps some of the other more successful closures, should be assessed against a wider range of Test Cases which successively introduce other effects likely to be encountered in real engineering flows. Apart from other free-stream turbulence intensities and length scales (with varying degrees of anisotropy), and variable pressure gradients, the effect of alternative leading edge geometries should also be examined. (The present test plate leading edge geometry was not in fact representative of internal blading and the additional complexity of an initial laminar separation bubble was deliberately avoided).

However it should also be remembered that a large number of other turbulence models have already been proposed, at many different levels of closure, which may be capable of predicting transition, perhaps better than the relatively small selection of models considered in this study. The alternative low-Re Reynolds stress closures of Launder & Shima [27], Shih & Lumley [28], and Lai & So [29] should certainly be tested. At the same time the two-equation models of Nagano & Tagawa [23], Muong & Kasagi [24] and Speziale et al.[25] clearly merit evaluation in view of their successful prediction of a range of fully turbulent flow cases, and the fact that they would appear to build on the success of the 'best' present models satisfying assymptotic wall limits for more quantities. Unfortunately they may all suffer from defficiencies associated with using y-dependent damping fuctions when applied to low-Reynolds number transitional flows.It might be that these could be overcome by including some specific x-dependent damping fuction, perhaps in the manner proposed by Grundmann and colleagues (eg.see [30]) for extending the MacDonald integrated k-1 model of [5], but it would probably be advantageous to attempt to devise a closure satisfying the same new constraints; avoiding any y-dependence. If this approach were to be adopted it would be logical to try to ensure that all the individual terms of the transport equations satisfy the correct wall limiting behaviour which is certainly not yet the case in any model. For example Speziale et al.[25] have demonstrated that the Diffusion term in the k-ε and k-τ models is not assymptotically correct unless pressure-diffusion is zero (as assumed in most turbulence models), but this has been shown not be the case near walls from analyses of the NASA Ames Direct Simulation data bases for turbulent boundary layers and channel flows by researchers working at the Centre for Turbulence Research. Such Direct Simulations are however now aiding the further refinement of low-Reynolds number near-wall modeling at both the k-ε [31] and Reynolds stress level [32]. Since the DNS of Yang & Voke [10] proved so successful in reproducing both the T3A and T3B experiments it is clear that this could be used in a similar manner to refine the models specifically for low-Reynolds number transitional flow regions. It may be of course, for the reasons discussed above, that the two resulting closure approximations are not the same or even mutually compatible, and although some type of zonal approach could presumably be devised to overcome any such difficulty, it would clearly be more satisfactory if a clearer distiction could be made between the two types of low-Re behaviors. One way to achieve this may be to utilise the new '2D/2C-limit' modelling approach (to handling both Pressure-strain and Dissipation within Reynolds stress closures) which has been adopted by Launder and coworkers [33,34], Reynolds [35] and Speziale et al.[36], since this recognises that the primary effect of a wall boundary is not to induce low-Reynolds number condition, but to impose a two-dimensional or more strictly a two-component limitation on the near-wall flow.

There are of course a other refinements that could be made to any of the above models. For example some consideration might be given to replacing the dissipation transport equation by the alternative 'structural' dissipation length scale prescription which has been proposed by Hunt et al.[37], also on the basis of analyses of the NASA data bases at CTR. Cho & Chung [38] have also recently shown how intermittency may quite simply be added in to an existing k-ε scheme. While this has not yet been applied to transitional flows, Vancoille & Dick [39] have had some success modelling such flows with a k-l-intermittency model.

It has also to be remembered of course that elliptic versions of the models will be required to make predictions for real flows in order to include leading edge effects and to predict any separated flow regions, unless parabolic models are to be patched in to some form of leadfing edge treatment. Certainly more elliptic computations of even the present Test cases, ideally with local grid refinement, would be useful.

For the future it may well be that some form of corrected LES can provide useful predictions of transition, for the present it appears this can more usefully be used to examine the effect of turbulence impinging on the leading edge region (as Mortensses & Albraten demonstrated at the Workshop). Since Roach [3] found that development of turbulent activity in the pseudo-laminar boundary layer on the present test plate did not corrlate with the local free-stream conditions, but instead appeared to depend on the conditions established at the origin of the layer at the leading edge, further information is required on the anisotropy set up in the impinging free-stream turbulence due to the distortion imposed on this by the leading edge geometry. This can probably best be provided by a Rapid Distortion Theory analysis. In addition, since Roach [3] has also shown that the initial u' peak within the pseudo-laminar layer continues to grow as the fress-stream intensity decays it way well be that some allowance must be made for non-local, counter-gradient transport effects, possibly in the manner discussed by Savill [40]. Certainly it is likely that such 'Diffusion-controlled' transitional flows will require some improvements to current Diffusion approximations. Up to now rather more emphasis has been placed on refining Pressure-strain and Dissipation closure simply because in the majority of other flows any weaknesses in the Diffusion modelling are far less important than defficiencies in either of these. Such developments could again be aided by Direct Simulations. The present DNS need to be repeated at higher resolution and with a more realistic treatment of the free-stream turbulence and Yang & Voke have already made some progress towards this using a 'precursor' Simulation.

{NB. In view of the clear need to carry the T3 Test Case work forwards, and to build on the successes of the Workshop evaluation, a 'T3 Follow-up' project has already been initiated. This has been accepted as one of the activities of the new ERCOFTAC Special Interest Group on Transition/Re-transition launched by Professor D.I.A.Poll through the UK North ERCOFTAC Pilot Centre at UMIST. In response to invitations sent out following the Workshop a far larger number of research groups (from more countries) are now participating in this project, which is again being co-ordinated by the author. Progress reports are being published as Poster papers at the RAeS Transition and Boundary Layer Control Conference, and the 1st European Fluid Mechanics Conference, in 1991, and a further Workshop is planned for 1992}

Acknowledgements
Rolls-Royce plc have actively encouraged my Supervision of the T3 Test Case problem and supported this as part of University Research Brochure PVA3-120D.

References
[1] P.E.Roach & D.H.Brierley (1990) The Influence of A Turbulent Free-Stream on Zero Pressure Gradient Transitional Boundary Layer Development including the T3A & T3B Test Case Conditions. In this Proceedings Volume.
[2] P.E.Roach (1987) The generation of nearly isotropic turbulence by means of grids. Int. J. Heat Fluid Flow, 8 (2), 82-92.

[3] P.E.Roach (1990) A Correlation Analysis Approach to the T3Test Case. In this Proceedings Volume.

[4] N.T.Birch, Y.K.Ho & P.Stow (1990) Calculations for the T3 Transition Test Cases. In this Proceedings Volume.

[5] J.Bernard, J.M.Fougeres & J.L.Mandlier (1990) Prediction of the Effect of Free Stream Turbulence on a Laminar Boundary Layer and Transition. In this Proceedings Volume.

[6] G.Theodoridis P.Prinos & A.Goulas (1990) ERCOFTAC PROBLEM T3 The effect of free-stream turbulence on the laminar boundary layer transition. Original submission to the Lausanne Workshop.

[7] W.Rodi, N.Fujisawa, & B.Shonung (1990) Calculation of Transitional Boundary Layer under the Influence of Free-Stream Turbulence. In this Proceedings Volume.

[8] F.Tarada (1990) A new version of the k-ε Model of Turbulence applied to Boundary-layer Transition. In this Proceedings Volume.

[9] A.M.Savill (1990) Application of a low-Reynolds number Reynolds Stress Transport model to the Prediction of Transition under the Influence of Free-Stream Turbulence. In this Proceedings Volume.

[10] H.E.Mortenssen, L.-E.Eriksson & P.J.Albraten (1990) Contribution to the ERCOFTAC Workshop at EPFL, March 26-28,1990 Flat Plate Transition Test Case T3. In this Proceedings Volume.

[11] Z.Y.Yang & P.R.Voke (1990) Numerical Simulation of Boundary Layer Transition in the Presence of Free-Stream Turbulence. In this Proceedings Volume.

[12] V.C. Patel, W.Rodi & G. Scheuerer (1985) Turbulence Models for Near-Wall and low Reynolds Number Flows: A Review. AIAA J. 23 (9) p.1308.

[13] Y.Nagano & M.Hishida (1987) Improved Form of the k- ε Model for Wall Turbulent Shear Flows. Trans ASME J.F.E. 109 p.156.

[14] B.J. Abu Ghannam & R.Shaw (1980) Natural Transition of Boundary Layers, the Effects of Turbulence, Pressure Gradients, and Flow History. J. Mech. Eng. Sci. 22 (5) p.213.

[15] W.Rodi & G.Scheuerer (1985) Calculation of laminar-turbulent boundary layer transition on turbine blades. Proc. AGARD-PEP 65th Symposium, Bergen.

[16] C.A.Stephens & M.E.Crawford (1990) An Investigation Into the Numerical Prediction of Boundary Layer Transition Using the K.Y.Chien Turbulence Model. NASA-CR-185252.

[17] N.Fujisawa (1990) Calculations of Transitional Boundary-Layers With a Refined Low Reynolds Number Version of a k-ε Model of Turbulence. In Engineeering Turbulence Modelling and Experiments (Ed. Rodi & Ganic; Elsevier) p.23.

[18] P.Prinos & A.Goulas(1990) Presentation at the ERCOFTAC General Assembly Meeting, UK North Pilot Centre, Manchester.

[19] G.Theodoridis P.Prinos & A.Goulas (1990) Test Case - Free-Stream Turbulence. In this Proceedings Volume.

[20] A.M.Savill (1991) Turbulence Model Predictions for Transition under Free-Stream Turbulence. Poster Paper for RAeS Transition and Boundary Layer Control Conference, Cambridge.

[21] D.Arnal (1990) Transition Description and Prediction.In this Proceedings Volume.

[22] D.R.Chapman & G.D.Kuhn (1986) The Limiting Behaviour of Turbulence Near the Wall. JFM 170, p.265.

[23] Y.Nagano & M.Tagawa (1990) An Improved k-ε Model for Boundary Layer Flows. Trans ASME J.F.E. 112 p.33.

[24] H.K.Myong & N.Kasagi (1990) A New Approach to the Improvement of k-ε Turbulence Model for Wall-Bounded Shear Flows. JSME J. Ser.II 33 (1) p.63.

[25] C.G.Speziale, R.Abid & E.C.Anderson (1990) A Critical Evaluation of Two-Equation Models for Near Wall Turbulence. AIAA-90-1481.

[26] U.Piomelli, T. A.Zang, C.Speziale & M.Y.Hussaini (1989) On The Large-Eddy Simulation of Transitional Wall-Bounded Flows. NASA-CR-181883.

[27] B.E.Launder & N.Shima (1989) Second-Moment Closure for The Near-Wall Sublayer: Development and Application. AIAA J. 27 (10) p.1319.

[28] T.-H.Shih & J.L.Lumley (1986) Second-Order Modeling of Near-Wall Turbulence. Physics of Fluids 29, p.971.

[29] Y.G.Lai & R.M.C.So (1991) Int. J. Heat & Mass Transf. (To appear).
- see also: Y.G.Lai, R.M.C.So, M.Anwer & B.C.Hwang (1990) Modelling of Turbulent Curved pipe Flows. In Engineeering Turbulence Modelling and Experiments (Ed. Rodi & Ganic; Elsevier) .

[30] R.Grundmann & U.Nehring (1984) Berechnung von zweidimensionalen, inkompressiblen, transitionellen Grenzschichten an gekrummten Oberflachen. Z. Flugwiss, Wetraumforsch 8, Heft 4, p.249.

[31] E.W.Miner, T.F.Swean, R.A.Handler & J.Leighton (1989) Evaluation of the enar-wall turbulence model by comparison with direct simulations of turbulent channel flow. NRL Memo Rep. 6499.

[32] T.H.Shih & N.N.Mansour (1990) In Engineeering Turbulence Modelling and Experiments (Ed. Rodi & Ganic; Elsevier) .

[33] S.Fu, B.E.Launder & D.P.Tselepidakis (1987) Accomodating the effects of high strain rates in modelling the pressure strain-correlation. UMIST Mech. Eng. Dept. Rep. TFD/87/5.

[34] B.E.Launder & D.P.Tselepidakis (1991) Directions in Second-Moment Modelling of Near-Wall Turbulence. AIAA-91-0219.

[35] W.C.Reynolds (1987) Fundamentals of turbulence modeling and simulations. Lecture Notes for Von Karman Istitiute AGARD Lecture Series No.86.

[36] C.G.Speziale, S.Sarkar & T.B.Gatski (1990) Modelling The Pressure-Strain Correlation Of Turbulence - An Invariant Dynamical Systems Approach. ICASE Report No. 90-5.

[37] J.C.R.Hunt, W.Weng, K.J.Richards & D.J.Caruthers (1989) New formulae for Dissipation and 'Mixing' Lengths. Proc. GAMNI-SMAI-IMA-ERCOFTAC MiniSymposium on Turbulence Modelling, Antibes (see also ERCOFTAC Bulletin III).

[38] J.R.Cho & M.K.Chung (1990) Intermittency modelling based on interaction between intermittency and mean velocity gradients. In Engineeering Turbulence Modelling and Experiments (Ed. Rodi & Ganic; Elsevier) p.101.

[39] G.Vancoille & E.Dick (1988) A Turbulence Model for the Numerical Simulation of the Transition Zone in a Boundary Layer. J. Eng. Fluid Mech.p.28.

[40] A.M.Savill (1991) Recent refinements to stress transport turbulence models. Proceedings 2nd ERCOFTAC Summer School, Oxford (CUP To appear).

Invited Lecture

ON RECENT NUMERICAL SIMULATIONS OF COMPRESSIBLE NAVIER-STOKES FLOWS

M.O. Bristeau, INRIA, B.P. 105, F-78153 Le Chesnay Cedex, France.

R.Glowinski, University of Houston, Houston, TX 77004, U.S.A. and INRIA.

L. Dutto, J. Périaux , G. Rogé, Dassault Aviation, B.P. 300, F-92214 St Cloud Cedex, France.

Summary

We discuss in this paper the numerical simulation of compressible viscous flows by a combination of finite element methods for the space approximation, implicit second order multistep scheme for the time discretization and GMRES iterative methods for solving the nonlinear problems encountered at each time step. Numerical results corresponding to flow around airfoils and 2D and 3D air intakes illustrate the possibility of these methods.

I. Introduction

In [1], we have discussed the numerical solution of the compressible Navier-Stokes equations by operator splitting methods. In this paper, we consider the solution of the same problem by methods which are in a sense more implicit since they are based on a time discretization by an implicit second order multistep scheme. This scheme is combined to finite element methods for the space discretization, and to a GMRES algorithm with preconditioning to solve the nonlinear problems encountered at each time step.

An important issue which is discussed here is the necessity (at least with the centered space approximations used here) to use different finite element approximations for velocity and density. The necessity of such compatibility condition which is well known in the incompressible case, has been discussed for a simple compressible case in [13] and more recently in [24], [25] ; actually, from the numerical experiments of Section 6, this compatibility condition seems to be also required for more complicated compressible flow.

In addition to the experiments of Section 6, the methods discussed in this paper are used to simulate flow around airfoils and 2D and 3D air intakes.

2. The compressible Navier-Stokes equations

Let $\Omega \subset R^N$ (N=2,3 in practice) be the flow domain and Γ be its boundary. The non-dimensional conservative form of the equations is given below by :

$$\frac{\partial \rho}{\partial t} + \nabla \cdot \rho \mathbf{u} = 0, \qquad (2.1)$$

$$\frac{\partial \rho \mathbf{u}}{\partial t} + \mathbf{\nabla} \cdot (\rho \mathbf{u} \otimes \mathbf{u}) + \mathbf{\nabla} p = \frac{1}{Re}[\Delta \mathbf{u} + \frac{1}{3}\mathbf{\nabla}(\mathbf{\nabla} \cdot \mathbf{u})]. \tag{2.2}$$

$$\frac{\partial e}{\partial t} + \mathbf{\nabla} \cdot (e + p)\mathbf{u} = \frac{1}{Re}[\mathbf{\nabla} \cdot [\mathbf{u}(-\frac{2}{3}\mathbf{\nabla} \cdot \mathbf{u} + \mathbf{\nabla}\mathbf{u} + \mathbf{\nabla}\mathbf{u}^t)] + \frac{\gamma}{Pr}\Delta \varepsilon]; \tag{2.3}$$

with $\rho, \mathbf{u}, T$ the density, velocity and temperature variables, respectively.

The pressure obeys the ideal gas law :

$$p = (\gamma - 1)\rho \varepsilon \tag{2.4}$$

and for the total energy e, we have

$$e = \rho \varepsilon + \rho |\mathbf{u}|^2/2. \tag{2.5}$$

The above equations express the conservation of mass, momentum and energy. We normalize the temperature T by $|\mathbf{u}_r|^2/c_v$, implying that

$$T = \varepsilon. \tag{2.6}$$

The constants Re, Pr and γ are the Reynolds number, the Prandtl number and the ratio of specific heats, respectively ($\gamma = 1.4$ in air).

From (2.1)-(2.6), we can deduce the following non-conservative form of the Navier-Stokes equations :

$$\frac{\partial \rho}{\partial t} + \mathbf{u} \cdot \mathbf{\nabla}\rho + \rho \mathbf{\nabla} \cdot \mathbf{u} = 0, \tag{2.7}$$

$$\frac{\partial \mathbf{u}}{\partial t} + (\mathbf{u} \cdot \mathbf{\nabla})\mathbf{u} + (\gamma - 1)(\frac{T}{\rho}\mathbf{\nabla}\rho + \mathbf{\nabla}T) = \frac{1}{Re\rho}[\Delta \mathbf{u} + \frac{1}{3}\mathbf{\nabla}(\mathbf{\nabla} \cdot \mathbf{u})], \tag{2.8}$$

$$\frac{\partial T}{\partial t} + \mathbf{u} \cdot \mathbf{\nabla}T + (\gamma - 1)T\mathbf{\nabla} \cdot \mathbf{u} = \frac{1}{Re\rho}[\frac{\gamma}{Pr}\Delta T + F(\mathbf{\nabla}\mathbf{u})], \tag{2.9}$$

where (2.4)-(2.6) still hold.

For three-dimensional flows, we have $\mathbf{u} = \{u, v, w\}$ and $F(.)$ in (2.9) has the following expression :

$$F(\mathbf{\nabla}\mathbf{u}) = \frac{4}{3}\left[(\frac{\partial u}{\partial x})^2 + (\frac{\partial v}{\partial y})^2 + (\frac{\partial w}{\partial z})^2\right] + (\frac{\partial v}{\partial x} + \frac{\partial u}{\partial y})^2 + (\frac{\partial w}{\partial x} + \frac{\partial u}{\partial z})^2 + (\frac{\partial v}{\partial z} + \frac{\partial w}{\partial y})^2$$

$$-\frac{4}{3}(\frac{\partial u}{\partial x}\frac{\partial v}{\partial y} + \frac{\partial u}{\partial x}\frac{\partial w}{\partial z} + \frac{\partial v}{\partial y}\frac{\partial w}{\partial z}) \tag{2.10}$$

In this paper, we will consider mainly the non conservative form (2.7)-(2.10) of the equations written in function of the primitive variables ; in this case, the expression of the different terms is much simpler.

Boundary and initial conditions have to be added.

We consider external flows ; the domain of computation is described in Fig. 2.1. Let Γ_∞ be the far-field boundary of the domain ; we introduce then

$$\Gamma_\infty^- = \{x | x \in \Gamma_\infty, \mathbf{u}_\infty \cdot \mathbf{n} < 0\}, \qquad (2.11)$$

$$\Gamma_\infty^+ = \Gamma_\infty \setminus \Gamma_\infty^-, \qquad (2.12)$$

where $\mathbf{u}_\infty$ denotes the free stream velocity and $\mathbf{n}$ the unit vector of the outward normal to Γ.

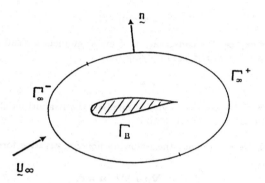

Figure 2.1.

We assume the flow to be uniform at infinity, and the corresponding variables to be normalized by the free stream values ; we require on the upstream boundary Γ_∞^- the following conditions :

$$\mathbf{u} = \mathbf{u}_\infty = \begin{pmatrix} cos\alpha \\ sin\alpha \end{pmatrix}, \qquad \alpha \quad \text{angle of attack}, \qquad (2.13)$$

$$\rho = 1, \qquad (2.14)$$

$$T = T_\infty = 1/\gamma(\gamma - 1)M_\infty^2, \qquad (2.15)$$

where M_∞ denotes the free stream Mach number.

On the downstream boundary Γ_∞^+, we require Neumann boundary conditions on $\mathbf{u}$, and T, i.e.

$$\frac{\partial \mathbf{u}}{\partial n} = 0, \qquad (2.16)$$

$$\frac{\partial T}{\partial n} = 0, \tag{2.17}$$

and if $M_\infty < 1$, we prescribe also on Γ_∞^+

$$\rho = 1. \tag{2.18}$$

On the rigid boundary Γ_B, we shall use the following conditions :

$$\mathbf{u} = 0, \quad \text{no-slip condition,} \tag{2.19}$$

$$T = T_B = T_\infty(1 + (\gamma - 1)M_\infty^2/2), \quad \text{free stream total temperature.} \tag{2.20}_1$$

or

$$\frac{\partial T}{\partial n} = 0, \text{ adiabatic condition.} \tag{2.20}_2$$

Finally, since we consider time dependent equations (even if we are looking for steady solutions), initial conditions have to be added ; we shall take

$$\rho(x, 0) = \rho_o(x), \tag{2.21}$$

$$\mathbf{u}(x, 0) = \mathbf{u}_o(x), \tag{2.22}$$

$$T(x, 0) = T_o(x). \tag{2.23}$$

3. Time discretization

With $\mathbf{U} = (\rho, \mathbf{w}, e)$ for the system (2.1)-(2.5), (we denote $\mathbf{w} = \rho\mathbf{u}$) or $\mathbf{U} = (\rho, \mathbf{u}, T)$ for the system (2.7)-(2.9), the unsteady problems to be solved are of the form :

$$\frac{\partial \mathbf{U}}{\partial t} + G(\mathbf{U}) = 0. \tag{3.1}$$

We introduce fully implicit schemes, either the Euler backward scheme which is first order accurate in time, or a Gear scheme which is second order accurate. Let $\Delta t(> 0)$ be a time discretization step.

The value $\mathbf{U}_o$ being given by the initial data (2.21)-(2.23), the Euler backward scheme is described by :

$$\mathbf{U}^o = \mathbf{U}_o \tag{3.2}$$

then, for $n \geq 0$, knowing $\mathbf{U}^n$, we compute $\mathbf{U}^{n+1}$ by :

$$\frac{\mathbf{U}^{n+1} - \mathbf{U}^n}{\Delta t} + G(\mathbf{U}^{n+1}) = 0. \tag{3.3}$$

On the other hand, the Gear scheme is given by :

$$\mathbf{U}^o = \mathbf{U}_o, \tag{3.4}$$

$$U^1 \quad \text{computed by (3.3)},$$

then for $n \geq 1$, knowing U^{n-1} and U^n we compute U^{n+1} by :

$$\frac{3U^{n+1} - 4U^n + U^{n-1}}{2\Delta t} + G(U^{n+1}) = 0. \tag{3.5}$$

If a steady solution is computed, a local time step is used.

4. Variational Formulation

We consider the non conservative formulation of the Navier-Stokes equations. At each time step of scheme (3.2)-(3.3) or (3.4)-(3.5), we have to solve a nonlinear problem of the following form :

$$\alpha \rho + \mathbf{u} \cdot \nabla \rho + \rho \nabla \cdot \mathbf{u} = g, \tag{4.1}$$

$$\alpha \mathbf{u} + (\mathbf{u} \cdot \nabla)\mathbf{u} + (\gamma - 1)(\frac{T}{\rho}\nabla \rho + \nabla T) - \frac{1}{Re\rho}[\Delta \mathbf{u} + \frac{1}{3}\nabla(\nabla \cdot \mathbf{u})] = \mathbf{f} \tag{4.2}$$

$$\alpha T + \mathbf{u} \cdot \nabla T + (\gamma - 1)T\nabla \cdot u - \frac{1}{Re\rho}[\frac{\gamma}{Pr}\Delta T + F(\nabla \mathbf{u})] = h, \tag{4.3}$$

where α is a positive parameter and where $\mathbf{f}, g, h$ are given functions ; the variables $\rho, \mathbf{u}, T$ satisfying the boundary conditions (2.13)-(2.20).

We introduce the following functional spaces of Sobolev's type :

$$R_r = \{\varphi | \varphi \in H^1(\Omega), \quad \varphi = r \quad \text{on} \quad \Gamma_r\}, \tag{4.4}$$

with

$$\Gamma_r = \Gamma_\infty \quad \text{if} \quad M_\infty < 1,$$

$$\Gamma_r = \Gamma_\infty^- \quad \text{if} \quad M_\infty \geq 1,$$

$$W_z = \{\mathbf{v} | \mathbf{v} \in (H^1(\Omega))^N, \mathbf{v} = \mathbf{z} \quad \text{on} \quad \Gamma_B \cup \Gamma_\infty^-\}, \tag{4.5}$$

$$V_s = \{\theta | \theta \in H^1(\Omega), \theta = s \quad \text{on} \quad \Gamma_t\}. \tag{4.6}_1$$

with

$$\Gamma_t = \Gamma_B \cup \Gamma_\infty^- \text{ if } (2.20)_1 \text{ holds},$$

$$\Gamma_t = \Gamma_\infty^- \text{ if } (2.20)_2 \text{ holds}.$$

If $r(resp. z, s)$ is sufficiently smooth, then $R_r(resp. W_z, V_S)$ is non-empty (the above choice for the space of the densities R_r is motivated by the fact that ρ will be approximated by continuous functions and that the restriction of ρ to Γ_∞^- makes sense ; of course, this supposes implicitly that ρ has some regularity).

Then an equivalent variational formulation of equations (4.1)-(4.3) is

$$\alpha \int_\Omega \rho \varphi dx + \int_\Omega (\mathbf{u} \cdot \nabla \rho)\varphi + \int_\Omega \rho(\nabla \cdot \mathbf{u})\varphi dx = \int_\Omega g \varphi dx \tag{4.7}$$

$$\alpha \int_\Omega \mathbf{u} \cdot \mathbf{v} dx + \int_\Omega (\mathbf{u} \cdot \nabla)\mathbf{u} \cdot \mathbf{v} dx + (\gamma - 1) \int_\Omega (\frac{T}{\rho}\nabla\rho + \nabla T) \cdot \mathbf{v} dx$$

$$-\frac{1}{Re} \int_\Omega \frac{1}{\rho}[\Delta\mathbf{u} + \frac{1}{3}\nabla(\nabla \cdot \mathbf{u})] \cdot \mathbf{v} dx = \int_\Omega \mathbf{f}\mathbf{v} dx \qquad (4.8)$$

$$\alpha \int_\Omega T\theta dx + \int_\Omega (\mathbf{u} \cdot \nabla T)\theta dx + (\gamma - 1) \int_\Omega T(\nabla \cdot \mathbf{u})\theta dx$$

$$-\frac{1}{Re} \int \frac{1}{\rho}[\frac{\gamma}{Pr}\Delta T + F(\nabla\mathbf{u})]\theta dx = \int_\Omega h\theta dx \qquad (4.9)$$

$$\forall\{\varphi, \mathbf{v}, \theta\} \in R_o \times W_o \times V_o, \quad \{\rho, \mathbf{u}, T\} \in R_r \times W_z \times V_s$$

where the value of r, z, s are precised by the boundary conditions (2.13)-(2.15), (2.18)-(2.20).

5. Solution of the nonlinear problems by preconditioned GMRES algorithms

Using appropriate finite element methods described in Section 6, the above nonlinear system of equations is reduced to a nonlinear system in finite dimension.

Among the various numerical methods which can be used for solving nonlinear problems of large dimension, let's mention nonlinear least squares methods, since these methods coupled to conjugate gradient algorithms have been successfully applied to the solution of complicated problems arising from fluid mechanics (cf. e.g. [1], [2], [3] for these applications).

One of the major drawbacks of the above methods is that they require an accurate knowledge of the gradient of the cost function ; for some problems, this knowledge is very costly in itself (for example this seems to be the case for the compressible Navier-Stokes equations). Recently, several investigators have introduced variants of the above methods which do not require the exact knowledge of the gradient. Among these methods, GMRES (Generalized Minimal Residual) algorithm (cf. [4], [5], [6]) has shown interesting possibilities for nonlinear problems (cf. [6]-[8] for the theory and some applications of GMRES in Fluid Dynamics).

One of the first applications of the (nonlinear) GMRES algorithm to nonlinear problems was done by Wigton, Yu and Young (cf. [6]). They have introduced an algorithm which seems quite efficient for solving some aerodynamics problems.

Saad and Brown (cf. [5]) have formulated this preliminary algorithm in the more general context of the Inexact Newton's methods, and they have proposed an automatic adjustment of the parameters.

In this section we shall use the framework of abstract Hilbert spaces to describe a generalization of GMRES whose finite dimensional variants can be seen as preconditioned variations of the original algorithm.

Let's consider an Hilbert space V, whose scalar product and corresponding norm are denoted by $(.,.)$ and $\|.\|$ respectively. We denote by V' the dual space of V, by $< .,. >$ the duality pairing between V' and V, and finally, by S the duality isomorphism between V and V', i.e. the isomorphism from V onto V' satisfying :

$$< Sv, w >= (v, w), \forall v, w \in V,$$

$$< Sv, w >=< Sw, v >, \forall v, w \in V,$$

$$< f, S^{-1} >=< f, g >_* \forall f, g \in V'.$$

With F a (possibly nonlinear) operator from V to V' we consider the following problem

$$F(u) = 0 \tag{5.1}$$

Description of a GMRES algorithm for the solution of (5.1)

$$u^o \in V \text{ is given ;} \tag{5.2}$$

Then, for $n \geq 0, u^n$ being known, we obtain u^{n+1} as follows :

$$r_1^n = S^{-1} F(u^n), \tag{5.3}$$

$$w_1^n = r_1^n / \|r_1^n\|. \tag{5.4}$$

Then for $j = 2, \ldots, k$ we compute r_j^n and w_j^n by

$$r_j^n = S^{-1} DF(u^n; w_{j-1}^n) - \sum_{i=1}^{j-1} b_{ij-1} w_i^n, \tag{5.5}$$

$$w_j^n = r_j^n / \|r_j^n\|. \tag{5.6}$$

in (5.5), $DF(u^n; w)$ is defined by either

$$DF(u^n; w) = F'(u^n).w \tag{5.7}_1$$

(where $F'(u^n)$ is the derivative of F at u^n), or, if the calculation of $F'(u^n)$ is too costly by

$$DF(u^n; w) = \frac{F(u^n + \varepsilon w) - F(u^n)}{\varepsilon} \tag{5.7}_2$$

with $\varepsilon(> 0)$ sufficiently small ; we define then b_{il}^n by

$$b_{il}^n =< DF(u^n; w_l^n), w_i^n > (= (S^{-1} DF(u^n; w_l^n), w_i^n)). \tag{5.8}$$

Then we solve

$$\begin{cases} Find\ a^n = \{a_j^n\}_{j=1}^k \in \mathbb{R}^k \text{ such that} \\ \forall c = \{c_j\}_{j=1}^k \in \mathbb{R}^k, \text{ one has} \\ \|F(u^n + \sum_{j=1}^k a_j^n w_j^n)\|_* \leq \|F(u^n + \sum_{j=1}^k c_j w_j^n)\|_* \end{cases} \qquad (5.9)$$

and obtain u^{n+1} by

$$u^{n+1} = u^n + \sum_{j=1}^k a_j^n w_j^n. \qquad (5.10)$$

Do $n = n + 1$ and go to (5.3).

In algorithm (5.2)-(5.10), k is the dimension of the so-called Krylov space.

Remark 5.1. We can easily show that

$$(w_j^n, w_l^n) = 0, \forall 1 \leq l, j \leq k, j \neq l. \qquad (5.11)$$

Remark 5.2. To compute $DF(u^n; w)$ instead of $(5.7)_2$, we can use the following second order accurate approximation of $F'(u^n).w$

$$DF(u^n; w) = \frac{F(u^n + \varepsilon w) - F(u^n - \varepsilon w)}{2\varepsilon} \qquad (5.12)$$

the choice of ε in $(5.7)_2$ and (5.12) can be done automatically by checking the variation of $DF(u^n, w)$ (cf. e.g. [29]).

Remark 5.3. The $\|f\|_*$ form in (5.9) satisfies the following relations

$$\|f\|_* = \|S^{-1}f\|, \forall f \in V', \qquad (5.13)$$

$$\|f\|_* = <f, S^{-1}f>^{\frac{1}{2}}, \forall f \in V'; \qquad (5.14)$$

in practice, one uses (5.13) to evaluate the various $\|.\|_*$ norms occuring in (5.9).

In order to improve the convergence of the algorithm, different choices for preconditioner S have been tested and compared ([15]). As an example, we show on Figure 5.1, the efficiency obtained with S chosen as the incomplete LDU factorization of the Jacobian matrix $A = \frac{\partial F}{\partial u}$; this test case concerns a transonic flow around an elliptic body (cf. [30]).

Remark 5.4. To compute a^n from the solution of (5.9), it suffices (in general) to approximate (in the neighborhood of $c = 0$) the functional

$$c \rightarrow \|F(u^n + \sum_{j=1}^k c_j w_j^n)\|_*^2 \qquad (5.15)_1$$

by the quadratic one defined by

$$c \to \|F(u^n) + \sum_{j=1}^{k} c_j DF(u^n; w_j^n)\|_*^2 \tag{5.15$_2$}$$

Minimizing $(5.15)_2$ is equivalent to solving a linear system with a symmetric and positive definite $k \times k$ matrix.

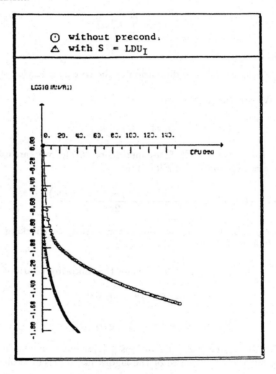

Figure 5.1. Convergence history of the algorithm.

6. Compatible Finite Element Approximations

It is well known that for the incompressible Navier-Stokes equations, pressure and velocity cannot be approximated independently (cf. e.g. [10] and the references therein).

Concerning now the compressible Navier-Stokes equations, there has been a natural tendancy to use the same approximation for all variables, the numerical viscosity of the scheme (due to upwinding, artificial viscosity, viscosity introduced via time discretization...) being generally sufficient to obtain solutions without oscillations.

Our aim was, in view of accuracy, to explore the possibility of using a centered Galerkin scheme without any viscosity added, at least for moderate Reynolds numbers and for solution without sharp shocks.

With the same piecewise linear continuous approximations, for all variables, we have obtained satisfying results using the Glowinski-Pironneau solver for the solution of the Generalized Stokes problem [20] which is a sub-problem of the Navier-Stokes equations ; this method implies more regularity on the density than, for instance, an Hood-Taylor method ; this explains why we obtain smooth solutions with the Glowinski-Pironneau solver and spurious oscillations if an Hood-Taylor method is used.

The same oscillations appear if we solve the Navier-Stokes system by the algorithm defined in Sec. 3.5 with the same approximations for all the variables.

Let T_h be a standard finite element triangulation of Ω, we introduce the following discrete spaces (with P_k = space of polynomials of degree $\leq k$) :

$$R_{rh} = \{\phi_h | \phi_h \in C^o(\bar{\Omega}), \quad \phi_{h|T} \in P_1, \forall T \in T_h, \quad \phi_h = r_h \quad \text{on } \Gamma_h\}, \qquad (6.1)$$

$$W_{zh} = \{v_h | v_h \in (C^o(\bar{\Omega}))^2, \quad v_{h|T} \in P_1 \times P_1, \forall T \in T_h, \quad v_h = z_h \quad \text{on } \Gamma_B \cup \Gamma_\infty^-\}, \qquad (6.2)$$

$$V_{sh} = \{\theta_h | \theta_h \in C^o(\bar{\Omega}), \quad \theta_{h|T} \in P_1, \forall T \in T_h, \quad \theta_h = s_h \quad \text{on } \Gamma_B \cup \Gamma_\infty^-\}. \qquad (6.3)$$

Then we can write for the discrete problem, the algorithm previously defined for the continuous one.

We have considered, as test case, the flow around a NACA0012 ; we have used first the triangulation (800 nodes, 1514 triangles) shown on Figure 6.1. The test case is a transonic calculation at $M_\infty = 0.8$, $Re = 73.$, at an angle of attack of $10°$, then the density contours exhibit spurious oscillations as shown on Figure 6.3.

Let $\tilde{T}_h$ be the triangulation deduced from T_h by joining the midpoints of the edges of $T \in T_h$; we have computed the same test case with T_h replaced by $\tilde{T}_h$ (as shown on Figure 6.2), then we have the same kind of oscillations on density contours (Figure 6.4).

These oscillations look like the checker board oscillations of pressure which appear when the same approximations are used for the pressure and velocity variables, to solve the incompressible Navier-Stokes equations.

We have considered the following equations :

$$\alpha \rho + \nabla \cdot u = g, \qquad (6.4)$$

$$\alpha\mathbf{u} - \mu\Delta\mathbf{u} + \nabla\rho = 0, \tag{6.5}$$

$$\mathbf{u} = 0 \quad \text{on} \quad \Gamma.$$

This system is a subproblem of (4.1)-(4.2) (or of (4.2)-(4.3) with ρ replaced by T) and is a "Generalized" Stokes problem.

If we solve this problem with the discrete spaces defined by (6.1), (6.2), we obtain also oscillating solutions (see Figure 6.5 for an example with $\alpha = 1, \mu = 0.01$ and all the variables, approximated on a triangulation of 3114 nodes ; the figure shows the density contours).

We define

$$\bar{W}_{zh} = \{\mathbf{v}_h | \mathbf{v}_h \in (C^\circ(\bar{\Omega}))^2, \quad \mathbf{v}_h|_{\tilde{T}} \in P_1 \times P_1, \quad \forall \tilde{T} \in \tilde{T}_h, \quad \mathbf{v}_h = \mathbf{z}_h \quad \text{on} \quad \Gamma_B \cup \Gamma_\infty^-\} \tag{6.6}$$

Then, if we choose,

$$\rho_h \in R_{rh}.$$

$$\mathbf{u}_h \in \bar{W}_{zh}$$

we obtain, for problem (6.4)-(6.5) the satisfying solution shown on Figure 6.6 (with for T_h the triangulation presented on Fig. 6.1).

This result tends to prove numerically that for problem (6.4), (6.5) a compatibility condition (some Inf-Sup condition (see [12])) has to be satisfied by the approximations of the different variables.

Concerning the compressible Navier-Stokes equations (Secs. 2 to 4), if we consider the test case previously defined with $\rho_h \in R_{rh}, T_h \in V_{sh}$ and $\mathbf{u}_h \in \bar{W}_{zh}$, we obtain also good results, as shown on Figure 6.7, for the density contours. We will denote this approximation by '$P_1, P_1 iso - P_2$' (the velocity has the same number of degrees of freedom as with a P_2 approximation).

We can also replace $\bar{W}_{zh}$ by (this space has been introduced in Ref. [16])

$$\bar{W}_{zh} = \{\mathbf{v}_h | \mathbf{v}_h \in (C^\circ(\bar{\Omega}))^2, \quad \mathbf{v}_h|_T \in P_{1T}^* \times P_{1T}^*, \quad \forall T \in T_h\} \tag{6.7}$$

In (6.7), P_{1T}^* is the subspace of P_3 defined as follows :

$$P_{1T}^* = \{q | q = q_1 + \lambda\phi_T, \quad \text{with} \quad q_1 \in P_1, \lambda \in R, \quad \text{and}$$

$$\phi_T \in P_3, \quad \phi_T = 0 \quad \text{on} \quad \partial T, \quad \phi_T(G_T) = 1\}, \tag{6.8}$$

where in (6.8), G_T is the centroid of T. A function like ϕ_T is usually called a bubble-function.

If we choose, $\rho_h \in R_{rh}, T_h \in V_{sh}, \mathbf{u}_h \in \bar{W}_{zh}$, a good solution is also computed as shown on Figure 6.8. We will denote this approximation by '$P_1, P_1 + \text{bubble}$'.

Concerning the conservative formulation (2.1)-(2.5), some first results prove also the interest of compatible approximations to avoid oscillations.

Some other results will be presented in Sec. 7, they prove numerically that accurate solutions can be computed by a centered scheme as soon as a compatibility condition is satisfied by the approximations of the different variables.

Some theoretical results have been proved on simplified model problems by Pironneau-Rappaz [13] for adiabatic stationary flows, by Bernardi-Pironneau [24], Bernardi et al [25] and by Fortin, Soulaimani [14].

7. Numerical Results

In this section, we will present some other 2D and 3D results to assess the accuracy and the efficiency of the methodologies introduced in the previous sections.

7.1. Flows around airfoils

The following results concern two dimensional transonic, subsonic and supersonic flows around a NACA0012 airfoil. For the two first cases, the temperature is prescribed on the body $(2.20)_1$ and for the third one the adiabatic condition $(2.20)_2$ is prescribed.

The first simulation consists of a transonic flow at $M_\infty = 0.85, Re = 2000$, a test case of the compressible Navier-Stokes workshop held in Nice, 1985 [18]. For this example we have used the $P_1 - P_1 iso P_2$ and the P_1, P_1+ bubble approximations of Sec. 6. With the first type of approximation and the mesh shown on Fig. 7.1, for the density and the temperature, we have computed the solution described on Figures 7.2-7.5, by the density contours, the Mach contours, the pressure coefficient on the body and the skin coefficient, respectively.

For the second approximation, four calculations [17] have been performed on successively improved meshes. Starting from the mesh of Fig. 7.1, the final mesh shown of Fig. 7.6 has been obtained by local enrichment according to physical criteria (cf. [21]-[23]) and by mesh deformation relying on a spring analogy ([31]).

The Figure 7.7 shows the Mach contours and the Figure 7.8 shows the skin friction coefficient.

With the two types of approximations, the results are in good agreement with the results issued from the workshop and which can be considered as references ; the adapted mesh used with the second type of approximation allows a better computation of the extremal values of the skin friction coefficient.

The following example concerns a supersonic computation ; it has been computed on an adapted mesh obtained also by enrichment (Fig. 7.9), the physical criteria used to choose the area to be refined is the gradient of temperature ([21]-[23]). The test case is $M_\infty = 2., Re = 500$, Figure 7.10 shows the Mach contours and Figure 7.11, the pressure contours.

The third case deals with a subsonic flow at $M_\infty = 0.5, Re = 5000., \alpha = 0$. This test case has been computed by different authors and recently on unstructured meshes by Mavriplis, Jameson and Martinelli [26]. The interest of this problem relies in the Reynolds number which is near to the upper limit for steady laminar flows ; a small recirculation bubble appears at the trailing edge.

For this simulation, we have tested an other adaptation technique based on regeneration of stretched triangles according to anisotropic information issued from a previous computation ([27], [28]). A first simulation has been done with the triangulation T_{1h} (1127 nodes, 2182 triangles) of Figure 7.12 for the density and temperature ; the triangulation $\tilde{T}_{1h}$ associated having 4426 nodes and 8728 triangles. The Figure 7.13 compares the skin friction coefficient obtained on this coarse mesh with the one obtained by Mavriplis et al [26].

For the computation of the skin friction coefficient, we use the fact that the velocity field is defined at the same nodes as in a quadratic approximation ; so in the post processing computing the skin friction coefficient, the velocity derivatives on the body are deduced by simulating a quadratic approximation of the velocity ; this implies an important improvement of the accuracy of the skin friction coefficient.

An other computation has been done on a mesh T_{2h} deduced from the previous one just by dividing the triangles into four sub-triangles near to the body ; this mesh T_{2h} (2907 nodes, 5706 triangles) is shown on Figure 7.14 (associated $\tilde{T}_{2h}$: 11520 nodes, 22825 triangles).

The Mach number contours are shown on Figure 7.15, and the recirculation bubble is depicted on the Figure 7.16 by the velocity vectors.

The skin friction distribution on these meshes is also presented on Figure 7.17.

Finally Table 1 gives for the two results, the values of pressure drag, viscous drag and separation point.

For the two computations, we have a good agreement with the results of Mavriplis et al and the other references given in their paper and this proves the accuracy of the solutions obtained on rather coarse meshes.

Meshes	CD_p	CD_v	Separation Point
T_{1h} : 1127 nodes			
$\tilde{T}_{1h}$: 4426 nodes	0.0220	0.0319	81.9 %
T_{2h} : 2907 nodes			
$\tilde{T}_{2h}$: 11520 nodes	0.0223	0.0320	82.1 %

Table 1. Pressure and viscous drags, separation point.
$M_\infty = 0.5, Re = 5000., \alpha = 0$.

7.2. Flow around and inside a 2D air intake at high incidence

The next computation concerns an unsteady flow around and inside a two-dimensional air-intake at $M_\infty = 0.6, Re = 750$. and $\alpha = 40°$; the characteristic length is taken as the distance between the walls of the nozzle. Concerning the time discretisation, we use the Gear scheme (3.4)- (3.5) with $\Delta t = 0.05$. The $P_1, P_1 isoP_2$ approximation is used ; the Figures 7.18 (resp. 7.19) shows the detail of the triangulation T_h (resp. $\tilde{T}_h$) close to the air intake. The velocity fields at different time steps ($t = 0.05, 2., 4., 6., 8., 10.$) are presented in Figures 7.20 to 7.25. We can observe the creation of the vorticity at the leading edge and the growing structures which propagate around the inlet.

7.3. Flow around and inside a 3D air intake

Finally a 3D computation around and inside an idealized air intake at $M_\infty = 0.9$ and a Reynolds number of 100 (based on the internal diameter) has also been performed with the 'P_1, P_1+ bubble' approximation. Figures 7.26 and 7.27 are views of the mesh and of the iso Mach lines in a plane of symmetry. Figure 7.28 is a view of the mesh on the inlet and Figure 7.29 represents the pressure coefficient levels.

8. Conclusion

We have discussed in this paper the numerical simulation of compressible viscous flow by a methodology combining finite element for the space discretization, an implicit second order multistep scheme for the time discretization, and a preconditioned GMRES iterative algorithm for the solution of the finite dimensional system encountered at each time step. Numerical experiments show that using similar space approximations for velocity and density leads to spurious oscillations, which dissapear if one employs the same type of elements than these used in the incompressible case for velocity and pressure. Numerical experiments show that the methodology described here provides a good basis for compressible viscous flow calculations ; indeed there is still room for progress and we are presently working at various improvements concerning the preconditioning, the approximations and the control of the oscillations close to the sharp layers and shocks.

Acknowledgment

We would like to acknowledge B. Mantel and M.G. Vallet for their collaboration and also O. Pironneau for fruitful discussions.

Special thanks are due to C. Demars for digilently processing this paper and last, but not least, to I. Ryhming and T.V. Truong for their quasi infinite patience with us. This work is partly supported by a grant from DRET (Grant $n^O 88.103$).

References :

[1] M.O. Bristeau, R. Glowinski and J. Périaux, Numerical Methods for the Navier-Stokes Equations. Applications to the simulation of compressible and incompressible viscous flows. *Computer Physics Report 6*, (1987) North-Holland, Amsterdam, pp. 73-187.

[2] R. Glowinski, *Numerical Methods for Nonlinear Variational Problems*, Springer-Verlag, New-York, 1984.

[3] M.O. Bristeau, O. Pironneau, R. Glowinski, J. Périaux, P. Perrier and G. Poirier, On the numerical solution of nonlinear problems in Fluid Dynamics by least squares and finite element methods (II). Application to transonic flows simulations, *Comp. Meth. in Appl. Mech. Eng.* 51, (1985), pp. 363-394.

[4] Y. Saad and M.H. Schultz, GMRES : A generalized minimal residual algorithm for solving nonsymmetric linear systems, *SIAM J. Sci. Stat. Comp.* 7 (1986), pp. 856-89.

[5] P.N. Brown, Y. Saad, Hybrid Krylov Methods for Nonlinear Systems of Equations, Lawrence Livermore National Laboratory Research Report UCLR-97645, Nov. 1987.

[6] L.B. Wigton, N.J. Yu and D.P. Young, GMRES Acceleration of Computational Fluid Dynamics Codes, *AIAA 7th Computational Fluid Dynamics Conference*, Cincinnati, Ohio, July 1985, Paper 85-1494, pp. 67-74.

[7] C. Bègue, M.O. Bristeau, R. Glowinski, B. Mantel, J. Périaux, Acceleration of the convergence for viscous flow calculations, in *Numeta 87*, Vol. 2, C.N. Pande J. Middleton eds., Martinus Nighoff Publishers, Dordrecht, 1987, pp. T4/1-T4/20.

[8] M. Mallet, J. Périaux, B. Stoufflet, On fast Euler and Navier-Stokes solvers. *Proceedings of the 7th GAMM Conference on Numerical Methods in Fluid Mechanics*, Notes on Numerical Fluid Mechanics, Vieweg, Vol. 20, pp. 199-210.

[9] P.N. Brown, A local convergence theory for combined inexact-Newton/finite difference projection methods, *SIAM J. Numer. Anal.*, 24 (1987), pp. 407-434.

[10] V. Girault and P.A. Raviart, *Finite Element Methods for Navier-Stokes equations*, Springer-Verlag, Berlin, 1986.

[11] M.O. Bristeau, R. Glowinski, B. Mantel, J. Périaux, G. Rogé, Self-adaptive finite element method for 3D compressible Navier-Stokes flow simulation in Aerospace Engineering, *Proceedings of the 11th Int. Conf. on Num. Meth. in Fluid Dynamics*, Williamsburg, U.S.A. June 1988, Springer-Verlag.

[12] F. Brezzi, On the existence, uniqueness and approximation of saddle point problems arising from Lagrangian multiplier, *RAIRO*, Série Analyse Numérique, R2, 1974, pp. 129-151.

[13] O. Pironneau, J. Rappaz, Numerical Analysis for compressible viscous adiabatic stationary flows, *IMPACT of computing in Science and Engineering*, Academic Press, Boston 1, 1989.

[14] M. Fortin, A. Soulaimani, Finite Element approximation of compressible viscous flows, Proc. *Computational Methods in Flow Analysis*, Vol. 2, H. Niki and M. Kawahara eds., Okayama University of Sciences Press, 1988, pp. 951-956.

[15] L. Dutto, Etude de préconditionnements pour la résolution par la méthode des éléments finis des équations de Navier-Stokes pour un fluide compressible, Thèse Université Pierre et Marie Curie, Paris 6, Nov. 1990.

[16] D.N. Arnold, F. Brezzi and M. Fortin, A stable finite element for the Stokes equations, *Calcolo* 21, (1984) 337.

[17] G. Rogé, Sur l'approximation et l'accélération de la convergence lors de la simulation numérique en éléments finis d'écoulements de fluides visqueux compressibles. Thèse, Université Pierre et Marie Curie, Paris 6, Mai 1990.

[18] M.O. Bristeau, R. Glowinski, J. Périaux, H. Viviand Eds., *Numerical Simulation of Compressible Navier-Stokes Flows*, A GAMM Workshop, Notes on Numerical Fluid Dynamics, Vieweg, Vol. 18, 1987.

[19] R. Glowinski, J. Périaux, G. Terrasson, On the coupling of Inviscid and Viscous Models for Compressible flows via Domain Decomposition, *Finite Element Analysis in Fluids*, T.J. Chung, G.R. Karr eds., Univ. of Alabama in Huntsville Press, 1989, pp. 444-452.

[20] M.O. Bristeau, R. Glowinski, J. Périaux, Acceleration procedures for the numerical simulation of compressible and incompressible viscous flows, *Advances in Computational Nonlinear Mechanics*, J. S. Doltsinis ed., CISM Courses no 300, Springer-Verlag, 1989.

[21] B. Palmerio, Self adaptive F.E.M. algorithms for the Euler equations, *Rapport de Recherche INRIA Sophia-Antipolis*, n°338, 1985.

[22] C. Pouletty, Génération et Optimisation de maillages éléments finis. Application à la résolution de quelques équations en Mécanique des Fluides. Thèse de Docteur-Ingénieur, Ecole Centrale, Paris, Déc. 1985.

[23] M.O. Bristeau and J. Périaux, Finite Element Methods for the calculation of compressible viscous flows using self-adaptive mesh refinements, *Lectures Notes in Computational Fluid Dynamics*, Von Karman Institute for Fluid Dynamics, Rhode-St-Genèse, Belgium (March 1986).

[24] C. Bernardi, O. Pironneau, On the shallow water equations at small Reynolds number. Internal report, Université Paris 6, 1989.

[25] C. Bernardi, F. Laval, B. Metivet, B. Thomas, Projet N3S de mécanique des fluides : Approximation par éléments finis de fluides visqueux avec masse volumique variable, Rapport EDF, DER, DMM, HI 72/7044, 1990.

[26] D. Mavriplis, A. Jameson, L. Martinelli, Multigrid solution of the Navier-Stokes equations on triangular meshes, Icase report 89-11, Feb. 1989.

[27] M.G. Vallet, F. Hecht, B. Mantel, Anisotropic control of mesh generation based upon a Voronoi type method, Proceedings of the Third Int. Conf. on Numerical Grid Generation in Computational Fluid Dynamics, June 1991, A.S. Arcilla ed., Elsevier Science Pub. (to appear).

[28] M.G. Vallet, Génération de maillages anisotropes adaptés, Application à la capture de couches limites, Rapport INRIA N , 1991.

[29] J.E. Dennis, R.B. Schnabel, Numerical Methods for Unconstrained Optimization and Nonlinear Equations, Prentice Hall, Englewood Cliffs, N.J. 1983.

[30] M.O. Bristeau, R. Glowinski, L. Dutto, J. Périaux, G. Rogé, Compressible Viscous Flow Calculations using compatible finite element approximations, Int. Journal for Num. Methods in Fluids, Vol. 11, 1990, pp. 719-749.

[31] B. Palmerio, A. Dervieux, 2D and 3D unstructured mesh adaption relying on physical analogy, Proceedings of the Sec. Int. Conf. on Numerical Grid Generation in Computational Fluid Dynamics, Miami, S. Sengupta, J. Hauser, P.R. Eiseman, J.F. Thompson eds., Pineridge Press, 1988.

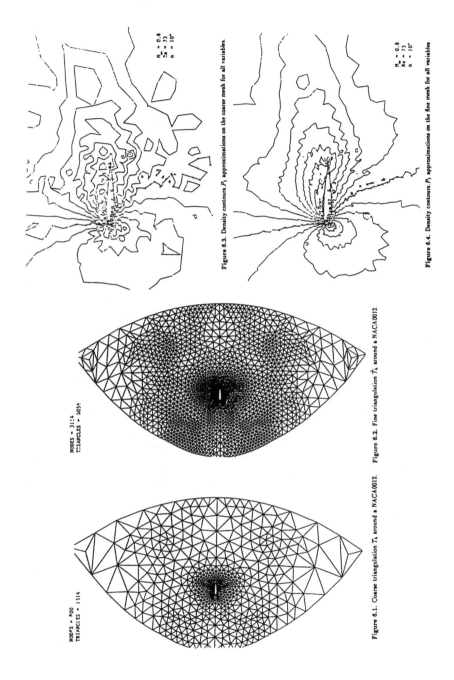

Figure 6.3. Density contours P_1 approximations on the coarse mesh for all variables.

Figure 6.4. Density contours P_1 approximations on the fine mesh for all variables.

$M_\infty = 0.8$
$Re = 73$
$\alpha = 10^\circ$

$M_\infty = 0.8$
$Re = 73$
$\alpha = 10^\circ$

NODES = 3114
TRIANGLES = 5054

NOEPS = 800
TRIANGLES = 1514

Figure 6.2. Fine triangulation $\hat{\mathcal{T}}_h$ around a NACA0012.

Figure 6.1. Coarse triangulation $\mathcal{T}_h$ around a NACA0012.

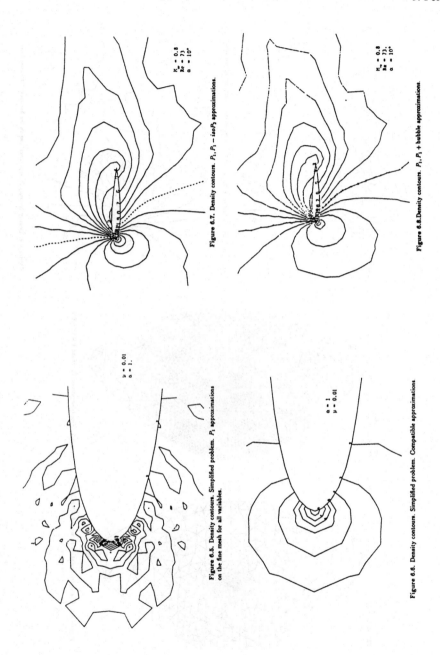

Figure 6.7. Density contours. $P_1, P_1 - iso-P_2$ approximations.

Figure 6.8. Density contours. $P_1, P_1 +$ bubble approximations.

Figure 6.5. Density contours. Simplified problem. P_1 approximations on the fine mesh for all variables.

Figure 6.6. Density contours. Simplified problem. Compatible approximations.

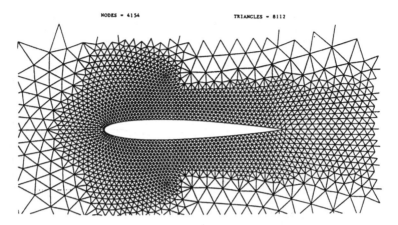

Figure 7.1. Enlargment of a triangulation around a NACA0012.

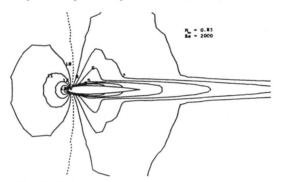

Figure 7.2. Density contours. $P_1, P_1 - isoP_2$ approximations.

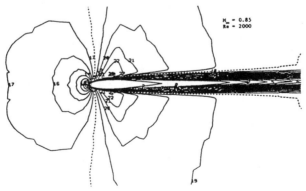

Figure 7.3. Mach contours. $P_1, P_1 - isoP_2$ approximations.

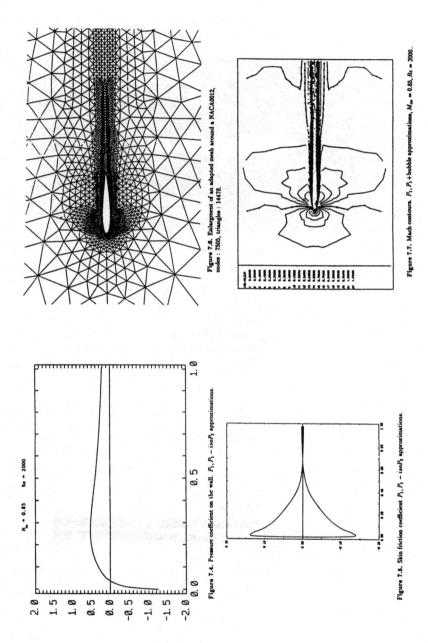

Figure 7.6. Enlargement of an adapted mesh around a NACA0012, nodes : 7505, triangles : 14478.

Figure 7.7. Mach contours. P_1, P_1 + bubble approximations, $M_\infty = 0.85$, $Re = 2000$.

Figure 7.4. Pressure coefficient on the wall. P_1, $P_1 - isoP_2$ approximations.

Figure 7.5. Skin friction coefficient P_1, $P_1 - isoP_2$ approximations.

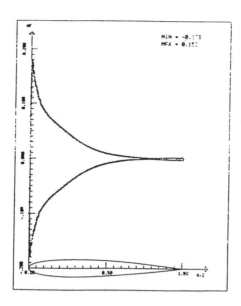

Figure 7.8. Skin friction coefficient. P_1, P_1 + bubble approximations, $M_\infty = 0.85, Re = 2000.$.

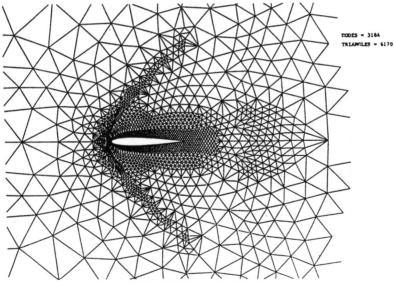

Figure 7.9. Enlargment of an adapted mesh around a NACA0012.

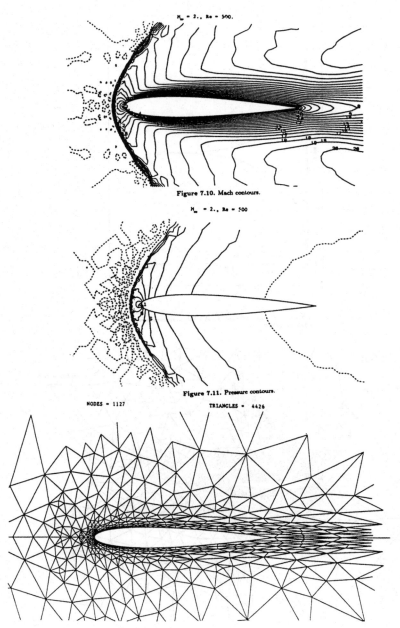

$M_\infty = 2., Re = 500.$

Figure 7.10. Mach contours.

$M_\infty = 2., Re = 500$

Figure 7.11. Pressure contours.

NODES = 1127   TRIANGLES = 4426

Figure 7.12. An enlargment of the adapted mesh T_{14}.

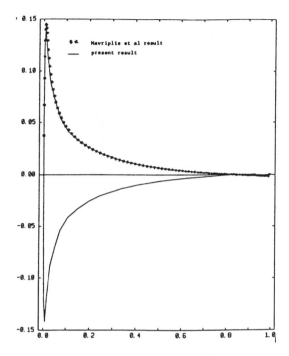

Figure 7.13. Skin friction distribution computed on $T_{1A}/\hat{T}_{1A}$ mesh. Comparison with Mavriplis et al results. $M_\infty = 0.5$, $Re = 5000.$, $\alpha = 0$.

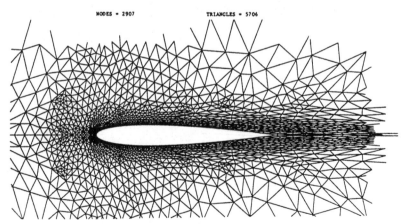

Figure 7.14. An enlargment of the adapted mesh T_{2A}.

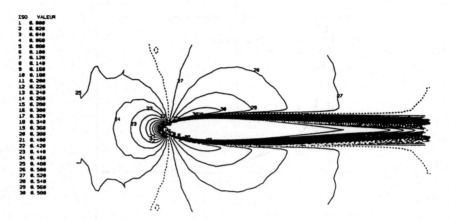

Figure 7.15. Mach contours of the solution computed on $T_{2h}/\dot{T}_{2h}$,
$M_\infty = 0.5, Re = 5000., \alpha = 0$.

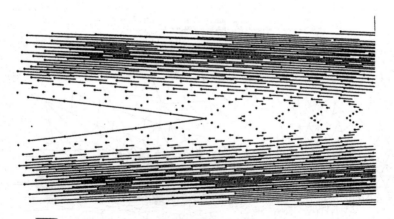

Figure 7.16. Velocity field near the trailing edge, recirculation bubble.
$\overline{M}_\infty = 0.5, Re = 5000., \alpha = 0$.

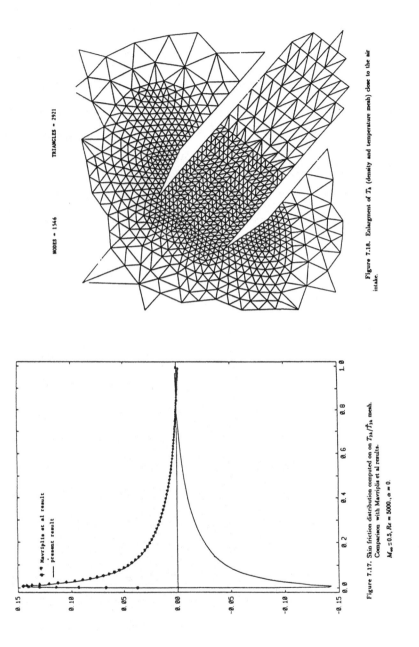

Figure 7.18. Enlargement of T_h (density and temperature mesh) close to the air intake.

NODES = 1546   TRIANGLES = 2921

Figure 7.17. Skin friction distribution computed on on T_{2h}/T_{3h} mesh. Comparison with Mavriplis et al results.

$M_\infty = 0.5, Re = 5000, \alpha = 0.$

● ● Mavriplis et al result

— present result

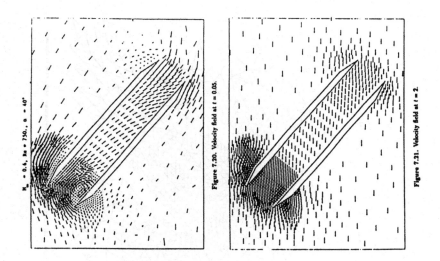

M∞ = 0.6, Re = 750., α = 40°

Figure 7.20. Velocity field at t = 0.05.

Figure 7.21. Velocity field at t = 2.

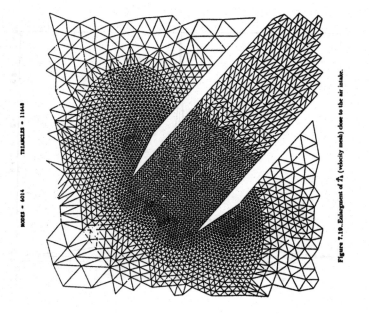

NODES = 6014   TRIANGLES = 11648

Figure 7.19. Enlargment of $\mathcal{T}_h$ (velocity mesh) close to the air intake.

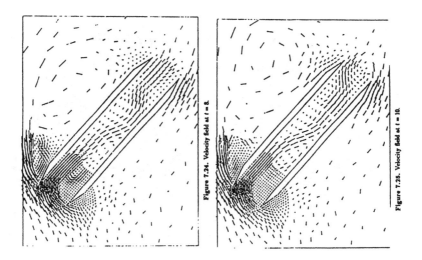

Figure 7.24. Velocity field at $t = 8$.

Figure 7.25. Velocity field at $t = 10$.

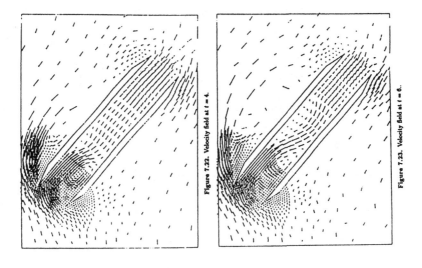

Figure 7.22. Velocity field at $t = 4$.

Figure 7.23. Velocity field at $t = 6$.

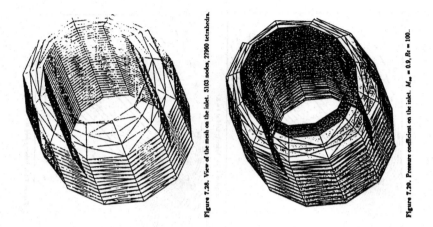

Figure 7.28. View of the mesh on the inlet. 5103 nodes, 27960 tetrahedra.

Figure 7.29. Pressure coefficient on the inlet. $M_\infty = 0.9$, $Re = 100$.

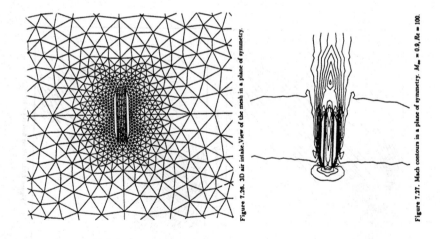

Figure 7.26. 3D air intake. View of the mesh in a plane of symmetry.

Figure 7.27. Mach contours in a plane of symmetry. $M_\infty = 0.9$, $Re = 100$.

Test Cases T4, T5:
Confined Axisymmetric Jet

Introduction of Test Case T4

W. RODI and J. ZHU

Institute for Hydromechanics, University of Karlsruhe, D-7500 Karlsruhe 1
F.R.Germany

1. INTRODUCTION

The axisymmetric confined jet flow in a slightly divergent duct studied experimentally by Binder and Kian (1983) was chosen as test case T4 for the ERCOFTAC Workshop on Numerical Simulation of Unsteady Flows, Transition to Turbulence and Combustion held in Lausanne, March 1990. This type of flow is of great practical importance as it occurs in a number of engineering applications, particularly in ejectors and combustion chambers. From the computational standpoint, confined jets present several interesting features such as the presence of an adverse pressure gradient, recirculation with unfixed separation and reattachment points as well as the coexistance of both strong and weak shear regions which are sensitive to turbulence modelling and to the numerical solution procedure. All these make the flow a suitable test case for assessing calculation methods for turbulent flows.

2. FLOW CHARACTERISTICS

The flow configuration is shown in Figure 1, together with the coordinate system with the origin being at the centre of the entrance plane. Craya and Curtet developed an approximate theory of confined jets (Curtet, 1960 and Rajaratnam, 1976). By analytically solving simplified equations of motion, they found that confined jets in constant area ducts can be characterized by a parameter C_t which is defined by

$$C_t = \frac{1}{\sqrt{m}}, \qquad m = \frac{S}{Q^2} \int_S (U^2 + \frac{p}{\rho}) dS - \frac{1}{2} \qquad (1)$$

where m is the total momentum flux non-dimensionalized with the flow rate Q and the duct area S. Under the conditions $d_o << D_o$ and $U_a << W_{oo}(W_{oo} = U_J - U_a)$, C_t can be simplified to

$$C_t \approx \frac{U_a}{W_{oo}} \frac{D_o}{d_o} \qquad (2)$$

Although the Craya-Curtet number C_t is not constant in variable area ducts, its value at the entrance can still be used to characterize the inlet flow conditions. Experiments

474

have shown that recirculation occurs in cylindrical ducts when $C_t \leq 0.96$ (Barchilon and Curtet, 1964) and in a conical duct with 5° divergence when $C_t \leq 1.1$ (Binder and Kian, 1983). It follows from Eq.(2) that, in a given duct, recirculation may be generated or intensified by reducing the ambient flow velocity while keeping the jet velocity constant.

According to experiments of Binder and Kian (1983), the flow field can be divided into the following four different regions:

(1) Potential core in which the jet discharging at a uniform velocity U_J from the nozzle remains undisturbed. The length of the potential core is about $6d_o$.

(2) Ambient region occupied by the secondary flow issuing with a uniform velocity U_a at the entrance. The flow velocity decreases downstream but remains constant radially.

(3) Boundary layer developing at the duct wall. However, this plays only a minor role in most cases, because the shear stress at the wall is one order of magnitude lower than that in the jet.

(4) Mixing region which is characterized by an initial high-shear layer and a continuous entrainment process until complete mixing of the jet with the ambient flow is achieved. Depending on the inlet value of C_t, different flow regimes occur in this region: at large C_t, the jet reaches the duct wall before it has consumed the ambient fluid and the flow does not separate, while at small C_t, the jet has entrained all the ambient fluid before it reaches the wall, thereby creating reverse flow to satisfy the total mass flux conservation. After the mixing is complete, the flow degenerates eventually to the fully developed regime, if the duct is long enough and becomes cylindrical.

In regions (1) and (2) the turbulence levels are very low and the flows can be treated as potential.

3. EXPERIMENTAL DATA

The experiments are described in detail in Kian (1981) and Binder and Kian (1983). The fluid used was water. The jet velocity was 6.5m/sec for flow visualization studies as well as wall-pressure measurements and 40cm/sec for velocity measurements. Figure 2 shows the separation and reattachment points obtained from the flow visualization. Only a range was given for the reattachment points because it was difficult to localize them with some accuracy due to the instability of the reattachment. The wall-pressure distribution was measured for different values of C_t ranging from 0.57 to 2.28. Detailed measurements of the mean velocity and turbulence fields were carried out with LDA for three values of C_t, namely 1.23 for non-separation, 0.775 for mild separation, and 0.59 for strong separation situations. Apart from the wall-pressure distribution, the profiles of the mean axial velocity and of the Reynolds stresses $\overline{uu}$,

$\overline{vv}$ and $\overline{uv}$ were measured at different x-locations. Since no experimental data were available for $\overline{ww}$, the turbulent kinetic energy k was assumed as $(\overline{uu} + 2\overline{vv})/2$. The selected experimental data are given in the authors' summary paper where they are compared with the predictions of different computor groups.

4. SPECIFICATION OF TEST CASE

Only the case with C_t=0.59 was chosen for the workshop for which previous calculations (Zhu, 1986) produced the worst agreement with the experiment.

4.1 Geometry

(a) Jet diameter do=1.6cm
(b) Initial ambient flow diameter Do=16cm
(c) Outflow section diameter De=21.8cm
(d) Length of the divergent duct L=64cm

4.2 Flow Conditions

Density ρ=0.9982gm/cm^3
Viscosity μ=0.01gm/cm·sec

The inflow boundary of the calculation is placed at x=0. The velocity parallel to the axis at this place is given the constant values U_J in the jet and U_a in the Ambient flow, where:

Jet velocity U_J=40cm/sec
Ambient velocity U_a=2.33cm/sec.

The radial velocity at the inflow boundary is put to zero. The way of prescribing turbulence at the inflow boundary is left to the user.

The outflow section is assumed to continue as an infinitely long cylindrical duct with constant diameter De. The outflow boundary of the calculations is placed in this section sufficiently far downstream so that the exact conditions used there have no influence on the solution in the divergent section. The exact location of this boundary is left to the user.

4.3 Results Required

(1) Separation and reattachment locations: x_s and x_r
(2) Pressure distribution C_p=[p(x)-p(x=0)-0.5ρU_a^2]/(0.5ρU_J^2) along the duct wall
(3) Centreline velocity U_0/U_m as function of x(cm), where U_m is the mean velocity of the section at x
(4) Axial velocity profiles U/U_m at 4 sections x=10, 20, 30 and 40cm

(5) Profiles of the turbulent kinetic energy k/U_0^2 at 4 sections x=10, 20, 30 and 40cm

(6) Profiles of the Reynolds shear stress $\overline{uv}/U_0^2$ at 4 sections x=10, 20, 30 and 40cm

5. REFERENCES

Barchilon, M., and Curtet, R., 1964, "Some details of the structure of an axisymmetric confined jet with backflow," *J. Basic Eng.*, **86**, 777-787.

Binder, G., and Kian, K., 1983, "Confined jets in a diverging duct," *Proc. 4th Symp. on Turbulent Shear Flows*, Karlsruhe, F.R.Germany, 7.18-7.23.

Curtet, R., 1960, "Sur l'écoulement d'un jet entre parois," *Publications Scientifiques et Techniques du Ministère de l'Air*, Paris.

Kian, K., 1981, "Jets confinés dans un divergent," Thèse de Docteur-Ingénieur de l'Université Scientifique et Médicale et l'Institut National Polytechnique de Grenoble, France.

Rajaratnam, N., 1976, *Turbulent jets*, Elsevier, Chapter 8.

Zhu, J., 1986, "Calcul des jets turbulents confinés avec recirculation," Thèse de Docteur, l'Institut National Polytechnique de Grenoble, France.

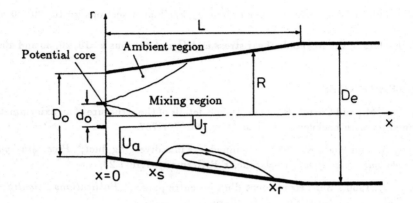

Figure 1. Flow configuration

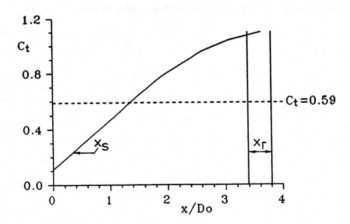

Figure 2. Separation and reattachment points

Problem T4: Predictions Of An Axisymmetric Jet Flow In A Diverging Duct.

D.S.Clarke

CFD Department,
Harwell Laboratory,
Oxon, OX11 0RA, UK

1 NUMERICAL METHOD

The numerical method used for these calculations is the same as that used for Problem T1.

2 CALCULATIONS

2.1 Grids

For this problem two sizes of non-orthogonal axisymmetric grid were used: 36×25×3 (coarse) and 54×37×3 (fine). The coarse grid in the x–r plane is shown in Figure 1.

2.2 Turbulence Models and Differencing Schemes

Calculations have been carried out on both grids with both the k–ε and DSM turbulence models using upwind and higher upwind differencing.

Further results have been obtained using the DSM with the addition of wall reflection terms. These terms are included to account for pressure reflections at the wall and have a significant effect on the pressure-strain term in the Reynolds stress equations. Their effect is to enhance the component of normal stress parallel to the wall whilst that perpendicular to the wall is well damped.

2.3 Initial and Boundary Conditions

At the inlet of the duct a step function was used for the axial velocity. Preliminary calculations were made to predict the turbulence values for the inlet. Given the mass fluxes at the inlet, fully-developed profiles of the turbulence quantities were then used.

Along the wall on the outer edge of the duct logarithmic boundary conditions were used. The values of the Reynolds stresses were calculated by linear extrapolation from the values in control volumes interior to the flow. At the symmetry axis, the normal gradient of all quantities with symmetrical behavior were taken to be zero, whereas the normal velocity component and shear stresses were themselves taken to be zero.

479

At the outlet of the duct Neumann boundary conditions were imposed.

3 RESULTS

3.1 Comparison of Predictions with Experimental Data

Predictions using the k–ε model with higher upwind differencing on the different grids are in close agreement. All results using the DSM are in reasonable agreement.

Figures 2 and 3 show comparisons between predictions and experimental data. The predictions are those made using the k–ε and DSM turbulence models on the fine grid with higher upwind differencing. Figure 2 shows profiles of the centerline velocity U_0/U_m as a function of x (cm). Figure 3 shows profiles of the axial velocity $U(x,r)/U_0$ as a function of r/R at the location x=30cm. (where r and $R(x)$ are as defined in the problem specification).

Table 1 gives the dimensionless locations of separation x_s/D_0 and reattachment x_r/D_0 for the k–ε and DSM when using higher upwind differencing on the fine grid.

Model	x_s/D_0	x_r/D_0
k–ε	1.46	3.21
DSM	2.23	3.77

Table 1. Dimensionless Separation and Reattachment lengths

3.2 Cpu Time

Table 2 gives the running times for each fine grid calculation on a Cray 2. It should be noted that no attempts were made to optimize these times.

Model	Differencing	CPU (secs)
k–ε	UW	80
k–ε	HUW	+49
DSM	UW	300
DSM	HUW	+168

Table 2. Running times on a Cray 2

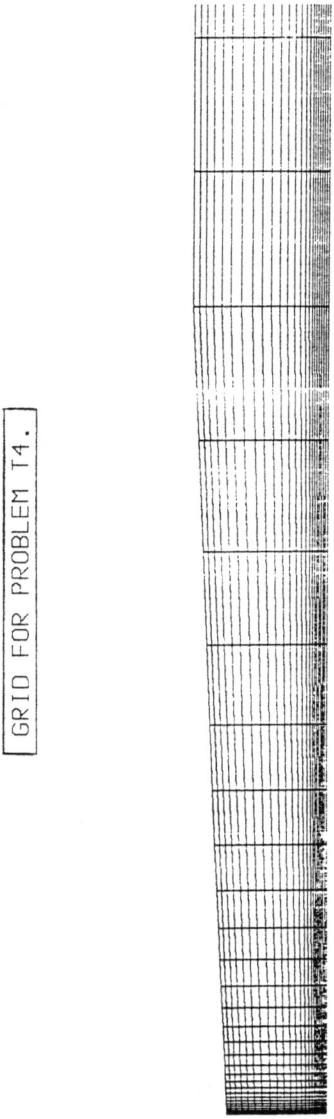

GRID FOR PROBLEM T4.

Figure 1

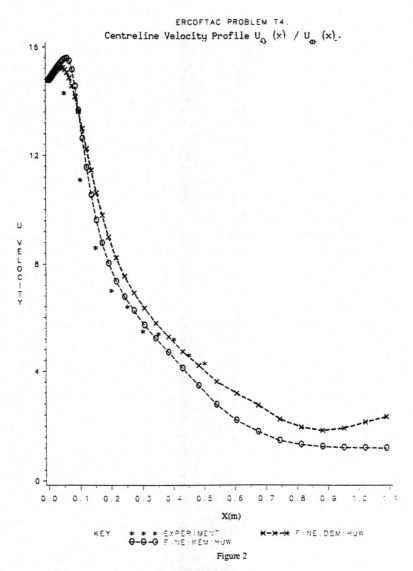

ERCOFTAC PROBLEM T4.
Centreline Velocity Profile U_Ω (x) / U_m (x).

KEY   * * * EXPERIMENT   x-x-x FINE:DSM:HUW
        O-O-O FINE:KEM:HUW

Figure 2

COMPARING TURBULENCE MODELS AND EXPERIMENTAL DATA

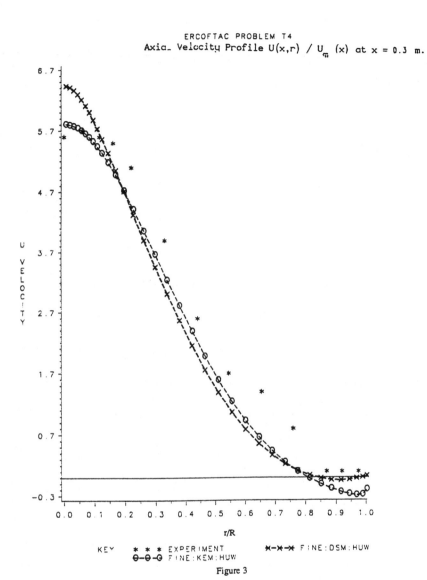

ERCOFTAC PROBLEM T4
Axial Velocity Profile U(x,r) / U_m (x) at x = 0.3 m.

Figure 3

COMPARING TURBULENCE MODELS AND EXPERIMENTAL DATA

Problem T5: Predictions Of The Turbulent Flow In An Axisymmetric Geometric Configuration.

D.S.Clarke

CFD Department,
Harwell Laboratory,
Oxon, OX11 0RA, UK

1 NUMERICAL METHOD

The numerical method used for these calculations is the same as that used for Problem T1.

2 CALCULATIONS

2.1 Grids

For this problem a rectangular axisymmetric grid of 32×28×3 was used.

2.2 Turbulence Models

Calculations have been carried out with both the $k–\varepsilon$ and differential stress (DSM) turbulence models using higher upwind differencing.

2.3 Initial and Boundary Conditions

The borders $C2_\infty$ and $C3_\infty$ (as defined in the problem specification) were modelled as boundaries on which the pressure was equal to atmospheric pressure.

To close the problem the following additional boundary conditions were imposed. On $C2_\infty$ and $C3_\infty$ normal derivatives of all the velocity components were set to zero except for the tangential velocity V on $C2_\infty$ which we set

$$\frac{1}{r}\frac{\partial(rV)}{\partial r} = 0.$$

The border $C1_\infty$ was modelled as a wall to prevent unphysical excess flow entering through it.

3 RESULTS

3.1 Comparison of Turbulence Models

Figures 1 and 2 show comparisons between the predictions using the $k–\varepsilon$ and DSM turbulence models using higher upwind differencing and with experimental data.

484

3.2 Comparison with Experimental Data

The predictions have been compared with experimental data (Chatou, Electricité de France). Figure 1 shows an axial profile of the axial velocity and Figure 2 shows an axial profile of the turbulent kinetic energy. Both sets of predictions are in reasonable agreement with the experimental data.

3.3 Cpu Time

Table 1 gives the running times for each calculation on a Cray 2. It should be noted that no attempts were made to optimize these times.

Model	Differencing	CPU (secs)
$k{-}\varepsilon$	UW	15
$k{-}\varepsilon$	HUW	+20
DSM	UW	52
DSM	HUW	+63

Table 1. Running times on a Cray 2

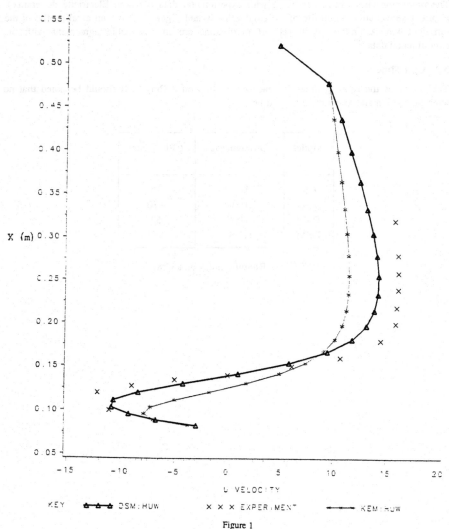

ERCOFTAC PROBLEM T5.
VALUES AT Y(M)=0

COMPARING TURBULENCE MODELS WITH HUW DIFFERENCING

Figure 1

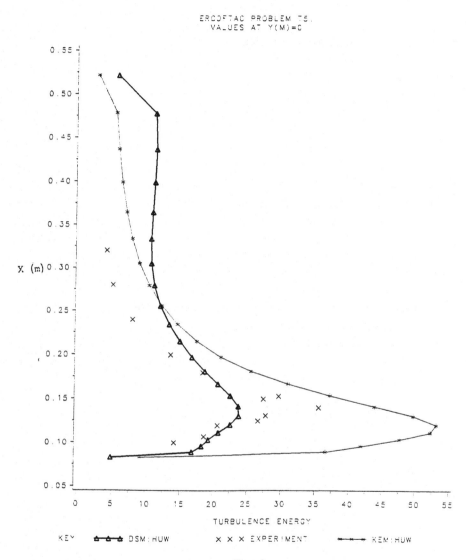

Figure 2
COMPARING TURBULENCE MODELS WITH HUW DIFFERENCING

TEST CASE T4: CONFINED AXISYMMETRIC JETS

P. Di Martino
Alfa Romeo Avio S.p.A.
I-80038 Pomigliano d'Arco (Na)
Italy

1. INTRODUCTION

In this study a finite volume method to solve time-averaged steady incompressible 2D Navier-Stokes equations for axisymmetric turbulent elliptic flows, employing a body-fitted curvilinear and non orthogonal grid, has been used, with a k-ε model for turbulence closure.

2. TRANSFORMATION OF THE TRANSPORT EQUATIONS TO A CURVILINEAR COORDINATE SYSTEM

In a transformed plane ξ, η the transport equation of a general dependent variable ϕ for a steady 2D incompressible flow can be written in the form:

$$\frac{\partial}{\partial \xi}(r\rho U\phi) + \frac{\partial}{\partial \eta}(r\rho V\phi) = rJS(\xi,\eta) +$$

(convection)   (source)

$$\frac{\partial}{\partial \xi}\left[\frac{r\Gamma}{J}\left(q_1\frac{\partial \phi}{\partial \xi} - q_2\frac{\partial \phi}{\partial \eta}\right)\right] + \frac{\partial}{\partial \eta}\left[\frac{r\Gamma}{J}\left(-q_2\frac{\partial \phi}{\partial \xi} - q_1\frac{\partial \phi}{\partial \eta}\right)\right] \qquad (1)$$

(diffusion)

The terms q_i and J are determined by geometrical derivatives of the curvilinear grid in the cartesian domain; q_2 accounts for the distortion of the grid and disappears for orthogonal grids along with the mixed derivatives of the diffusion terms in (1). U and V are the contravariant velocity components. A turbulence model of the k-ε type [1] has been used for the closure of the system of equations.

488

3. DISCRETIZATION OF THE EQUATIONS

The equations of type (1) are transformed into difference equations by integrating over each control volume with $\Delta\xi=\Delta\eta=1$. The diffusion terms are discretized with central difference scheme, while for convective terms a first order upwind scheme is used. The source terms are linearized and added to the mixed derivatives of diffusion terms.

The discretized transport equations can be cast in the following form:

$$a_P\phi_P = a_E\phi_E + a_W\phi_W + a_N\phi_N + a_S\phi_S + S_u; \qquad (2)$$

$$a_P = a_E + a_W + a_N + a_S - S_P; \qquad (3)$$

The resulting system of algebraic equations is solved by a line Gauss-Seidel method.

4. CONTINUITY EQUATION

The well established method SIMPLE [2] is used to solve the continuity equation, which, after applying the coordinate transformation, is integrated over the control volume, yielding:

$$[(r\rho U)_e - (r\rho U)_w]\Delta\eta + [(r\rho V)_n - (r\rho V)_s]\Delta\xi = 0 \qquad (4)$$

The non-staggered calculation grid used in this study has requested a particular procedure of interpolation of the components of velocity, suggested by Rhie and Chow [3], to avoid the checker-board splitting.

For example the velocity u in a point P is:

$$u_P=\alpha_u[H_P{}^u+D_{1P}{}^u(P_w-P_e)+D_{2P}{}^u(P_S-P_n)]+(1-\alpha_u)u_P{}^o; \qquad (5)$$

with $\quad H_P{}^u = (\Sigma a_{nb}{}^u + S_u{}^u)/a_P{}^u;$

and $D_1p^u = ry_\eta \Delta\eta/a_p{}^u$; $D_2p^u = - ry_\xi \Delta\xi/a_p{}^u$;

α_u is the relaxation factor for u.

The velocity on the face of the volume is interpolated in the following way:

$$u_w = \alpha_u[\overline{Hp^u + D_2p^u}(P_S - P_n) + \overline{D_1w}(P_W - P_p)] + (1-\alpha_u)u_w{}^o; \tag{6}$$

where overbars represent the average of the same quantities evaluated at the nodes adjacent to the face. Similar relations are obtained for the other components of velocity. Substituting these relations in the equation (4), a system of equations for the pressure-correction is obtained.

5. BOUNDARY CONDITIONS

To avoid detailed calculations in the near-wall regions a wall-function is used. Velocity distributions for the central jet and surrounding flow are known. The values of K and ϵ are prescribed in the following way:

$$K = \alpha_k\, u_{in}{}^2 \qquad \epsilon = C_\mu\, \frac{K\,{}^{3/2}}{\alpha_L Y} \tag{7}$$

where u_{in} is the velocity in the inlet plane and α_k and α_L are input coefficients. Along the axis of symmetry the gradients in the radial direction of all the variables are equal to zero. The axial gradient for all the variables are set to zero.

6. SOME DETAILS OF COMPUTATION

A non-uniformly distributed grid of 101x42 was used in the computation. There were 5 cells covering the central-jet inlet. The domain of calculation was prolonged downstream the exit of the divergent duct up to the position x=1.5 m in order to satisfy the condition of zero axial gradients.

The inlet conditions for the turbulence parameters k-ε were imposed in the hypothesis of low turbulence with $\alpha_k = \alpha_L = 10^{-5}$. The under-relaxation factors were fixed to 0.5 for all the variables except the pressure for which a factor 1. was used. Calculations stopped when the mass residual sources reached a specified minimum value, 10^{-6}. The number of iterations required for the present grid was about 1300 with a CPU time of 30' on a computer VAX 9410.

7. RESULTS

The dimensionless separation and reattachment points are:
$$x_s/D_o = 1.622 \qquad x_r/D_o = 3.660$$
The other required results are presented in the order:
Fig. 1: Centerline velocity $U_o(x)/U_m$ as function of x(cm).
Fig. 2: Wall static pressure $C_p(x) = (p(x)-p(0)-0.5\rho U_a^2)/$
$(0.5\rho U_j^2)$ as function of x/D_o.
Fig. 3: Axial velocity profiles $U(x,r)/U_m(x)$ at sections
x= 10, 20, 30, 40 cm.
Fig. 4: Profiles of the turbulent kinetic energy $k(x,r)/U_o^2$
at sections x=10, 20, 30, 40 cm.
Fig. 5: Profiles of the Reynolds shear stress $uv(x,r)/U_o^2$
at sections x=10, 20, 30, 40 cm.

REFERENCES

[1] Launder B.E., Spalding D.B., "The Numerical Computation of Turbulent Flows", Computer Methods Applied Mech. Eng., Vol. 3, pp. 269-289, 1974.

[2] Patankar S.V., "Numerical Heat Transfer and Fluid Flow", Hemisphere Publishing Co., New York, 1980.

[3] C.M. Rhie and W.L. Chow, " A Numerical Study of the Turbulent Flow Past an Isolated Airfoil with Trailing-edge Separation", AIAA J. Vol. 21, No. 11 (1983), pp. 1525-1532.

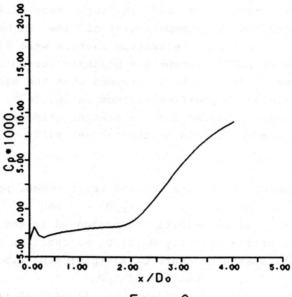

Fig. 2

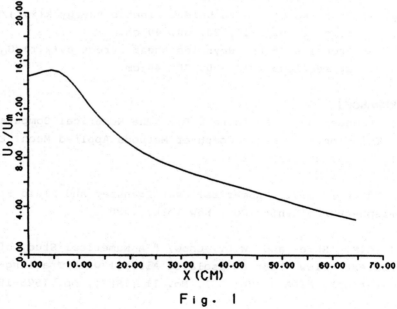

Fig. 1

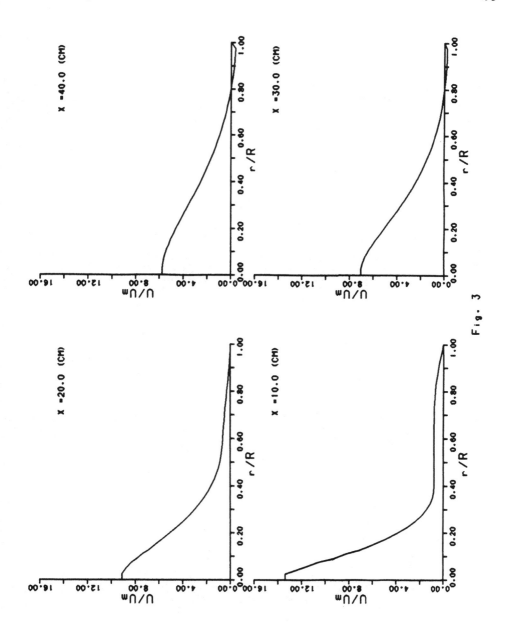

Fig. 3

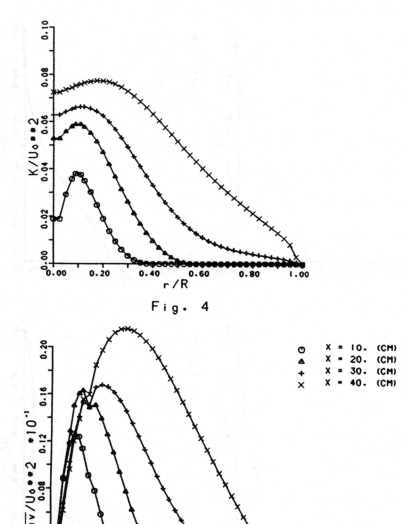

Fig. 4

Fig. 5

Modelling ERCOFTAC Test Case T4

J.R.NOYCE

SD-Scicon UK Ltd, Wavendon Tower, Wavendon, Milton Keynes, MK17 8LX, UK

1   INTRODUCTION

In 1989, ERCOFTAC specified a variety of test cases for computational modelling and for comparison with experimental results.   Results were presented at the workshop in Lausanne in March 1990.   This paper describes one approach which has been adopted to model test case T4, axisymmetric jet flow in a slightly divergent duct.   The modelling work has been carried out using the PHOENICS CFD code and the FEMVIEW post-processing package.

2   MODEL DESCRIPTION

2.1 System Configuration

The system to be modelled comprises a round jet with a diameter d_0 of 1.6 cm emerging into a tapered axisymmetric duct with an inlet diameter D_0 of 16.0 cm, a length L of 64.0 cm and a divergence half-angle θ of 2.59°.   The tapered duct is followed by a cylindrical duct with a diameter D_e of 21.8 cm.   At the inlet plane, the jet is surrounded by a co-flowing stream with a velocity 0.0583 times the jet inlet velocity, leading to an inlet Craya-Curtet number C_t of 0.59.   This configuration has been investigated experimentally by Binder and co-workers (1983, 1987), using water at the working fluid.

2.2 CFD Software

The calculations have been carried out using the PHOENICS computational fluid dynamics code from CHAM Limited.   This is a general-purpose finite-volume package for the solution of single or multi-phase flow problems in one, two or three dimensions (Rosten *et al* (1987)).   In its standard form, the package contains facilities to model a wide variety of flow types.   There is also a capability to add additional software in order to represent special flow phenomena.

2.3 Computational Grids

Body-fitted computational grid meshes have been generated for the analysis, extending from the jet inlet plane to one or two pipe diameters along the parallel outlet duct.

Several computational grids have been used in order to investigate the grid dependency of the solution. These contained 30 x 48, 30 x 96, 45 x 72 and 60 x 96 computational cells in the axial and radial directions respectively. The grid has been refined particularly within one jet diameter of the inlet, where the work of Zhu *et al* (1987) has indicated the importance of obtaining accurate results in order to predict adequately conditions further downstream. With the finest mesh, the radial and axial grid spacings in this area are of the order of $d_0/32$ and $d_0/16$.

2.4 Equations Solved

The equations which have been solved in the analysis are for the conservation of mass, momentum and turbulence quantities in the general form

$$\text{div}(\rho u \phi - \Gamma_\phi \text{grad} \phi) = S_\phi \qquad (1)$$

where ϕ is the solved quantity, or unity for the mass conservation equation, ρ is the density, u is the velocity vector, Γ_ϕ is the diffusive exchange coefficient and S_ϕ is the net source term.

The momentum equations are solved in terms of the resolutes of the total velocity vector in the axial and radial grid directions. Further details of the solution procedure for this type of grid are given by Malin *et al* (1985). The turbulence quantities solved are the turbulence kinetic energy (k), and the turbulence dissipation rate (ε).

2.5 Boundary Conditions

The inlet velocities which have been used in the model are taken from the report on the experimental results by Binder and Kian (1983):

Jet inlet velocity = U_j = 40.00 cm/s   Duct inlet velocity = U_a = 2.33 cm/s

The inlet velocities have been assumed to be uniform across both the jet and the duct. Since no inlet turbulence levels were provided in the problem specification, three jet inlet turbulence levels were employed in order to investigate the importance of this parameter on the predictions. These corresponded to inlet turbulence levels, k_{in}, of 0.2%, 1.0% and 5.0% of the jet inlet velocity. The inlet values of turbulence dissipation, ε_{in}, have been calculated from the relationship

$$\varepsilon_{in} = \frac{0.1643k_{in}^{1.5}}{0.045d_0} \tag{2}$$

A zero pressure boundary condition has been imposed at the outlet plane to ensure mass continuity between inlet and outlet.

Initial conditions for all solved variables except the axial velocity have been set close to zero (10^{-10}). The initial axial velocity field has been set to U_a throughout the computational domain.

2.6 Solution Procedure
The facilities of PHOENICS software which have been activated to complete the solution of this problem are summarised below:

- single-phase, incompressible, isothermal flow;
- full elliptic solution;
- upwind interpolation;
- modified SIMPLE algorithm for the hydrodynamic solution;
- standard or modified k-ε turbulence model;
- logarithmic law of the wall on the duct walls.

2.7 Turbulence Model Modifications
As well as the standard k-ε turbulence model, the simulation has also been run using the modifications to the standard model suggested by Rodi (1984) and Malin (1982) which improve predictions for free axisymmetric jets. In this case, the standard turbulence model constants C_μ and $C_{2\varepsilon}$ are replaced by variable values calculated from

$$C_\mu = 0.09 - 0.04f \tag{3}$$

$$C_{2\varepsilon} = 1.92 - 0.0667f \tag{4}$$

where

$$f = \left| \frac{\delta}{2\Delta U_{max}} \left(\frac{\partial U_{CL}}{\partial x} - \left| \frac{\partial U_{CL}}{\partial x} \right| \right) \right|^{0.2} \tag{5}$$

where $\partial U_{CL}/\partial x$ is the centreline deceleration, δ is the jet width, and ΔU_{max} is the maximum velocity difference across the jet.

2.8 Computational Details
The calculations were carried out on a MicroVAX II computer with 16 Mbyte of main memory. Calculations with the finest computational grid required out-of-core operation.

Convergence was deemed to have been attained when the outlet mass flow was within ±0.05% of the sum of the inlet mass flows. The number of iterations, or computational 'sweeps', of the domain required to achieve convergence depended on the grid size and the initial conditions assumed. Starting the 60 x 96 grid analysis from scratch required of the order of 300 iterations to reach a converged solution, and took approximately 16 hours of CPU time.

3   RESULTS

The ERCOFTAC-specified results of this analysis are presented elsewhere, together with those provided by other contributors. These results were obtained using the 60 x 96 grid with a 1% jet inlet turbulence level, an outlet duct length of $2D_e$, and with both the standard and modified versions of the k-ε turbulence model.

The influence of the grid density on the centreline normalised velocities is shown in Fig. 1, and on the wall static pressure coefficients in Fig. 2. These results are for the standard k-ε turbulence model. Although full grid independence is not claimed, the results from the 45 x 72 and 60 x 96 grids are similar (eg centreline normalised velocities within 3%).

Increasing the outlet duct length from $1D_e$ to $2D_e$ has a negligible influence on the results in the tapered duct section.

Varying the jet inlet turbulence level between 0.2% and 1% of the jet inlet velocity has a negligible influence on the jet centreline velocities (Fig. 3) or on the wall static pressure coefficients (Fig. 4). Raising the turbulence level to 5% has increased the spreading rate of the jet, reducing the centreline velocities by up to 7%.

4   REFERENCES

Binder, G., Kian, K. (1983).   Confined Jets in a Diverging Duct. *Proceedings of 4th Symposium on Turbulent Shear Flows*, Karlsruhe, FRG, pp 7.18 - 7.23.

Zhu, J., Binder, G., Kueny, J.L. (1987). Improved Predictions of Confined Jets with a Parabolic Computation of the Entrance Region, *AIAA Journal*, pp 1141 - 1142.

Rosten, H.I., Spalding, D.B. (1987).   The PHOENICS Reference Manual, *CHAM TR/200* Revision 06.

Malin, M.R., Rosten, H.I., Spalding, D.B., Tatchell, D.G. (1985).   Application of PHOENICS to Flow Around Ship's Hulls, *2nd International Symposium on Ship Viscous Resistance*, Goteborg.

Rodi, R. (1984). Turbulence Models and their Application in Hydraulics, *IAHR*.

Malin, M.R. (1982). Prediction of the Hydrodynamic and Thermal Characteristics of Self-Preserving Free Jets by Use a Revised k-W-g Model of Turbulence, *CHAM PDR/UK/18*.

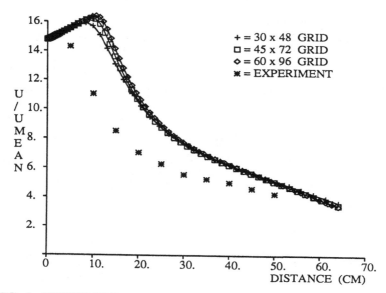

FIG. 1 CENTRELINE VELOCITIES FOR DIFFERENT GRIDS

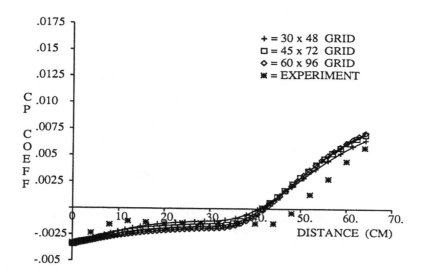

FIG. 2 WALL CP COEFFICIENTS FOR DIFFERENT GRIDS

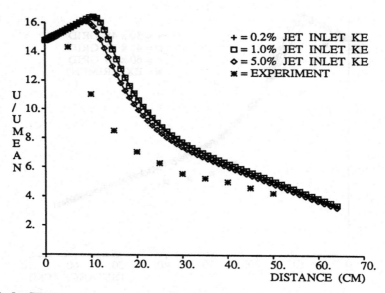

FIG. 3 CENTRELINE VELOCITIES FOR VARYING KE LEVELS

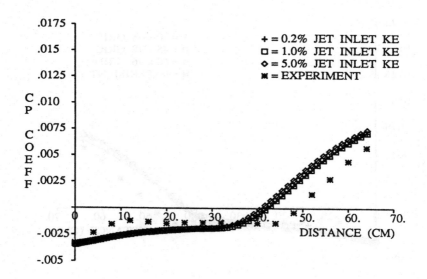

FIG. 4 WALL CP COEFFICIENTS FOR VARYING KE LEVELS

Calculation of a Coaxial Jet in a Diffuser

W. RODI and J. ZHU

Institute for Hydromechanics, University of Karlsruhe, D-7500 Karlsruhe 1
F.R.Germany

1. INTRODUCTION

The paper reports on the calculation of the test problem T4 - axisymmetric turbulent jet confined in a conical duct with a 5° divergence. The calculation was carried out using a two-dimensional axisymmetric version of the general, 3D vectorized finite-volume method described briefly in the authors' paper for test case T1. Since a synthesis of all the computational results from different participants is given in the authors' summary paper for test case T4, this paper concentrates on the computational details particular to the present problem and presents only a few results. It should be mentioned here that the authors, being also the test case supervisors for T4, had an advantage over other contributors in that they knew the experimental results. Also the experience gathered by one of the authors (Zhu, 1986) in previous calculations of the same flow was helpful.

2. GOVERNING EQUATIONS

The partial differential equations used to solve the flow can be written in the following form:

$$\frac{\partial}{\partial x_i}(C_i\phi - D_{i\phi}) = r^\alpha J S_\phi, \qquad i = 1, 2 \qquad (1)$$

where for different variables ϕ, the convective coefficients C_i, the diffusion terms $D_{i\phi}$ and the source terms S_ϕ are given in Table 1. J is the Jacobian of coordinate transformation between the general curvilinear system (x_i) and a reference rectangular system (y_i). Equation (1) includes both plane $(\alpha=0)$ and axisymmetric $(\alpha=1)$ forms. For the latter, y_1=x, y_2=r, and V_1=U, V_2=V, being the axial and radial components of the rectangular coordinate system and flow velocity vector, respectively. The second-order and bounded SOUCUP scheme (Zhu and Rodi, 1991) was used to discretize the convection terms of all the transport equations, including those for turbulence quantities k and ϵ.

501

3. BOUNDARY CONDITIONS

In spite of the uniform velocity profiles at the inlet, as declared in the experiment, there always exists a thin shear layer with high gradients between the jet and ambient flow near the nozzle. Therefore, outside the initial jet shear layer ($r < r_1$ or $r > r_2$, where r_1 and r_2 are the coordinates of the inner and outer edge of the shear layer at the inlet, taken respectively as $0.25d_o$ and $0.75d_o$ in this calculation), the axial velocities were given the experimental values, i.e., jet velocity $U_J = 40$cm/sec and ambient velocity $U_a = 2.33$cm/sec, and the turbulent kinetic energy k and its dissipation rate ϵ both set to a very small value 10^{-5}. Inside the shear layer ($r_1 \leq r \leq r_2$), the following Schlichting profile (Razinsky and Brighton, 1971) was assumed

$$\frac{U - U_a}{U_J - U_a} = \left[1 - \left(\frac{r - r_1}{r_2 - r_1} \right)^{3/2} \right]^2 \tag{2}$$

and k and ϵ were calculated from

$$k = \frac{|\nu_t \partial U / \partial r|}{C_\mu^{1/2}}, \qquad \epsilon = \frac{C_\mu k^2}{\nu_t} \tag{3}$$

where the turbulent viscosity ν_t was determined from

$$\nu_t = C^2 (r_2 - r_1)^2 |\partial U / \partial r| \tag{4}$$

and C is an empirical coefficient taken from the experimental data of Rajaratnam and Pani (1972)

$$C^2 = 0.0042 + 0.004 U_a / U_J, \qquad 0 \leq U_a / U_J \leq 0.2 \tag{5}$$

The outflow boundary was placed at x=$7.5D_o$ which is sufficiently far away from the end section of the divergent duct. At this boundary zero gradient conditions were assumed.

4. COMPUTATIONAL DETAILS AND RESULTS

The grid-dependence of the solution was tested using two computational grids, one with 68×50 and the other with 102×82 nodes. Both are non-uniform, with the grid lines being refined particularly in the initial jet shear layer. Figure 1 shows the upstream part of the coarse 68×50 grid (the total length of the solution domain is 120cm). The results from these two grids have been found to be very similar, as can be seen in Figure 2 for the decay of centreline velocity.

The iteration solution procedure was started from the fully developed pipe flow conditions and stoped when the maximum normalised residue of all the dependent variables was below 5×10^{-3}. The calculation with the fine 102×82 grid required 2924 iterations, which took 4.3 minutes of CPU-time on the Siemens VP400-EX vector computer. Figure 3 shows the streamlines calculated with the 102×82 grid.

5. REFERENCES

Rajaratnam, N., and Pani, B.S., 1972, "Turbulent compound annular shear layers," *J. Hydraulics Division, Proceedings of ASCE*, **98**, 1101-1115.

Razinsky, E., and Brighton, J.A. 1971, "Confined jet mixing for non-separating conditions," *J. Basic Eng.*, **93**, 333-349.

Zhu, J., 1986, "Calcul des jets turbulent confinés avec recirculation," Thèse de Docteur, l'Institut National Polytechnique de Grenoble, France.

Zhu, J., and Rodi, W., 1991, "A low dispersion and bounded convection scheme," to appear in *Comput. Methods. Appl. Mech. Eng.*

Table 1: Form of terms in the individual equations

ϕ	$D_{1\phi}$	$D_{2\phi}$	S_ϕ
1	0	0	0
V_1	$\dfrac{r^\alpha \mu_e}{J}(D_1 + \beta_1^1\omega_1^1 + \beta_2^1\omega_2^1)$	$\dfrac{r^\alpha \mu_e}{J}(D_2 + \beta_1^2\omega_1^1 + \beta_2^2\omega_2^1)$	$-\dfrac{1}{J}[\dfrac{\partial}{\partial x_1}(p\beta_1^1) + \dfrac{\partial}{\partial x_2}(p\beta_1^2)]$
V_2	$\dfrac{r^\alpha \mu_e}{J}(D_1 + \beta_1^1\omega_2^1 + \beta_2^1\omega_2^2)$	$\dfrac{r^\alpha \mu_e}{J}(D_2 + \beta_1^2\omega_1^1 + \beta_2^2\omega_2^2)$	$-\dfrac{1}{J}[\dfrac{\partial}{\partial x_1}(p\beta_2^1) + \dfrac{\partial}{\partial x_2}(p\beta_2^2)]$ $-2\mu_e\alpha\dfrac{V_2}{r^2}$
k	$\dfrac{r^\alpha \mu_e}{J\sigma_k}D_1$	$\dfrac{r^\alpha \mu_e}{J\sigma_k}D_2$	$G - \rho\epsilon$
ϵ	$\dfrac{r^\alpha \mu_e}{J\sigma_\epsilon}D_1$	$\dfrac{r^\alpha \mu_e}{J\sigma_\epsilon}D_2$	$(C_{1\epsilon}G - C_{2\epsilon}\rho\epsilon)\dfrac{\epsilon}{k}$

$$C_1 = \rho r^\alpha(V_1\beta_1^1 + V_2\beta_2^1), \quad C_2 = \rho r^\alpha(V_1\beta_1^2 + V_2\beta_2^2)$$

$$D_1 = B_1^1\frac{\partial \phi}{\partial x_1} + B_2^1\frac{\partial \phi}{\partial x_2}, \quad D_2 = B_1^2\frac{\partial \phi}{\partial x_1} + B_2^2\frac{\partial \phi}{\partial x_2}$$

$$\beta_j^i = \text{cofactor of } \partial y_j/x_i \text{ in } J$$

$$B_j^i = \beta_1^i\beta_1^j + \beta_2^i\beta_2^j, \quad \omega_j^i = \beta_j^1\frac{\partial V_i}{\partial x_1} + \beta_j^2 + \frac{\partial V_i}{\partial x_2}, \qquad J = \begin{vmatrix} \dfrac{\partial y_1}{\partial x_1} & \dfrac{\partial y_2}{\partial x_1} \\ \dfrac{\partial y_1}{\partial x_2} & \dfrac{\partial y_2}{\partial x_2} \end{vmatrix}$$

$$\mu_e = \mu + \mu_t, \quad \mu_t = \rho C_\mu k^2/\epsilon$$

$$G = \frac{\mu_t}{J^2}[2(\beta_1^1\frac{\partial V_1}{\partial x_1} + \beta_1^2\frac{\partial V_1}{\partial x_2})^2 + 2(\beta_2^1\frac{\partial V_2}{\partial x_1} + \beta_2^2\frac{\partial V_2}{\partial x_2})^2 + 2\alpha J^2(\frac{V_2}{r})^2$$
$$+ (\beta_2^1\frac{\partial V_1}{\partial x_1} + \beta_2^2\frac{\partial V_1}{\partial x_2} + \beta_1^1\frac{\partial V_2}{\partial x_1} + \beta_1^2\frac{\partial V_2}{\partial x_2})^2]$$

$$C_\mu = 0.09, \quad C_{1\epsilon} = 1.44, \quad C_{2\epsilon} = 1.92, \quad \sigma_k = 1.0, \quad \sigma_\epsilon = 1.32.$$

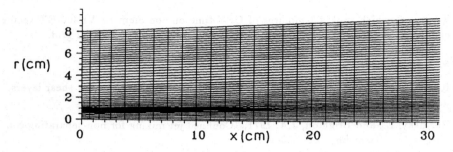

Figure 1. Computational grid

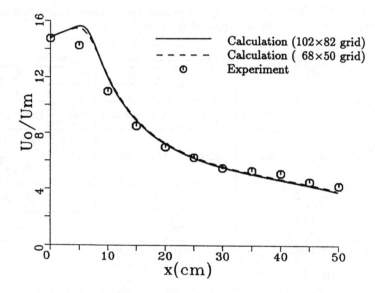

Figure 2. Decay of centreline velocity

Figure 3. Calculated streamlines

Summary of Predictions for a Coaxial Jet in a Diffuser (Test Case T4)

W. RODI and J. ZHU

Institute for Hydromechanics, University of Karlsruhe, D-7500 Karlsruhe 1
F.R.Germany

1. INTRODUCTION

Predictions for test case T4 have been submitted by four computor groups. This paper summarises the calculation methods employed and compares the main calculation results with each other as well as with the experimental data. The specification of the test case and the main flow characteristics have been given in the authors' introduction to this test case also published in these proceedings.

2. SUMMARY OF CALCULATION METHODS

All the calculations were carried out with finite-volume methods and on boundary-fitted non-orthogonal grids, three with non-staggered and one with staggered variable arrangement. Besides the standard k-ε turbulence model, which was used by all the contributors, a modification of this model devised by Rodi (1972) for round jets (labelled here as k-ε(R)) was used by Noyce, and a differential stress model (DSM) by Clarke and Wilkes. Three different types of discretization of the convection terms were used: upwind (UP), higher upwind (HUP) and second-order bounded SOUCUP schemes. Table 1 lists the main elements of each method. The details of the methods are described in the papers of the individual contributors. It should be noted that the computer codes FLOW3D used by Clarke and Wilkes and PHOENICS used by Noyce are commercially available CFD codes.

The wall-function approach (Launder and Spalding, 1974) bridging the viscous sub-layer was used in all the calculations to formulate the boundary conditions at the duct wall. It is of interest to note that contrary to heat-transfer and friction coefficients, the overall flow behaviour is not very sensitive to the near-wall modelling because in this case the maximum shear stress occurs in a region far away from the wall (Chieng and Launder, 1980 and Zhu, 1986). Problematic and critical is the specification of the inlet boundary conditions for which no detailed experimental data are available. The

505

previous work of Zhu et al. (1987) has indicated already a fairly strong influence of these conditions on the calculations. Table 2 lists the inlet boundary conditions used by the various computor groups, and Figure 1 compares the inlet turbulent kinetic energy (k) and eddy viscosity (μ_t) distributions resulting from these inlet conditions. It should be noted that Rodi and Zhu were the only ones who did not use an exact step profile for the mean velocity U, but a gradual transition of U over a shear layer originating from the nozzle with an estimated thickness of 0.5do. After the Workshop, Rodi and Zhu repeated the calculations using also a step profile, and these results are included in this paper to show the sensitivity to inlet profiles.

Table 1: Main features of different methods

name of contributor	turbulence model	convection scheme	variable arrangement
Clarke & Wilkes	k-ϵ, DSM	HUP	non-staggered
Di Martino	k-ϵ	UP	non-staggered
Noyce	k-ϵ, k-ϵ(R)	UP	staggered
Rodi & Zhu	k-ϵ	SOUCUP	non-staggered

Table 2: Inlet boundary conditions

contributor	U	k, ϵ
Clarke & Wilkes	step profile	BC1: fully-developed profiles
Di Martino	step profile	BC2: $k=\alpha_k U^2$, $\epsilon = C_\mu k^{1.5}/(\alpha_\epsilon r)$ with $\alpha_k = \alpha_\epsilon = 10^{-5}$
Noyce	step profile	BC3: $k=\alpha U_j^2$, $\epsilon = 3.65 k^{1.5}/d_o$ with $\alpha=0.01$
Rodi & Zhu	nearly step profile (gradual transition over thin shear layer centred at the nozzle wall)	BC4: derived from a mixing length formulation

Information on computational details of each contributor are given in Table 3. It can be seen that reasonably fine grids were used by all groups, with Clarke and Wilkes having the coarsest and Rodi and Zhu the finest grid. The latter contributors have later verified that their calculations obtained with a higher-order discretization scheme are grid-independent (Zhu and Rodi, 1990). This is not certain for the calculations on coarser grids performed with lower-order schemes.

The number of iterations depends on the convergence criterion employed, and it increases with grid fineness. As the convergence criteria were either not clearly stated or could not be translated, the degree of convergence of the individual solutions cannot be compared. However, the number of iterations cited by Noyce is much lower than that reported by the other contributors and there is some question whether his calculations were carried out to the same convergence level as the others. The computing time depends of course strongly on the computer used. Clarke and Wilkes, and Rodi and Zhu carried out their calculations on fast vector computers having comparable speed. Their calculations took only a few minutes of CPU time so that high efficiency has been achieved in their methods. In terms of CPU time per grid point and iteration, the method of Rodi and Zhu appears fastest.

Table 3: Computational details

	Clarke & Wilkes	Di Martino	Noyce	Rodi & Zhu
number of grid points	54×37	101×42	60×96	102×82
number of iterations	1200 (k-ϵ) 1500 (DSM)	1300	300	2924
CPU time in seconds	129 (k-ϵ) 468 (DSM)	6600	57600	258
CPU time per iteration and grid point	5.4×10^{-5} (k-ϵ) 1.6×10^{-4} (DSM)	1.2×10^{-3}	3.3×10^{-2}	1.1×10^{-5}
computer used	Cray 2	Vax 8830	MicroVAX II	VP400-EX

3. COMPARISON OF RESULTS

3.1 Recirculation Zone

Figure 2 shows the streamlines predicted by Rodi and Zhu and illustrates the size of the separation bubble obtained in their calculations. The separation and reattachment points predicted by the various methods are compared in Table 4 with experimental results (Kian, 1981 and Binder and Kian, 1983). It can be seen from the table that the various k-ϵ model calculations lead to separation and reattachment locations which differ by up to 20%, while the length of the separation zone differs by even more. Results from the authors' sensitivity study are also included and show that the use of an initial step profile for U instead of a more gradual transition causes earlier separation and later reattachment. The use of a numerically less accurate scheme has the same effect. Rodi and Zhu's original calculations are more or less in the middle of the field and agree fairly well with experiments. The differential stress model (DSM) and the modified k-ϵ(R) lead to lower shear stresses (see Figure 7 below), and this delays both separation and reattachment. In the case of the modified k-ε model, reattachment is now predicted considerably too late.

Table 4: Separation and reattachment points

	x_s (cm)	x_r (cm)	length (cm)	
Binder & Kian	21.6	54.2 ~ 60.8	32.6 ~ 39.2	experiment
Clarke & Wilkes	23.4	51.4	28	k-ε
	35.7	60.3	24.6	DSM
Di Martino	25.9	58.6	32.7	k-ε
Noyce	21.9	63.0	41.1	k-ε
	24.2	83.2	59	k-ε(R)
Rodi & Zhu	22.1	59.2	37.1	k-ε
results from	20.6	61.3	40.7	inlet step U-profile
sensitivity study	19.4	62.6	43.2	Noyce's conditions

3.2 Wall Static Pressure Distribution

Figure 3 displays the distribution of the static pressure along the diffuser wall. The behaviour of the pressure along the wall is closely related to that of the separation bubble. The pressure rises first due to the duct divergence and the entrainment of ambient fluid into the jet (reducing the mass flow and hence the velocity in the ambient stream). As can be seen from Figure 2, the separation bubble leads to an effective contraction of the flow when the bubble is approached and also in the upstream part of the bubble. This counteracts the mechanisms mentioned above and hence in this region the pressure is virtually constant, a behaviour often observed in the presence of separation bubbles. On the downstream side of the bubble, the flow diverges and the opposite effect comes into play, leading to a steep increase in pressure in this region. The pressure distribution is therefore mainly governed by the size and the location of the separation bubble.

To various degrees, the k-ε model calculations miss the observed levelling off of the pressure and predict the onset of the pressure rise too early. Both discrepancies point to an underprediction of the width of the separation bubble (confirmed by the velocity profiles in Figure 5), and the start of the pressure rise is clearly correlated with the reattachment point. In Clarke and Wilkes' calculations the pressure rises much too early because reattachment is predicted too early while Noyce comes closest to the experimental distribution because his reattachment is predicted latest. The DSM and the modified k-ε(R) model leading to smaller shear-stress levels and later reattachment predict the onset of the pressure rise in much better agreement with the experiments, but the modified k-ε(R) model leads to an underprediction of the following increase in pressure. It appears that these two models produce a thicker separation bubble.

The pressure distribution calculated by Rodi and Zhu after the Workshop with a step function for the inlet profile of U is also shown in Figure 3. This distribution was virtually insensitive to the grid fineness and discretization scheme used. It is in fairly close agreement with the distributions of Di Martino and Noyce. The original pressure distribution of Rodi and Zhu is lower at larger x because the momentum influx is somewhat smaller when an inlet U profile with gradual transition is used.

3.3 Axial Velocity

The decay of the streamwise velocity on the centerline is shown in Figure 4. Apart from the distributions submitted by the various contributors to the Workshop, additional calculations of the authors are presented in which they tested the sensitivity of the results to inlet conditions. Figure 4(b) presents results obtained with an initial velocity profile having a gradual transition over a finite shear-layer thickness and various turbulence conditions given in Table 2. Figure 4(c) presents results with a sharp step in the U-profile and various turbulence conditions while Figure 4(d) displays results of a calculation obtained with the inlet condition, upwind differencing and a 60×96 grid as used by Noyce.

From Figures 4(b) and (c) it is clear that the differences in the k-ε model predictions in Figure 4(a) between Clarke and Wilkes, Di Martino and Rodi and Zhu can be explained by the different initial conditions used. The better agreement with experiments obtained by Rodi and Zhu is mainly due to the use of initial conditions based on the shear-layer with finite width. Figure 4(d) shows that the difference between Noyce's k-ε model calculation and the others is only partly due to different initial conditions. As was already surmised above, the calculations of Noyce may not have been fully converged.

The DSM model can be seen to predict the decay of the centerline velocity fairly well and the results could probably be improved further by using more realistic inlet conditions. The modification to the k-ε model tested by Noyce makes the predictions worse. This modification was designed to reduce the turbulent shear-stress level and hence the spreading rate of round jets in unconfined ambient. This model therefore leads to a slower decay of the centerline velocity, which is not wanted in this case.

Calculated velocity profiles are compared with measured ones at four downstream locations in Figure 5. The behaviour near the centerline is covered already by the discussion given above on the decay of the centerline velocity. Away from the centerline, there is generally good agreement between predictions and measurements. At the station x=40cm, which is just beyond the middle of the separation zone, the thickness of the region with negative velocities is predicted too small by all models, particularly so by the unmodified k-ε model. Also, the maximum of negative velocity is considerably underpredicted. It appears therefore that in the experiment a thicker recirculation bubble with a stronger recirculating motion exists.

3.4 Turbulent Kinetic Energy

Figure 6 compares calculated and measured turbulent kinetic energy profiles at four downstream stations. Here the differences in the k-ϵ model predictions can be traced mainly to the inlet conditions of the turbulence quantities. Di Martino using very small initial values of both kinetic energy and eddy viscosity predicts the correct level in kinetic energy, while in the calculations of Rodi and Zhu using estimated values in the finite-width shear layer, the kinetic energy level is initially somewhat too high.

3.5 Turbulent Shear Stress

The corresponding shear stress distributions are shown in Figure 7. Similar observations can be made as for the kinetic energy distribution, i.e. here also the differences in the k-ϵ model predictions are due primarily to different inlet conditions for turbulence quantities. Of course, further downstream, numerical diffusion and convergence differences may also be partly responsible for the different results. Generally, the shear stress distribution, both with respect to shape and levels, are well predicted by the k-ϵ model except in the separation region. In the measurements, the shear stress tends to go negative quite a distance from the wall, indicating again a relative thick separation bubble. The k-ϵ model predictions do not reproduce this behaviour as the shear stress remains positive until very near the wall (regions with negative shear stress are barely recognisable). Hence, the shear stress is considerably too high in the recirculation region. The DSM model appears to perform better in this region, even though the predicted shear stress does not actually go negative, but it has a very low level in the recirculation region.

4. CONCLUSIONS

Four calculations with the same k-ϵ turbulence model were found to produce considerable differences in the results. These could be traced mainly to differences in the specified inlet conditions, but in one case they may have been caused also by insufficient convergence of the solution. Also, not all results may have been entirely grid-independent. The k-ϵ model has been found capable of predicting well some features of the test flow, such as the distribution of mean velocity and turbulence quantities outside the recirculation region and also the separation and reattachment points. The mean flow quantities are predicted best with inlet profiles based on the existence of a finite-width shear layer as used by Rodi and Zhu who, it must be stated for fairness' sake, kwew the experimental results. However, their inlet turbulence quantities estimated with a mixing-length formula led to an initial overprediction of kinetic energy and shear stress levels. Although little experimental evidence is available on this, it appears that the thickness of the separation bubble is underpredicted by the k-ϵ model, leading to a relatively poor wall-pressure distribution in flow regions affected by the separation. The shear stress in the separation region is predicted considerably too high. Its distribution is simulated much better with the differential stress model and so is the onset and the rate of pressure rise, but the other features are not predicted markedly better by this model. More detailed experiments, particularly in the separation region, would be required in order to shed further light on the relative performance of the various models.

5. REFERENCES

Binder, G., and Kian, K., 1983, "Confined jets in a diverging duct," *Proc. 4th Symp. on Turbulent Shear Flows*, Karlsruhe, F.R.Germany, 7.18-7.23.

Chieng, C.C., and Launder, B.E., 1980, "On the calculation of turbulent heat transport downstream from an abrupt pipe expansion," *Num. Heat Transfer*, **3**, 189-207.

Launder, B.E., and Spalding, D.B., 1974, "The numerical computation of turbulent flows," *Comput. Methods Appl. Mech. Eng.*, **3**, 269-289.

Kian, K., 1981, "Jets confinés dans un divergent," Thèse de Docteur-Ingénieur, l'Université Scientifique et Médicale de Grenoble, France.

Rodi, W., 1972, "The prediction of free turbulent boundary layers by use of a two-equation model of turbulence," Ph.D. Thesis, University of London.

Zhu, J., 1986, "Calcul des jets turbulents confinés avec recirculation," Thèse de Docteur, l'Institut National Polytechnique de Grenoble, France.

Zhu, J., Binder, G., and Kueny, J.L., 1987, "Improved predictions of confined jets with a parabolic computation of the entrance region," *AIAA J.*, **25**, 1141-1142.

Zhu, J., and Rodi, W., 1990, "Computation of axisymmetric confined jets in a diffuser," *Engineering Turbulence Modelling and Experiments*, eds. W. Rodi and E.N. Ganić, Elsevier, 237-247.

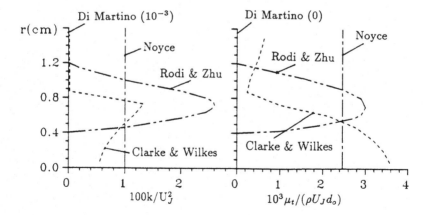

Figure 1. Inlet turbulent kinetic energy and eddy viscosity distributions

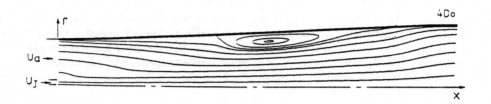

Figure 2. Calculated streamlines (Rodi and Zhu)

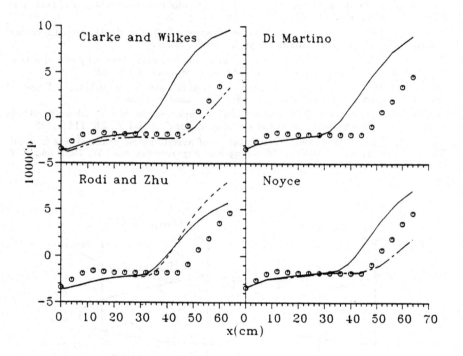

Figure 3. Wall static pressure: ———— k-ε; ——— - ——— k-ε(R);
——— - - ——— DSM; - - - - - - Inlet step U-profile; ⊙ Experiment

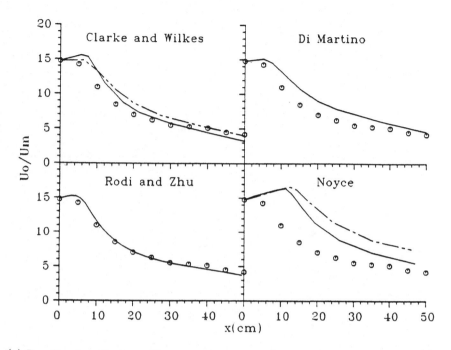

(a) Results of predictions submitted to workshop (key to symbols as in Figure 3)

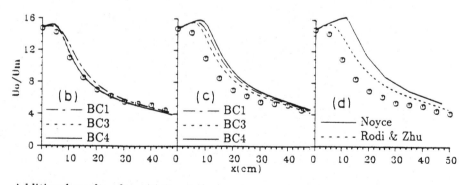

Additional results of sensitivity study:
(b) Finite-width shear-layer profile for U, various k and ϵ inlet conditions (Table 2)
(c) Step profile for U, various k and ϵ inlet conditions (Table 2)
(d) Calculations with Noyce's inlet conditions

Figure 4. Centerline velocity decay

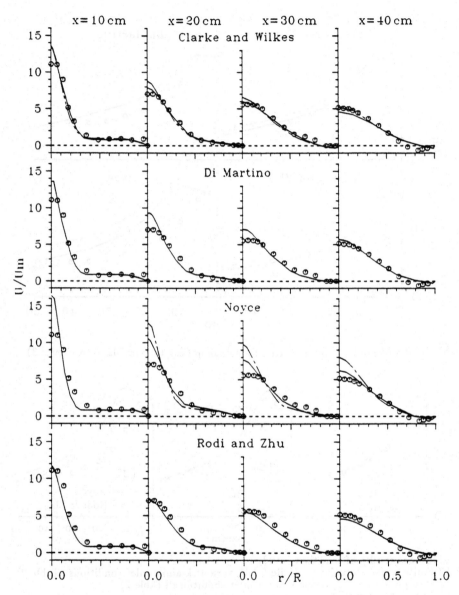

Figure 5. Axial mean velocity (key to symbols as in Figure 3)

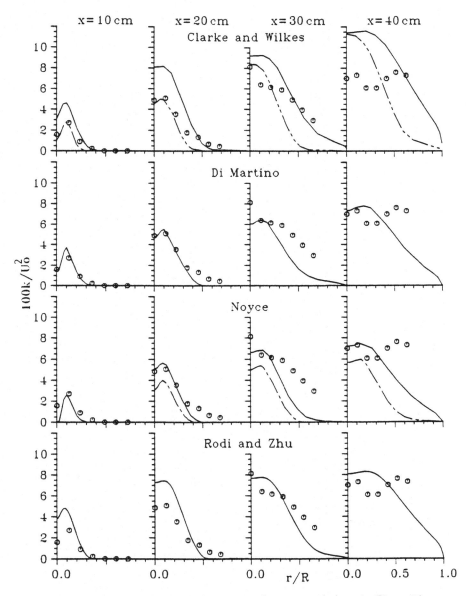

Figure 6. Turbulent kinetic energy (key to symbols as in Figure 3)

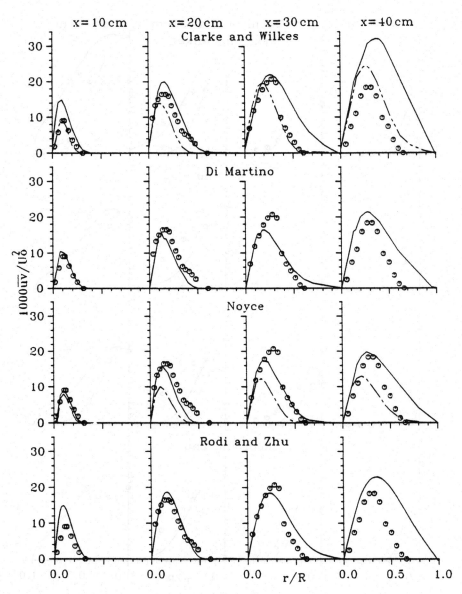

Figure 7. Turbulent shear stress (key to symbols as in Figure 3)